Polarimetric SAR Imaging

SAR Remote Sensing Series

Series Editor
Jong-Sen Lee

Imaging from Spaceborne and Airborne SARs, Calibration, and Applications
Masanobu Shimada

Polarimetric SAR Imaging
Theory and Applications
Yoshio Yamaguchi

For more information about this series, please visit: https://www.routledge.com/SAR-Remote-Sensing/book-series/CRCSRS

Polarimetric SAR Imaging

Theory and Applications

Yoshio Yamaguchi

CRC Press
Taylor & Francis Group
Boca Raton London New York

CRC Press is an imprint of the
Taylor & Francis Group, an **informa** business

First edition published 2020
by CRC Press
6000 Broken Sound Parkway NW, Suite 300, Boca Raton, FL 33487-2742

and by CRC Press
2 Park Square, Milton Park, Abingdon, Oxon, OX14 4RN

© 2021 Taylor & Francis Group, LLC

CRC Press is an imprint of Taylor & Francis Group, LLC

ISBN: 978-0-367-47831-5 (hbk)
ISBN: 978-0-367-50310-9 (pbk)
ISBN: 978-1-003-04975-3 (ebk)

Typeset in Times
by Lumina Datamatics Limited

Contents

Preface

This book is devoted to polarimetric radar remote sensing with an emphasis placed on the utilization of scattering matrix data. Radar polarimetry, that is, full utilization of the vector nature of electromagnetic waves, has been attracting attention in radar sensing for more than three decades. However, since the electromagnetic wave cannot be seen by the human eye, the treatment of polarimetric data has been difficult for general users. In order to fulfill a gap between expectation and complex treatment of polarimetry, we have been engaged in the data analysis, aiming at intuitive comprehension by everybody. Since power information is one of the most important and stable parameters in radar data, we tried to use power and assign polarimetric scattering powers to an RGB color-code for creating color images. Fully color-coded images generated by polarimetric powers became easy to understand for everybody. Recognition of polarization by color is one of the purposes. Based on this assigning methodology, it became possible to see the observation areas just like a color photograph based on polarimetric information.

The radar images in the microwave frequency are different from optical images. For the case of Earth observation from space, we see white clouds in the optical image, whereas we do not see any clouds in the microwave radar image. This is caused by the difference of propagation and scattering properties of the electromagnetic waves in optical and in microwave frequency regions. For the case of spaceborne radar observation, radar waves can penetrate clouds and reach to the ground, and then are scattered back to the radar. We can see terrain in the radar image clearly. Clouds are transparent for the radar waves, and therefore the echo from the clouds is zero. In optical sensing, the optical wave is scattered and absorbed by clouds; hence, it is impossible to see beneath the cloud. Even for cloud-free conditions, the scattered information is different. For example, bright and dark areas in each image are different. This means each piece of information is complementary. Although the information is different, radar imaging is the only one that is achievable in cloudy and rainy weather conditions or in rainforest areas where the sky is always covered with clouds. Among radar imaging, polarimetric radar provides a lot of information on observation objects and serves for accurate remote sensing by the scattering matrix information.

This book is intended to provide complete and consistent theory and applications of radar polarimetry. By Advanced Land Observing Satellite—Phased Array L-band Synthetic Aperture Radar (ALOS-PALSAR) by JAXA in 2006, a great number of scattering matrix data sets over the globe have been acquired, and much attention is paid to polarimetric radar remote sensing. The scattering matrix from space is available to users nowadays. It is expected to contribute to monitoring planet Earth in more detail. The contents and images in this book are mainly based on the ALOS/ALOS2 data, airborne PiSAR/PiSAR-2 data, international polarimetry community workshops, IEEE GARSS series, and other conferences. This book is a revised version of the Japanese edition in 2007, plus an addition of several recent observation results after 2007.

Acknowledgments

I would like to express sincere appreciations to the radar polarimetry community in the world, especially to my mentor Professor Emeritus Wolfgang M. Boerner at the University of Illinois at Chicago who for almost 30 years has always inspired me to work. I have learned polarimetric radar remote sensing during my visit to the University of Illinois at Chicago (1988–1989), and since then I have worked together for the advancement of polarimetric remote sensing for preserving a healthy earth. Through his diverse interactions with researchers and experts worldwide, I became acquainted with so many people in the polarimetry community. I would like also to thank Dr. Jong-Sen Lee for providing me the chance to write this book.

We are all "co-strugglers" of polarimetric radar sensing. We meet at IEEE IGARSS series and at International Polarimetric SAR Workshops almost every year. I thank all members for sharing important ideas: Dr. Thomas Ainsworth, NRL, USA; Dr. Jacob van Zyl and Dr. Scott Hensley, JPL, USA; Dr. Shane Cloude, AEL, UK; Prof. Eric Pottier, University of Rennes 1, France; Dr. Kostas Papathanassiou, DLR, Germany; Prof. Kun-Shan Chen, CAS, China; and Prof. Dhamendra Singh, IITR, India, among others.

A group photo was taken at the International Polarimetric SAR Workshop in Niigata 2012 and is displayed to thank all of colleagues. The workshop series and anechoic chamber with polarimetric measurement system were supported by the Ministry of Education (MEXT), Japanese Government, as a project of "International collaboration on space sensing—full utilization of polarimetric information." I am really thankful for the project support of Niigata University and the MEXT, Japan.

Group photo at the International Polarimetric SAR Workshop in Niigata 2012

The works in this book were carried out by my former students and colleagues in Niigata University, Japan: Prof. Hiroyoshi Yamada; Prof. Ryoichi Sato; Prof. Hirokazu Kobayashi; Prof. Toshifumi Moriyama, Nagasaki University; Prof. Jian Yang, Tsinghua University, China; Prof. Dhamendra Singh, Indian Institute of Technology Roorkee, India; Prof. Sang Eun Park, Sejong University, Korea; Dr. Cui Yi, Stanford University, USA; Prof. Gulab Singh, Indian Institute of Technology Bombay, India, and more than 100 students in our laboratory. The contributions of Prof. G. Singh to the recent multiple scattering-power decompositions such as G4U, 6SD, and 7SD, are highly appreciated. Many students conducted experimental verifications of PolSAR measurements in anechoic chambers and in outdoor sites, and many other students engaged in computer simulations and image processing of huge amounts of polarimetric SAR data sets. I also thank Prof. Emeritus Masakazu Sengoku (former vice president of Niigata University) and other members for supporting our international space sensing project (2010–2012) granted by the MEXT, Japan.

The data sets provided by JAXA and NICT are highly appreciated. The total number of PolSAR scenes exceeds more than 2,700. Without Dr. Masanobu Shimada, a former JAXA senior scientist and now Prof. at Tokyo Denki University, ALOS and ALOS2 polarimetric data analyses were impossible. Very high resolution and precious Pi-SAR-2 data have been provided by NICT, led by Dr. Seiho Uratsuka, Dr. Shoichiro Kojima, and his group. All polarimetric data are now processed to show colorful decomposition images on the website of Advanced Industrial Science and Technology (AIST), Japan. I would like to thank Dr. Ryosuke Nakamura and his colleagues.

I also appreciate the Institute of Electrical and Electronics Engineers (IEEE) and the Institute of Electronics, Information and Communication Engineers (IEICE) for allowing me to use figures that have appeared in the publications. Many figures are reused here again because this book is the second and revised version of *Radar Polarimetry from Basics to Applications* (in Japanese) IEICE, in 2007. The author thanks CRC Press and Lumina Datamatics Limited for editing the manuscript, and careful English corrections done by Ms. Suyun Wang of Tohoku University.

Yoshio Yamaguchi

Author

Yoshio Yamaguchi received his BE degree in electronics engineering from Niigata University, Niigata, Japan, in 1976 and his ME and Dr. Eng. degrees from Tokyo Institute of Technology, Tokyo, Japan, in 1978 and 1983, respectively. He is a Professor Emeritus and Fellow of Niigata University after his retirement in 2019. He is also an invited researcher at the National Institute of Advanced Industrial Science and Technology (AIST), Japan.

He had been with the faculty of engineering, Niigata University, during 1978–2019 as an assistant professor, associate professor, and professor, where he had engaged in the research of electromagnetic wave propagation in tunnels, in lossy media, and FMCW radar for short-range sensing. From 1988 to 1989, he was a research associate at the University of Illinois at Chicago, where he learned basic polarimetry under the guidance of Prof. Wolfgang M. Boerner. His interests moved toward the fields of radar polarimetry, microwave scattering, scattering power decomposition, and imaging. He has published more than 150 journal papers, 3 books, 4 book chapters, and more than 300 conference proceedings.

Dr. Yoshio Yamaguchi has served as chair of IEEE Geoscience & Remote Sensing Society (GRSS) Japan Chapter (2002–2003), chair of International Union of Radio Science Commission F Japan Committee (URSI-F) Japan (2006–2011), associate editor for *Asian Affairs of GRSS Newsletter* (2003–2007), and Technical Program Committee (TPC) co-chair of the 2011 IEEE International Geoscience and Remote Sensing Symposium (IGARSS). He is a Life Fellow of IEEE, and a Fellow of the Institute of Electronics, Information and Communication Engineers (IEICE), Japan. He is a recipient of the 2008 IEEE GRSS Education Award "in recognition of significant educational contributions to Geoscience and Remote Sensing Society" and 2017 IEEE GRSS Distinguished Achievement Award "for contributions to Polarimetric Synthetic Aperture Radar Sensing & Imaging and its Utilization."

1 Introduction

1.1 REMOTE SENSING

Remote sensing is defined as a technique for identifying, classifying, and determining objects, as well as obtaining information on their physical properties through analysis of the data on the objects collected by remotely located sensors that do not contact physically with objects. Usually, the data analyzed are those on the electromagnetic waves emitted from or reflected by the objects. Remote sensing, in a narrow sense, is defined as a method for retrieving the physical properties of objects around the earth by remote sensors on platforms such as aircraft and satellites. Representative references on remote sensing are listed in reference books [1–25] and websites.

There are roughly two kinds of remote sensing methods as shown in Figure 1.1. The first one is passive remote sensing in which sensors receive the electromagnetic spectrum emitted from the object. Radiometer operative in the microwave frequency and optical sensors such as a spectrometer or hyper-spectral meter in the optical frequency are typical passive remote sensors. Passive sensors do not emit any signals toward objects just like a camera without flash light. The second one is active sensing in which sensors emit signals and receive the echo. Radio detection and ranging (radar), synthetic aperture radar (SAR), and light detection and ranging (lidar) are typical sensors. A camera with a flash light can be considered an active sensor. These categories together with the frequency range are shown in Figure 1.1 [16,23].

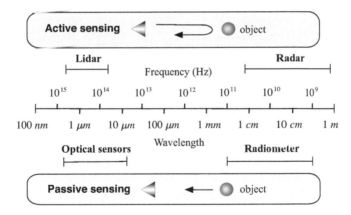

FIGURE 1.1 Passive and active remote sensing.

1.1.1 DIFFERENCE OF OPTICAL AND MICROWAVE SENSORS

Optical remote sensing has a longer history than microwave sensing. Images obtained by optical sensors, such as photographs, is rather familiar to our eyes and feelings. We can interpret optical images based on our visual perception. In addition to visible light bands, there are many spectral bands in the optical region, which have specific responses to physical quantity, such as infrared to thermal temperature and certain bands to chlorophyll. On the other hand, microwave images generated by SAR are not so familiar to everybody. Since we cannot see microwaves at all, the images generated by microwaves are out of our perception range. For example, if we see a forest image by a P-band microwave radar, the image looks completely different from that of an optical sensor.

P-band frequencies can partially penetrate forests. The reflection image from a forest area shows tree echoes and underlying ground information. In an optical image, we always see the top canopies of the forest. This makes the two images totally different. This difference comes from the propagation and scattering properties inherent to the frequency, and hence the information obtained by optical and microwave regions becomes totally different.

One example is shown in Figure 1.2 where Mt. Bromo, Indonesia, is imaged by an optical sensor and microwave radar (Polarimetric SAR, ALOS-PALSAR). In the optical image, we see a white cloud covering the mountain slope and smoke erupting from a volcano summit. It is impossible to see beneath the cloud of smoke because optical waves cannot penetrate. Since this image size is approximately 40 × 70 km, several optical subimages (taken at different times and conditions) are combined to make the whole image. On the other hand, it is possible to see all information on the ground objects by radar. There is no effect of clouds in the SAR image due to the penetration capability of microwaves. In addition, the information in the image looks different because color coding as well as scattering mechanisms are different. Polarimetric information is used to create the full-color image based on the scattering mechanisms.

In rainforest areas near the equatorial region, the land is covered by clouds almost 90% of the time. The penetration of clouds is one of the most important issues for remote sensing from the sky. The penetration capability depends on the attenuation characteristics of electromagnetic waves in the atmosphere. The attenuation constant (dB/km) shows how the wave is attenuated in 1000 m propagation. The value is less than 0.01 dB for 1 GHz and increases with frequency up to 4–5 dB for 10 GHz in heavy rain conditions [26]. For the optical region, the attenuation is more than 100 dB, which means the optical wave is not available for sensing. Water vapor or aerosol in the atmosphere causes the attenuation in the optical wave.

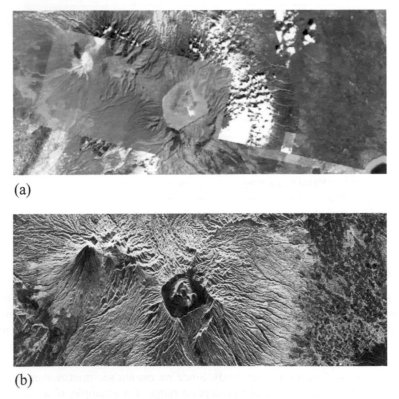

(a)

(b)

FIGURE 1.2 Volcano Bromo, Indonesia, observed by an optical sensor and microwave PolSAR: (a) Google Earth optical image, and (b) PolSAR image.

1.1.2 RADAR REMOTE SENSING

In this book, we deal with microwave SAR. In the microwave frequency region, frequencies are further divided into some bands as shown in Table 1.1. Each band has special frequency characteristics for targets with respect to target size and material constants having electrical properties. It is known the C-band, which is suited for crop monitoring, and the L-band is suitable for forest, land cover, disaster monitoring, etc. The lower frequency can propagate with low attenuation and penetrate the medium deeper inside. The P-band wave can penetrate not only forests but also dry ground, snow, ice, and the soil surface at several centimeters. The higher frequency above the X-band scatters at the surface of objects and does not penetrate objects. The scattering nature at higher frequency becomes close to those of optical waves. Therefore, these bands are suited for high-resolution imaging of terrain.

In addition to penetration (reflection, attenuation) characteristics, there is another physical effect, that is, resonance by the wavelength. Since the wavelength of microwave frequency ranges approximately from 1 to 100 cm, many object sizes can be in the same order of the wavelength or multiple of half wavelength. Resonance or Bragg scattering can occur when the phase condition matches. Sometimes we see strong Bragg scattering in crop fields.

Synthetic aperture radar is an active microwave instrument that performs two-dimensional high-resolution imaging under almost all weather conditions. The advantages of SAR are as follows:

1. Applicability in all weather conditions. Due to low attenuation characteristics of electromagnetic propagation, it can penetrate haze, clouds, and rain with very small losses. This allows SAR to operate under all weather conditions regardless of time. Day and nighttime operation is also available.
2. Quick data acquisition time over a wide area. Since the radar illumination area is wide, a vast area can be observed in a very short time. For example, an imaging area of width 40 km and length 70 km can be observed in less than 3 minutes by the spaceborne ALOS2-PALSAR2 system.
3. High-resolution two-dimensional imaging is achievable. The azimuth resolution is half the size of a radar antenna. The range resolution is inversely proportional to the bandwidth of a transmitting signal. The wider the bandwidth, the finer the resolution. Recent typical resolutions are 1–20 m for spaceborne SAR, and 0.3–3 m for airborne SAR. Range resolution is independent of radar platform height, which means the same resolution is obtained from space and from aircraft if the signal bandwidth is the same.

TABLE 1.1
Frequency Band

Band	Frequency (GHz)	Typical Wavelength (cm)
P	0.3–1.0	50
L	1–2	25
S	2–4	10
C	4–8	5
X	8–12.5	3
Ku	12.5–18	2
K	18–26.5	1.5
Ka	26.5–40	1
V	40–75	—
W	75–110	—

4. It provides complementary information to optical remotely sensed data. Since the scattering nature is completely different from the optical region, information by SAR and optical sensors should be used complementarily.
5. New image products such as 3D-mapping, tomographic mapping, and differential interferometry can be produced.

1.2 APPLICATIONS OF SAR TECHNOLOGY

Radar is a coherent system that transmits a pulse and receives the echo from target. With the advancement of technical development, scattered wave information (amplitude and phase) can be captured digitally and processed by computer very rapidly. Radar technologies are rich in mathematics, physics, electromagnetic theory, signal processing, image processing, and interpretation. It includes hardware technology, software technology, and IT-related technology.

SAR is an extension of the radar and plays an important role in the microwave imaging system. It utilizes the amplitude and phase of the received signal and the doppler frequency information to generate high-resolution images. In addition to these parameters, polarimetric information is now incorporated into SAR systems, which allow us to use the full vector nature of electromagnetic waves. A SAR system that can observe fully polarimetric information is called PolSAR or Quad PolSAR. Since polarimetric information is sensitive to the orientation of an object, it brings additional information from the object. SAR and PolSAR have vast application areas, and they are as follows:

1. Forest monitoring including ecology, management, ecosystem change, carbon cycle, illegal deforestation, and wild fire.
2. Soil moisture content, agriculture applications, crop monitoring, desertification, etc.
3. Snow and ice applications, water management, sea ice movement for safe vessel navigation, glacier monitoring, etc.
4. Ocean application, wind-speed retrieval, and oil spill detection.
5. Disaster motoring by earthquake, landslide detection, and volcano activity monitoring.
6. Wetland monitoring for environmental preservation and flood monitoring.
7. 2D mapping of terrain, urban area monitoring, land-use map, etc.

With interferometry and polarimetric interferometry, application areas by interferometric SAR (InSAR) or polarimetric interferometric SAR (PolInSAR) are further expanded to:

8. 3D mapping of terrain and high-resolution 3D height structure in urban area.
9. Tree height estimation for biomass estimation.
10. Precursor system for volcano eruption using high-resolution displacement with accuracy in the order of millimeters.

As an example of disaster monitoring, Figure 1.3 shows PiSAR-2 [27] images over Kumamoto, before and after the earthquake on December 12, 2015 and April 16, 2016, respectively. Just after the earthquake, the airborne PiSAR-2 system acquired scattering matrix data in the X-band over Minami-Aso, Kumamoto, Japan. The scattering power decomposition images of two acquisitions are compared to show the damages of the great earthquake. We can immediately recognize the landslide areas enclosed by white circles. Time series polarimetric images are very useful for identifying these landslide areas. In addition, full-color images obtained by scattering mechanisms better serves interpretation of the scene just like a photograph.

The radar application field has expanded very rapidly in recent years. The radar sensor is employed in our daily life, such as car collision avoidance systems, intelligent car systems, intrusion detection systems, ground-penetrating radar systems for detecting gas pipes, electric cables, or mine detections. Radar technology is expanding to our daily lives.

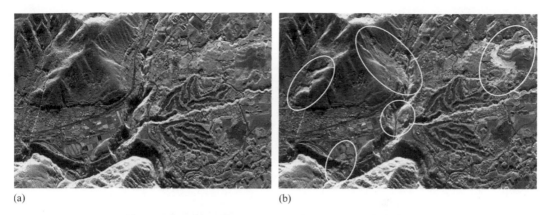

(a)   (b)

FIGURE 1.3 Time series fully PolSAR images near Minami-Aso before and after the Kumamoto earthquake in Japan [32]. Observation by airborne PiSAR-2 in the X-band with 30-cm resolution. White circles show landslide areas induced by the earthquake: (a) December 5, 2015 (before) and (b) April 17, 2016 (just after the earthquake).

1.3 IMPORTANCE OF POLARIMETRIC RESPONSE

Conventional radars use single polarization for transmitting and receiving. For example, suppose a target is an oblique dipole as shown in Figure 1.4. If a vertically polarized wave is incident on the dipole, a new horizontal component is created in the scattered wave in addition to the vertical component. A single polarization radar cannot obtain the new horizontally polarized component. Although the electric field has the vector nature, the single polarized radar receives one component only and misses the cross-polarized component. If the cross component is measured simultaneously, the vector nature of the scattered wave can be recovered. A fully polarimetric radar measures co-pol and cross-pol components of the scattered wave simultaneously so that the vector nature of the wave can be recovered.

One example of polarimetric response is shown in Figure 1.5. Metallic plate (green color), dihedrals (red), and wire structure (yellow) were placed as shown in the lower-left corner. A polarimetric combination of H (horizontal) and V (vertical) direction yields a polarimetric radar channel. The power of the channels is depicted in the HH, VV, and HV images.

We can see that co-pol HH and VV images are similar. The difference appears at the wire structure in the bottom, that is, the echo can be seen in the VV image, but it cannot be seen in the HH image. This is caused by the difference in the direction of metallic wires. The electric field parallel to the wire is reflected, and the field orthogonal to the wire is transmitted. This is due to the scattering property of the electromagnetic wave. Therefore, we have reflection in the VV channel. If we use single HH polarization radar, we miss this wire.

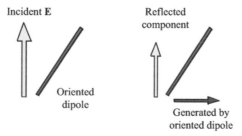

FIGURE 1.4 Vector nature of scattering phenomenon.

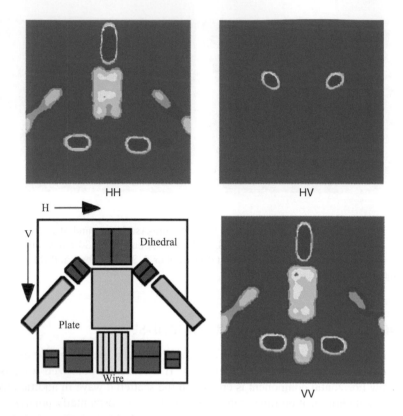

FIGURE 1.5   Polarimetric channel images of metallic plate, dihedrals, and wire.

Dihedrals oriented at 45° changes the polarization direction from *H* to *V*, and vice versa. Dihedrals placed at 0° or 90° return the signals in the same direction. These specific polarimetric characteristics can be recognized in all polarimetric images. There is no echo of 45°-oriented dihedrals in the *HH* and *VV* images due to polarization change. On the other hand, 45°-oriented dihedrals are clearly seen in the cross-polarized *HV* image.

From Figure 1.5, the polarimetric responses and the importance of fully polarimetric information are recognized. Polarization of the reflected wave is sensitive to the orientation of objects. Fully polarimetric radar measures all combinations of *H* and *V* (scattering matrix) and detects all responses, which serves the identification of objects.

1.4   BRIEF HISTORY OF RADAR POLARIMETRY

Optical polarization phenomena have been known a long time ago and is used in crystals for navigation by Vikings since AD 1000 [19]. It is known that some fishes in deep sea are sensitive to polarized light. There is a report that bees fly using information by the geomagnetic direction. We encounter a situation when we take photos of objects inside water by a camera, a polarization filter is helpful to suppress unwanted light from water surface. So far, animals, fishes, and humans have used optical polarization phenomena for many thousands of years.

These polarization phenomena have been theoretically investigated in the field of electromagnetic waves and now have been established as optical polarimetry in optic sensing and radar polarimetry in microwave radar sensing. The essential points are the vector nature of the electromagnetic waves and the interaction of material bodies and the propagation medium [19]. Radar polarimetry deals with and utilizes the full vector nature of electromagnetic waves in sensing.

According to [3, 35], the research on radar polarimetry started around 1950. George Sinclair (1950) treated the radar target as a polarization transformer and derived the 2×2 scattering matrix bearing his name [28]. E. M. Kennaugh at Ohio State University derived the 4×4 Kennaugh matrix suitable for radar measurement and the radar coordinate system [29]. In the radar measurement, the origin of the coordinate system is located in the radar system, whereas the origin resides in the target for optical sensing. The relation between forward scattering and backward scattering has been investigated in detail for the incoherent scattering scenario. By the redevelopment of the scattering theory in which the origin of the coordinate system locates on the radar, further advancements on the principle of radar polarimetry have been achieved by several researchers, like V. H. Rumsey [30], G. A. Deschamps [31] (1951), J. R. Copeland [32] (1960), S. H. Bickel (1965), and P. Beckman (1968). The Graves power matrix was derived for optimal power reception [33].

A great contribution in radar polarimetry was brought by J. R. Huynen [34] who had developed a unified theory on scattering matrix and its extension Stokes matrix and examined the target property and decomposition. He introduced Huynen parameters and the "polarization fork" on the Poincaré sphere with respect to characteristic polarization states. W. M. Boerner claimed the importance of polarization in the inverse problem [35] and investigated characteristic polarization states for target [3]. With the advancement of the hardware system [36,37], polarimetric radar became available and some test flights were demonstrated by NASA/JPL in the 1980s. The first polarimetric airborne system AIRSAR [38] conducted various polarimetric data takes and confirmed the principle. J. J. van Zyl (1987) showed the polarization signature of various targets, which is a graphical aid to understand the polarimetric scattering behavior [39–41]. W. M. Boerner continuously claimed the importance of polarimetry [42,43] and encouraged many people toward the advancement of radar polarimetry by holding numerous international workshops and creating a polarimetry community worldwide.

Once the basic principle of polarimetry was established, diverse application areas expanded in many ways. There was Russian research on polarization anisotropy (A. I. Kozlov), calibration (A. Freeman 1992; W. Wiesbeck 1991; S. Quegan 1994), and decomposition (S. R. Cloude 1985; E. Krogager 1993; A. Freeman 1993). From the 1990s, major research topics moved to retrieval of target information. For coherent scattering decomposition, E. Krogager (1990) proposed the KsKdKh method directly applied to the scattering matrix in the circular polarization basis. For incoherent scattering decomposition, S. R. Cloude and E. Pottier (1997) have a great contribution of the entropy/anisotropy/alpha-bar method based on the eigenvalue/eigenvector analysis using a coherency matrix. A. Freeman and Durden (1998) proposed a model-based scattering power decomposition method, which was further expanded to four-component, up to seven-component decompositions by Y. Yamaguchi and G. Singh.

Polarimetric filtering is also another attracting topic in polarimetric imaging. Since there are too many filters proposed, it is difficult to cite them all. Lee filter and its extensions [19] are standard polarimetric filters in many SAR image processing.

A combination of polarimetric SAR and interferometry yields PolInSAR, which can produce 3D structure of the objects such as a forest. Multiple data sets with slightly different vertical heights can be used to create 3D images by tomographic SAR. Such systems can be tested by airborne SAR in advance, and then can be achieved by steady state but slightly different repeat-pass observations of the satellite. The tutorial of SAR, polarimetric SAR (PolSAR), and new applications are well-documented by A. Moreira et al. in "A Tutorial on Synthetic Aperture Radar" [23].

Furthermore, a convenient analysis toolbox (software) is prepared in several research institutions. The representative one is the "PolSAR_Pro" series provided by ESA [61] and developed by E. Pottier and his colleagues. This open software deals with almost all new technologies in PolSAR remote sensing and has attracted the younger generation to work with it. A detailed history and development of radar polarimetry can be found in [3,7].

1.5   POLARIMETRIC SAR SYSTEMS

Radar system that can acquire a 2×2 scattering matrix is called PolSAR or Quad PolSAR. After the recognition of the effectiveness of PolSAR, various PolSAR systems have been developed all over the world [50–59]. There are two types of PolSAR systems depending on the platform: airborne SAR and spaceborne SAR. Airborne SAR is usually developed for conceptional verification of proposed SAR systems, checking the performance of future spaceborne systems, SAR technology developments, etc. It is designed for experimental purposes and not intended for routine operational purposes. Therefore, the performance of airborne SAR is usually superior to those of spaceborne systems.

1.5.1   Airborne SAR Systems

Airborne SAR has the following advantages:

1. It can be applied to any place, at any time, and for any observation direction, if the flight condition allows.
2. The signal-to-noise ratio is high compared to the spaceborne system because the SAR system is close to the target.
3. Spacious in the platform. This allows one to carry a high-powered battery and various instruments together for observation, which allows advanced radar measurement.

But it has the following disadvantages:

4. The aircraft platform suffers from air-flow disturbance and hence needs motion compensation for SAR signal processing.
5. Periodical observation is not easy. Even if the same area is observed, the flight condition is different. Therefore, data acquisition in the same condition is difficult.

There are so many airborne PolSAR systems in the world as listed in Table 1.2. Among them, the pioneering one is AIRSAR developed by NASA/JPL in the early 1980s, which has shown the effectiveness of polarimetric information. The successful results of AIRSAR brought the shuttle-based SIR-C/X-SAR mission into space (1994) [39]. E-SAR and the follow-up mission F-SAR by DLR conducted various radar measurements and confirmed the performances of conceptional designs. PiSAR-X/L and PiSAR-X2/L2 in Japan also contributed to the disaster monitoring and to the future design of spaceborne SAR missions. F-SAR and Pi-SAR are illustrated in Figure 1.6. There are many other excellent PolSAR systems, such as UAV-SAR/USA [50], EMISAR/Denmark, etc. The resolution of the recent SAR system is less than 1.5 m in the L-band, and less than 30 cm in the X-band. However, the high-resolution data is usually limited for domestic use.

TABLE 1.2
Airborne Quad PolSAR Systems

Name	Institution	Frequency Bands	Resolution (m)
AIRSAR	USA, NASA/JPL	P, L, C	Range: 7.5, 3.75 Azimuth: 1
UAVSAR	USA, NASA/JPL	L	1×1
EcoSAR	USA, NASA	P	0.75×0.5
SAR580	Canada, CCRS	C, X	6×6
F-SAR	Germany, DLR	P, L, C, X	X&C: 1.5×0.3, L: 2×0.4, P: 3×1.5
AER II-PAMIR	Germany, FHR	X	—
EMISAR	Denmark, DCRS	L, C	2.4×2.4, 8×8
Phraus	NL, TNO-FEL	C	3×3
SETHI	France, ONERA	P, L, X	X: 0.12×0.12
PiSAR-L2/X2	Japan, JAXA/NICT	L, X	1.6×1.6, 0.3×0.3

(a)   (b)

FIGURE 1.6 Photo images of airborne SAR systems: (a) F-SAR and (b) Pi-SAR.

1.5.2 SPACEBORNE PolSAR SYSTEMS

The first spaceborne SAR is SEASAT in 1978. Although the lifetime was 105 days, it demonstrated the capability as an imaging radar in the L-band under various weather conditions regardless of operation time. It opened the door for many follow-up spaceborne SAR systems up to now. Spaceborne SAR has the following advantages:

1. It works in all weather conditions and at any time.
2. Since the orbit is fixed, there is no need for motion compensation.
3. Observation from the same position is possible, which is suitable for time series data acquisition.

The disadvantages are:
4. Less flexibility compared to airborne SAR system. It is difficult to observe off-beam areas.
5. The revisit time is dependent on the radar system.
6. Resolution is usually less than those of airborne SAR.

Based on the experiments on the airborne SAR system, the first spaceborne and PolSAR measurement was carried out by SIR-C/X-SAR on the Space Shuttle by NASA/JPL (1994). This multiple frequency (L-, C-, X-band) mission proved the usefulness of polarimetry and interferometry, immediately after the launch, and became the path opener for the follow-up missions in the world. Usually, polarimetric data acquisition is added optionally to the single or dual polarization mode functions in spaceborne radar. Realization of PolSAR is really difficult, including, concept design, its approval, hardware development, launch, data take quality control, calibration, maintaining, continuous operation, and data management. Although PolSAR has been rather limited, ALOS PALSAR (2006–2011) in the L-band, RadarSAT2 (2007–) in the C-band, and TerraSAR-X/TanDEM-X (2007–) in the X-band, have been successfully developed to show the performance and usefulness. Table 1.3 lists some spaceborne PolSARs and Figure 1.7 illustrates some systems in space. Currently, ALOS2/PALSAR2 (2014–), TerraSAR/TanDEM-X, Gaofen-3 (2016–), RadarSAT 2, and SAOCOM are orbiting. Future missions with ALOS3/PALSAR3 (2021–) and NISAR (2022–) will be carrying the polarimetric mode. It is a golden age of SAR (Table 1.3).

It is interesting to see how the images over the same area are dependent on SAR systems and frequency. Figure 1.8 shows the L- and X-band images of Niigata prefecture acquired with ALOS PALSAR and TerraSAR-X, respectively. The radar illumination direction is from left to right for ALOS, and is opposite for TerraSAR-X. Although in the same area, the scattering nature is different, which results in a different image even if the same color-coding is applied.

TABLE 1.3

Spaceborne Quad PolSAR Systems

Name	Country	Launch	Frequency Band	Resolution (m)
SIR-C/X-SAR	USA	1994.4	L, C, X	10
	Germany/Italy	1994.9		
		2002.4		
ALOS-PALSAR	Japan	2006–2011	L	30
RADARSAT-2	Canada	2007–	C	11
TerraSAR-X	Germany	2007–	X	3
TanDEM-X		2010–	X	3
RISAT	India	2012–	C	—
ALOS2-PALSAR2	Japan	2014–	L	6
Gaofen-3	China	2016–	C	1
SAOCOM	Argentina	2017/2019	L	7
NovaSAR	UK	2018	S	30
Kompsat-6	Korea	2019	X	1
ALOS4-PALSAR3	Japan	2021	L	6
NISAR	USA/India	2022	L/S	4

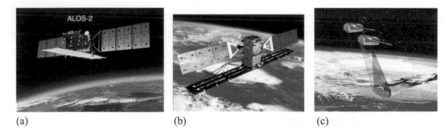

(a)   (b)   (c)

FIGURE 1.7 Photo images of spaceborne SAR systems: (a) ALOS2-PALSAR2, (b) RadarSAT-2, and (c) TerraSAR/TanDEM-X.

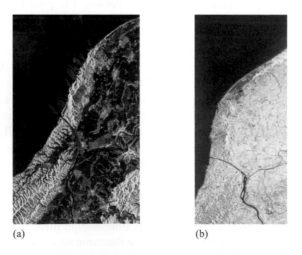

(a)   (b)

FIGURE 1.8 Niigata area observed by (a) ALOS PALSAR (ascending), and (b) TerraSAR-X (descending).

1.6   THE SCOPE OF THIS BOOK

This book is intended to provide the basic principles of radar polarimetry and its utilization using illustrations and figures as much as possible. The readers are assumed to know a little bit about electromagnetic waves. The mathematical expressions are frequently used for complete understandings. The contents are briefly categorized into the basics of polarization and its applications. Figure 1.9 shows the constitution of this book.

The basic principle of radar polarimetry and polarization matrices are defined from Chapter 1 to 4. Based on the scattering matrix in the coherent scattering, a 3×3 covariance or coherency matrix is used for data analyses in most of the incoherent scattering scenarios. Since these two matrices are positive semi-definite mathematically, they are convenient for various applications. We used coherency matrix formulation from Chapter 5 to 15, from the physical scattering point of view and from simple mathematical operations.

Chapter 2: Starting from Maxwell's equations, the solution of the Helmholtz wave equation is reviewed. The plane wave, which is the most fundamental solution of the wave equation, is described in detail to understand the property of the electromagnetic wave propagation. The expression of a polarized wave is represented in the form of the Jones vector, geometrical parameters (ellipticity angle and tilt angles), Stokes vector, and on the Poincaré sphere plot.

Chapter 3: The polarization state of the scattered wave from an object is different from that of the transmitted wave. The vector nature of wave scattering is defined by a 2×2 scattering matrix. The scattered wave is received by a radar antenna. This phenomenon is formulated in receiving antenna voltage and receiving power in the polarimetric radar channel. Taking into account the polarimetric characteristics, the basic principles of radar polarimetry are established. The polarization signatures of various objects are presented to show how the polarimetric response changes and how important it is for identifying objects. Polarimetric enhancement of a desired target against an undesired target can be achieved by polarimetric filtering. This chapter deals with scattering waves and receiving power in polarimetric radar channels.

Chapter 4: Once the scattering matrix is obtained, just like a snapshot in the coherent scattering case, there are five independent parameters in a relative scattering matrix form. For incoherent scattering, scattering matrix data are ensemble averaged to obtain the polarimetric information. There are nine independent parameters in the covariance matrix, coherency matrix, Kennaugh matrix, Mueller matrix, etc., in the form of the second-order statistics. This chapter introduces these matrices and their relations and indicates four important parameters in these polarization matrices.

Chapter 5: The H/A/alpha-bar method is the most frequently used decomposition method, developed by S. Cloude and E. Pottier. This method is based on the rigorous eigenvalue/eigenvector

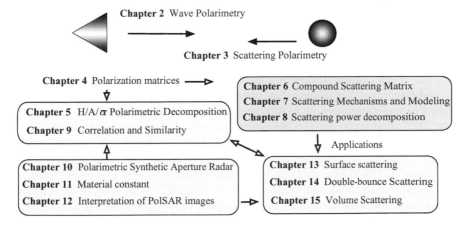

FIGURE 1.9   Book chapters.

expansion of the coherency matrix. Since the principle has been described in detail [19,20], only a brief review is provided.

Chapter 6: Any kind of scattering matrix can be created by a combination of dipoles aligned in the range direction. If dipoles with different orientation and appropriate spacing are used as a scatterer aligned in a radar range resolution, various kinds of scattering matrices can be generated. This is called a compound scattering matrix. These phenomena have been confirmed by finite difference time domain (FDTD) simulations and polarimetric measurements in a well-controlled anechoic chamber.

Chapter 7: Followed by the compound scattering matrix, scattering mechanisms and the modeling of coherency matrix elements are given. There are nine real-valued and independent parameters in a coherency matrix. For each parameter, an appropriate coherency matrix is derived based on the corresponding feasible compound scattering matrix. Scattering mechanisms and modeling with physically realizable situations are summarized.

Chapter 8: Using scattering models in Chapter 7, several scattering power decomposition schemes are presented. The measured coherency matrix is expanded as a sum of scattering models based on scattering mechanisms. The decomposition procedure and its algorithm, as well as the decomposition image of San Francisco, are displayed and compared. Starting from the original three-component decomposition by Freeman and Durden, the updated four-component decomposition (Y40, Y4R, S4R, G4U), the six-component decomposition (6SD), and up to seven-component decompositions are described in a unified manner.

Chapter 9: Correlation is one of the most important concepts in radar signal processing. This chapter deals with the correlation coefficient in various polarization bases and it shows the effectiveness of the circular polarization basis. Using the correlation coefficient in the circular polarization basis, a polar plot is used to classify target categories. A modified version of the correlation coefficient and the scattering power decomposition combination complementarily yielded excellent classification results. Furthermore, a similarity parameter is updated and redefined in order to make full use of nine polarimetric information more effectively. This new similarity parameter is used to extract the desired scattering objects successfully.

Chapter 10: In order to understand radar images, it is important to understand how the image is created. The range resolution as well as the azimuth resolution are the key parameters of radar images. This chapter explains the principle of SAR processing and image formulation using an example of frequency modulated and continuous wave (FMCW) radar. SAR is expanded to the PolSAR system with two transmitting antennas and two receiving antennas. Some applications of handmade FMCW PolSAR are explained with measured data. As an advanced application, the polarimetric Holo-SAR system is introduced to show its 3D imaging capability from 360° viewing angles. Concrete building models and trees are imaged by a polarimetric Holo-SAR system.

Chapter 11: The physical scattering nature from objects is determined by the boundary condition of electromagnetic waves on the object surface. The object has its own shape, orientation, and dielectric materials. This chapter summarizes the material constant of objects, that is, relative permittivity of various materials, using a well-known Debye model.

Chapter 12: When dealing with SAR and PolSAR images, we often encounter basic questions, such as foreshortening, layover, the effect of incidence angle, and resolution. Some notes on the interpretation of these questions on SAR images are presented for reference.

Chapter 13: Scattering power decompositions to PolSAR data sets yields the surface-scattering power P_s, the double-bounce scattering power P_d, and the volume-scattering power P_v, among other powers. This chapter shows the applications of the surface-scattering power P_s. Through the decomposition of ALOS/ALOS2 data sets, it is found that the surface-scattering power increases after disaster events, such as by mud, soil, landslide, and flooding. The power magnitude is also dependent on snow depth and volcano ashes, which might be effective for environmental monitoring.

Chapter 14: Right-angle structure induces the double-bounce scattering phenomenon. This scattering can be seen in man-made structures such as building walls, road surfaces, or tall tree/

vegetation stems on the ground surface. The scattering model is reviewed again, and some scattering measurement results on the incidence angle and squint angle characteristics are described. The applications to monitor tidal height, ship detection, and flooding are shown.

Chapter 15: Since the cross-polarized *HV* component is mainly generated from vegetation, this component can be a good indicator of the forest. This chapter is mainly devoted to monitoring forest, trees, and vegetation and seeks a possibility of classifying tree types using high-frequency and high-resolution PolSAR. In other applications, a reduction of the volume-scattering power P_v is successfully applied to detect a landslide area caused by an earthquake on a woody mountain area.

In addition, scattering power decomposition images are posted on the following websites:

https://gsrt.airc.aist.go.jp/landbrowser/index.html (Global ALOS quad pol images and data)
https://landbrowser.airc.aist.go.jp/polsar/index.html (Partial ALOS2 quad pol images)
http://www.wave.ie.niigata-u.ac.jp (which includes ALOS, ALOS2, PiSAR-L2, PiSAR-X2 data)
http://www.csre.iitb.ac.in/gulab/index.html (which includes ALOS2 data)

REFERENCES

1. F. T. Ulaby, R. K. Moore, and A. K. Fung, *Microwave Remote Sensing: Active and Passive*, vols. I–III, Artech House, Boston, 1986.
2. Y. Furuhama, K. Okamoto, H. Masuko, *Microwave Remote Sensing by Satellite*, IEICE, Tokyo, 1986.
3. W. M. Boerner et al., eds., Direct and Inverse Methods in Radar Polarimetry, *Proceedings of the NATO-ARW*, September 18–24, 1988, 1987–91, NATO ASI Series C: Math & Phys. Sciences, vol. C-350, Parts 1 & 2, Kluwer Academic Publication, the Netherland, 1992.
4. F. T. Ulaby and C. Elachi, *Radar Polarimetry for Geoscience Applications*, Artech House, Boston, 1990.
5. J. A. Kong, ed., *Polarimetric Remote Sensing*, PIER-3, Elsevier, New York, 1990.
6. H. J. Kramer, *Observation of the Earth and Its Environment: Survey of Missions and Sensors*, 3rd ed., Springer, 1996.
7. F. M. Henderson and A. J. Lewis, *Principles & Applications of Imaging Radar, Manual of Remote Sensing*, 3rd ed., vol. 2, ch. 5, pp. 271–357, John Wiley & Sons, 1998.
8. Y. Yamaguchi, *Fundamentals and Applications of Polarimetric Radar* (in Japanese), REALIZE Inc., Tokyo, 1998.
9. C. Oliver and S. Quegan, *Understanding Synthetic Aperture Radar Images*, Artech House, Boston, 1998.
10. C. H. Chen, *Information Processing for Remote Sensing*, World Scientific, 1999.
11. K. Okamoto, ed., *Global Environmental Remote Sensing: Wave Summit Course*, Ohmsha Press, Tokyo, 2001.
12. K. Ouchi, *Fundamentals of Synthetic Aperture Radar for Remote Sensing* (in Japanese), 2nd ed., Tokyo Denki University Press, 2004, Tokyo, 2009.
13. M. Takagi and H. Shimoda, eds., *New Handbook on Image Analysis*, Tokyo University Press, Tokyo, 2004.
14. I. Woodhouse, *Introduction to Microwave Remote Sensing*, CRC Press, Taylor & Francis Group, 2006.
15. C. Elachi and J. van Zyl, *Introduction to the Physics and Techniques of Remote Sensing*. John Wiley & Sons, Hoboken, NJ, 2006.
16. Y. Yamaguchi, *Radar Polarimetry from Basics to Applications: Radar Remote Sensing using Polarimetric Information* (in Japanese), IEICE, Tokyo, 2007.
17. Z. H. Czyz, *Bistatic Radar Polarimetry: Theory & Principles*, Wexford College Press, 2008.
18. D. Masonnett and J.-C. Souyris, *Imaging with Synthetic Aperture Radar*, EPFL/CRC Press, Taylor & Francis Group, 2008.
19. J. S. Lee and E. Pottier, *Polarimetric Radar Imaging from Basics to Applications*, CRC Press, 2009.
20. S. R. Cloude, *Polarisation: Applications in Remote Sensing*, Oxford University Press, Oxford, 2009.
21. J. van Zyl and Y. Kim, *Synthetic Aperture Radar Polarimetry*, Wiley, Hoboken, NJ, 2011.
22. Y.-Q. Jin and F. Xu, *Polarimetric Scattering and SAR Information Retrieval*, Wiley, IEEE Press, Singapore, 2013.

23. A. Moreira, P. Prats-Iraola, M. Younis, G. Krieger, I. Hajnsek, and K. Papathanassiou, "A tutorial on synthetic aperture radar," *IEEE Geosc. Rem. Sens. Magaz.*, vol. 1, no. 1, pp. 6–43, 2013.
24. S. W. Chen, X.-S. Wang, S.-P. Xiao, and M. Sato, *Target Scattering Mechanism in Polarimetric Synthetic Aperture Radar: Interpretation and Application*, Springer, Singapore, 2017.
25. M. Shimada, *Imaging from Spaceborne SARs, Calibration, and Applications*, CRC Press, 2019.
26. Recommendation of ITU-R p.527-5, "Electrical characteristics of the surface of the Earth," 2019.
27. https://directory.eoportal.org/web/eoportal/airborne-sensors/pi-sar2
28. G. Sinclair, "The transmission and reception of elliptically polarized waves," *Proc. IRE*, vol. 38, no. 2, pp. 148–151, 1950.
29. E. Kennaugh, Polarization properties of radar reflections, MSc Thesis, Ohio State University, 1952.
30. V. H. Rumsey, "Part I – Transmission between elliptically polarized antennas," *Proc. IRE*, vol. 38, pp. 535–540, 1951.
31. G. A. Deschamps, "Part 2, Geometrical representation of the polarization state of a plane magnetic wave," *Proc. IRE*, vol. 39, pp. 540–544, 1951.
32. J. D. Copeland, "Radar target classification by polarization properties," *Proc. IRE*, vol. 48, pp. 1290–1296, 1960.
33. C. D. Graves, "Radar polarization power scattering matrix," *Proc. IRE*, vol. 44, no. 5, pp. 248–252, 1956.
34. J. R. Huynen, Phenomenological theory of radar targets, PhD Thesis, University of Technology, Delft, the Netherlands, December 1970.
35. W.-M. Boerner, El-Arini, C. Y. Chan, and P. M. Mastoris, "Polarization dependence in electromagnetic inverse problems," *IEEE Trans. Antenna Propag.*, vol. AP-29, no. 2, pp. 262–271, 1981.
36. D. Giuli, "Polarization diversity in radar," *Proc. IEEE*, vol. 74, no. 2, pp. 245–269, 1986.
37. C. A. Wiley, Pulsed doppler radar methods and apparatus, U.S. Patent 3 196 436, 1954.
38. https://airsar.jpl.nasa.gov/index_detail.html
39. E. R. Stofan, D. L. Evans, C. Schmullius, B. Holt, J. J. Plaut, J. van Zyl, S. D. Wall, and J. Way, "Overview of results of Spaceborne Imaging Radar-C, X-band synthetic aperture radar (SIR-C/X-SAR)," *IEEE Trans. Geosci. Remote Sens*, vol. 33, no. 4, pp. 817–828, 1995.
40. J. J. van Zyl, H. A. Zebker, and C. Elachi, "Imaging radar polarization signatures: Theory and observation," *Radio Sci.*, vol. 22, no. 4, pp. 529–543, 1987.
41. D. L. Evans, T. G. Farr, J. J. van Zyl, and H. A. Zebker, "Radar polarimetry: Analysis tools and applications," *IEEE Trans. Geosci. Remote Sens.*, vol. 26, no. 6, pp. 774–789, 1988.
42. A. P. Agrawal and W.-M. Boerner, "Redevelopment of Kennaugh target characteristic polarization state theory using the polarization transformation ratio for the coherent case," *IEEE Trans. Geosci. Remote Sens.*, GE-27, pp. 2–14, 1989.
43. W.-M. Boerner, W. L. Yan, A.-Q. Xi, and Y. Yamaguchi, "On the basic principles of radar polarimetry: The target characteristic polarization state theory of Kennaugh, Huynen's polarization fork concept, and its extension to the partially polarized case," *Proc. IEEE*, vol. 79, no. 10, pp. 1538–1550, 1991.
44. E. Krogager, Aspects of polarimetric radar imaging, Doctoral Thesis, Technical University of Denmark, May 1993.
45. E. Luneburg, "Principles in radar polarimetry: The consimilarity transformation of radar polarimetry versus the similarity transformations in optical polarimetry," *IEICE Trans. Electron.*, vol. E-78C, no. 10, pp. 1339–1345, 1995.
46. A. Rosenqvist, M. Shimada, N. Ito, and M. Watanabe, "ALOS-PALSAR: A pathfinder mission for global-scale monitoring of the environment," *IEEE Trans. Geosci. Remote Sens.*, vol. 45, no. 11, pp. 3307–3316, 2007.
47. W. Pitz and D. Miller, "The TerraSAR-X satellite," *IEEE Trans. Geosci. Remote Sens.*, vol. 48, no. 2, pp. 615–622, 2010.
48. T. Kobayashi, T. Umehara, M. Satake, A. Nadai, S. Uratsuka, T. Manaba, H. Masuko, M. Shimada, H. Shinohara, H. Tozuka, and M. Miyawaki, "Airborne dual-frequency polarimetric and interferometric SAR," *IEICE Trans. Commun.*, vol. E83-B, no. 9, pp. 1945–1954, 2000.
49. Y. Yamaguchi, G. Singh, and H. Yamada, "On the model-based scattering power decomposition of fully polarimetric SAR data," *Trans. IEICE*, vol. J101-B, no. 9, pp.638–649, 2018.
50. http://www.jpl.nasa.gov/, https://uavsar.jpl.nasa.gov/
51. http://www.jaxa.jp/
52. http://www.restec.or.jp/
53. http://www.eorc.jaxa.jp/ALOS/about/palsar.htm

54. http://www.jaxa.jp/projects/sat/alos2/index_e.html
55. http://www.asc-csa.gc.ca/eng/satellites/radarsat2/
56. http://www.infoterra.de/terrasar-x.html, http://www.dlr.de/tsx/start_en.htm
57. http://www.dlr.de/eo/en/desktopdefault.aspx/tabid-5727/10086_read-21046/
58. http://www.eorc.jaxa.jp/ALOS/Pi-SAR-L2/index.html
59. http://www.isro.org/satellites/risat-1.aspx
60. http://www.wave.ie.niigata-u.ac.jp
61. https://step.esa.int/main/toolboxes/polsarpro-v6-0-biomass-edition-toolbox/

2 Wave Polarimetry

Polarization refers to the trace of the extremity of an electric field vector as a function of time in a fixed space as seen back from the propagation direction as shown in Figure 2.1. If the extremity of the electric field vector traces linear, it is called a linearly polarized wave. If the trace is circular, the wave is circularly polarized. The shape of the trace can be linear, circular, elliptical, or even oriented elliptical. In addition, the clockwise or anticlockwise direction of the rotation enriches the directional polarization pattern. Each one of them represents the polarization state of the wave. It should be noted that polarization is not a field vector. There are several ways to represent the polarization state of the wave. In this chapter, we deal with wave polarization. Starting from Maxwell's equation, we assess the fundamentals of the plane electromagnetic wave and how to represent polarization states using vector notation as well as geometrical parameters.

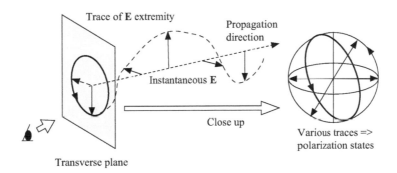

FIGURE 2.1 Trace of the extremity of time-varying electric field. (From Yamaguchi, Y., *Radar Polarimetry from Basics to Applications: Radar Remote Sensing Using Polarimetric Information [in Japanese]*, IEICE, 2007.)

2.1 PLANE WAVE

The electromagnetic wave radiated from the source (antenna or scattering point) spreads spherically in a three-dimensional space as shown in Figure 2.2. The spherical equiphase surface expands according to range *r*. If the observation point *P* is located at *r* far from the source, the wave front surface can be approximated by a tangential plane. We assume *r* is much larger than the wavelength λ. The wave near *P* can be considered as a plane wave, indicating that the equiphase surface is plane. Therefore, wherever $r \gg \lambda$ is satisfied, we may regard the wave as a plane wave.

Since the plane wave plays the essential role for presentation of the polarization state of electromagnetic waves and can be derived from Maxwell's equations, we start with Maxwell's equation.

2.1.1 Maxwell's Equation and Wave Equation

Position vector can be written as $\mathbf{r} = x\,\mathbf{a}_x + y\,\mathbf{a}_y + z\,\mathbf{a}_z$, where $\mathbf{a}_x, \mathbf{a}_y, \mathbf{a}_z$ are unit vectors in the *x*, *y*, and *z* directions, respectively. The electric field vector $\mathbf{E}(\mathbf{r}, t)$ and the magnetic field vector $\mathbf{H}(\mathbf{r}, t)$, as functions of position $\mathbf{r}$ and time *t*, satisfy the following Maxwell's equation,

$$\nabla \times \mathbf{E}(\mathbf{r},t) = -\frac{\partial}{\partial t}\mathbf{B}(\mathbf{r},t) \tag{2.1.1}$$

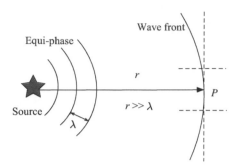

FIGURE 2.2 Plane wave approximation in the far field. (From Yamaguchi, Y., *Radar Polarimetry from Basics to Applications: Radar Remote Sensing Using Polarimetric Information [in Japanese]*, IEICE, 2007.)

$$\nabla \times \mathbf{H}(\mathbf{r},t) = \frac{\partial}{\partial t}\mathbf{D}(\mathbf{r},t) + \mathbf{J}(\mathbf{r},t) \tag{2.1.2}$$

$$\nabla \cdot \mathbf{D}(\mathbf{r},t) = \rho(\mathbf{r},t) \tag{2.1.3}$$

$$\nabla \cdot \mathbf{B}(\mathbf{r},t) = 0 \tag{2.1.4}$$

where, $\mathbf{D}$ is the electric flux density, $\mathbf{B}$ is the magnetic flux density, $\mathbf{J}$ is the electric current density, and ρ is the electric charge density. By taking the divergence of equation (2.1.2), the conservation law of charge can be obtained.

$$\nabla \cdot \mathbf{J}(\mathbf{r},t) + \frac{\partial}{\partial t}\rho(\mathbf{r},t) = 0 \tag{2.1.5}$$

In homogeneous media, the macroscopic electrical property can be expressed by the dielectric permittivity ε, the magnetic permeability μ, and the electric conductivity σ.

$$\mathbf{D} = \varepsilon\mathbf{E} \tag{2.1.6}$$

$$\mathbf{B} = \mu\mathbf{H} \tag{2.1.7}$$

$$\mathbf{J} = \sigma\mathbf{E} \tag{2.1.8}$$

where the medium is also assumed to be isotropic. Equations (2.1.6) through (2.1.8) are called as constitutive relations. If the medium is anisotropic such as active ionosphere, ε becomes the "tensor," which induces the Faraday rotation.

The electric current density $\mathbf{J}$ is expressed by the sum of the conductive current $\mathbf{J}_c$ and the source current $\mathbf{J}_s$.

$$\mathbf{J} = \mathbf{J}_c + \mathbf{J}_s = \sigma\mathbf{E} + \mathbf{J}_s \tag{2.1.9}$$

Introducing the vector operator identity $\nabla \times \nabla \times \mathbf{A} = \nabla(\nabla \cdot \mathbf{A}) - \nabla^2\mathbf{A}$, and operating the rotation operator on Equations (2.1.1) and (2.1.2), we obtain the vector wave equations for $\mathbf{E}$ and $\mathbf{H}$.

$$\nabla^2\mathbf{E} - \sigma\mu\frac{\partial \mathbf{E}}{\partial t} - \varepsilon\mu\frac{\partial^2 \mathbf{E}}{\partial t^2} = \mu\frac{\partial \mathbf{J}_s}{\partial t} + \frac{\nabla\rho}{\varepsilon} \tag{2.1.10}$$

$$\nabla^2 \mathbf{H} - \sigma\mu\frac{\partial \mathbf{H}}{\partial t} - \varepsilon\mu\frac{\partial^2 \mathbf{H}}{\partial t^2} = -\nabla \times \mathbf{J_s} \tag{2.1.11}$$

Equations (2.1.10) and (2.1.11) are the general forms of wave equation. It would be desirable to determine a time-dependent polarization state presentation satisfying equation (2.1.10) and (2.1.11); however, no simple useful presentation was found to exist. Instead, we will treat plane wave propagation in a source-free medium with time-harmonic oscillation fields. This will allow a frequency domain treatment of the wave equation for which a unique monochromatic presentation of the polarization state exists.

2.1.2 The Vector Wave Equation and Its Solution Using Phasor Representation

For a source-free medium ($\rho = 0, \mathbf{J_s} = 0$), the right-hand side of (2.1.10) becomes zero.

$$\nabla^2 \mathbf{E} - \sigma\mu\frac{\partial \mathbf{E}}{\partial t} - \varepsilon\mu\frac{\partial^2 \mathbf{E}}{\partial t^2} = 0 \tag{2.1.12}$$

In the following section and throughout the book, we shall adopt the harmonic time phasor definition. All the field quantity is assumed to have time-harmonic oscillation $e^{j\omega t}$ and expressed as,

$$\mathbf{A}(\mathbf{r},t) = \mathbf{A}(\mathbf{r})e^{j\omega t} \tag{2.1.13}$$

where $\mathbf{A}(\mathbf{r})$ is the phasor representation of $\mathbf{A}(\mathbf{r},t)$. $\mathbf{A}(\mathbf{r})$ is a complex-valued vector as a function of position $\mathbf{r}$, and independent of time t.

Now, we clarify here the relation of $\mathbf{A}(\mathbf{r},t)$ and $\mathbf{A}(\mathbf{r})$. If we let instantaneous field vector as $\mathbf{A}(\mathbf{r},t)$, it can be decomposed into

$$\mathbf{A}(\mathbf{r},t) = \mathbf{a}_x A_x(x,y,z,t) + \mathbf{a}_y A_y(x,y,z,t) + \mathbf{a}_z A_z(x,y,z,t) \tag{2.1.14}$$

On the other hand, the measurable quantity should be real-valued, and it can be written as

$$\mathbf{a}_x A_{mx}\cos(\omega t + \theta_{mx}) + \mathbf{a}_y A_{my}\cos(\omega t + \theta_{my}) + \mathbf{a}_z A_{mz}\cos(\omega t + \theta_{mz}) \tag{2.1.15}$$

$\theta_{mx}, \theta_{my}, \theta_{mz}$ are phases of the x-, y-, z- component, respectively. The subscript m refers to "measured." The x-component leads to the following relation:

$$A_{mx}\cos(\omega t + \theta_{mx}) = \mathrm{Re}\left\{A_{mx}e^{j(\omega t + \theta_{mx})}\right\} = \mathrm{Re}\left\{A_{mx}e^{j\theta_{mx}}e^{j\omega t}\right\}$$

$$= \mathrm{Re}\left\{(A_{mx}\cos\theta_{mx} + jA_{mx}\sin\theta_{mx})\right\}e^{j\omega t} = \mathrm{Re}\left\{(A_{rx} + jA_{ix})e^{j\omega t}\right\}$$

$$= \mathrm{Re}\left\{\dot{A}_x e^{j\omega t}\right\} \tag{2.1.16}$$

$\mathrm{Re}\{\bullet\}$ implies to take the real part of $\bullet$ Therefore, the measured value of the x-component can be expressed by the product of $\dot{A}_x$ and $e^{j\omega t}$. Similar expressions can be applied to the y- and z-components. Therefore, the vector $\mathbf{A}$ can be related to the following equation:

$$\mathbf{A}(\mathbf{r},t) = \mathrm{Re}\left\{\mathbf{A}(\mathbf{r})e^{j\omega t}\right\} \tag{2.1.17}$$

where

$$\mathbf{A}(\mathbf{r}) = \mathbf{a}_x\dot{A}_x + \mathbf{a}_y\dot{A}_y + \mathbf{a}_z\dot{A}_z \tag{2.1.18}$$

$$\dot{A}_x = A_{rx} + jA_{ix}, \quad \dot{A}_y = A_{ry} + jA_{iy}, \quad \dot{A}_z = A_{rz} + jA_{iz} : \text{complex scalar}$$

If we use phasor notation $\mathbf{A(r)}$ (2.1.13) in Maxwell's equation, the equation becomes simple, that is, independent of time. The field vector becomes a function of position $\mathbf{r}$ only. By assuming time harmonics $e^{j\omega t}$, the time derivative becomes $j\omega$. After calculating simple algebra and finding solution $\mathbf{A(r)}$, the actual field quantity can be obtained from $\mathbf{A(r},t) = \mathrm{Re}\{\mathbf{A(r)}e^{j\omega t}\}$. These are the advantages of using the phasor notation.

Now, the vector wave Equations (2.1.10) and (2.1.11) in the source-free space becomes in the phasor notation as

$$\nabla^2\mathbf{E(r)} + k^2\mathbf{E(r)} = 0 \qquad (2.1.19)$$

$$\nabla^2\mathbf{H(r)} + k^2\mathbf{H(r)} = 0 \qquad (2.1.20)$$

where k is called the "wave number" and is defined as

$$k^2 = \omega^2\varepsilon\mu - j\omega\mu\sigma \qquad (2.1.21)$$

The above equation is called the **wave equation** or **Helmholtz equation**.

2.1.2.1   Separation of Variables

Each scalar component of Equations (2.1.19) and (2.1.20) should satisfy the wave equation. For example, the equation for E_x becomes

$$\nabla^2 E_x + k^2 E_x = \frac{\partial^2 E_x}{\partial x^2} + \frac{\partial^2 E_x}{\partial y^2} + \frac{\partial^2 E_x}{\partial z^2} + k^2 E_x = 0 \qquad (2.1.22)$$

Now we try to solve this second-order partial differential equation. It is anticipated that the solution of this equation is a function of x, y, and z. Assuming the solution is in the form of $E_x = X(x)Y(y)Z(z)$, substitute it into (2.1.22) and divide by $X(x)Y(y)Z(z)$. Then the next equation comes out.

$$\frac{1}{X(x)}\frac{\partial^2 X(x)}{\partial x^2} + \frac{1}{Y(y)}\frac{\partial^2 Y(y)}{\partial y^2} + \frac{1}{Z(z)}\frac{\partial^2 Z(z)}{\partial z^2} + k^2 = 0$$

From this equation, we notice that each term should be independent of each other and must be a constant. Therefore, we put each term as

$$\frac{1}{X(x)}\frac{\partial^2 X(x)}{\partial x^2} = -k_x^2, \quad \frac{1}{Y(y)}\frac{\partial^2 Y(y)}{\partial y^2} = -k_y^2, \quad \frac{1}{Z(z)}\frac{\partial^2 Z(z)}{\partial z^2} = -k_z^2 \qquad (2.1.23)$$

The constants k_x, k_y, k_z must satisfy,

$$k^2 = k_x^2 + k_y^2 + k_z^2 \qquad (2.1.24)$$

For each variable, we have the second-order differential equation and its solution.

$$\frac{d^2 X}{dx^2} = -k_x^2 X \qquad \therefore X(x) = A_0 e^{-jk_x x} + A_1 e^{jk_x x} \qquad (2.1.25)$$

Therefore, $E_x = X(x)Y(y)Z(z)$ can be multiplied as

$$E_x = \left(A_0 e^{-jk_x x} + A_1 e^{jk_x x}\right)\left(B_0 e^{-jk_y y} + B_1 e^{jk_y y}\right)\left(C_0 e^{-jk_z z} + C_1 e^{jk_z z}\right) \qquad (2.1.26)$$

$A_0, ..., C_1$: amplitude coefficients

Since similar solutions can be obtained for E_y and E_z, the vector form $\mathbf{E}(\mathbf{r})$ can be written as

$$\mathbf{E}(\mathbf{r}) = \mathbf{E}_0 \exp(-j\mathbf{k}\cdot\mathbf{r}) + \mathbf{E}_1 \exp(+j\mathbf{k}\cdot\mathbf{r}) \qquad (2.1.27a)$$

$$\mathbf{H}(\mathbf{r}) = \mathbf{H}_0 \exp(-j\mathbf{k}\cdot\mathbf{r}) + \mathbf{H}_1 \exp(+j\mathbf{k}\cdot\mathbf{r}) \qquad (2.1.27b)$$

where

$$\mathbf{k} = k_x\mathbf{a}_x + k_y\mathbf{a}_y + k_z\mathbf{a}_z \qquad (2.1.28)$$

$$\mathbf{r} = x\mathbf{a}_x + y\mathbf{a}_y + z\mathbf{a}_z \qquad (2.1.29)$$

$$\mathbf{E}_0, \mathbf{E}_1, \mathbf{H}_0, \mathbf{H}_1 : \text{Amplitude vector}$$

Finally, the electric field vector $\mathbf{E}(\mathbf{r},t)$ as a function of space position and time can be obtained by the real part of the phasor multiplied by the time factor $e^{j\omega t}$,

$$\mathbf{E}(\mathbf{r},t) = \text{Re}\{\mathbf{E}(\mathbf{r})e^{j\omega t}\} = \mathbf{E}_+(\mathbf{r},t) + \mathbf{E}_-(\mathbf{r},t) \qquad (2.1.30a)$$

$$\mathbf{E}_+(\mathbf{r},t) = \text{Re}\{\mathbf{E}_0 \exp[j(\omega t - \mathbf{k}\cdot\mathbf{r})]\} \qquad (2.1.30b)$$

$$\mathbf{E}_-(\mathbf{r},t) = \text{Re}\{\mathbf{E}_1 \exp[j(\omega t + \mathbf{k}\cdot\mathbf{r})]\} \qquad (2.1.30c)$$

2.1.2.2 Physical Interpretation of the Solution

Now, let's check for a moment how the mathematical solution is related to physical phenomena. We know that the term $\omega t - \mathbf{k}\cdot\mathbf{r}$ in $\exp[j(\omega t - \mathbf{k}\cdot\mathbf{r})]$ is a phase. For simplicity, we assume $|\mathbf{E}_0| = 1$ and the wave is propagating toward the r direction so that $\mathbf{k}\cdot\mathbf{r} = kr$, and then (2.1.30b) becomes just a cosine function, $\cos(\omega t - kr)$.

Figure 2.3 shows how $\cos(\omega t - kr)$ changes with time t. When $t = 0$, it starts from 1. As time goes, the shape moves toward the right direction. We focus on a black dot ●, which has a phase $\theta_1 = \omega t - kr$. If black dot ● does not change its position with respect to time just like wind surfing, the next equation holds,

$$\frac{d\theta_1}{dt} = 0 = \omega - k\frac{dr}{dt} \qquad \therefore \frac{dr}{dt} = \frac{\omega}{k} = v\left[\frac{m}{s}\right]$$

Since $\frac{dr}{dt}$ represents velocity, the phase ● moves toward the positive r direction with the speed of $v = \frac{\omega}{k}$. Therefore, we can consider:

$\exp[j(\omega t - \mathbf{k}\cdot\mathbf{r})]$ represents a wave propagating toward $+r$ direction,
$\exp[j(\omega t + \mathbf{k}\cdot\mathbf{r})]$ represents a wave propagating toward $-r$ direction.

It is very important to pay attention to the sign in front of $\mathbf{k}\cdot\mathbf{r}$ because it plays the key role for polarization transformation or polarimetric analysis. Historically, in the field of optics, the expression

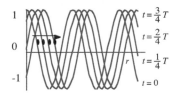

FIGURE 2.3   Constant phase with respect to time. (From Yamaguchi, Y., *Radar Polarimetry from Basics to Applications: Radar Remote Sensing Using Polarimetric Information [in Japanese]*, IEICE, 2007.)

exp $(i\mathbf{k} \cdot \mathbf{r})$ has been used from the outset. So, the wave expression $\exp[i(\mathbf{k} \cdot \mathbf{r} - \omega t)]$ represents wave propagating in the $+r$ direction [1]. On the other hand, $\exp(j\omega t)$ has been used in engineering field, and $\exp[j(\omega t - \mathbf{k} \cdot \mathbf{r})]$ represents wave propagating in $+r$ direction. Therefore, the sign in front of $\mathbf{k} \cdot \mathbf{r}$ is opposite between optics and engineering. In order to avoid confusion, it is convenient to assume that $i = -j$, or $j = -i$ for both academic communities.

The phase of the electric field vector (2.1.27) is constant if $\mathbf{k} \cdot \mathbf{r} = k_x x + k_y y + k_z z = $ const. Since the equation $\mathbf{k} \cdot \mathbf{r} = $ const. represents a two-dimensional plane, the phase is constant on the plane as shown in Figure 2.4. If the equiphase is a plane, it is called a plane wave. Therefore (2.1.30) represents a plane wave.

In Figure 2.4, $\mathbf{k}$ is taken in the propagation direction. The position vectors $\mathbf{r}_0$, $\mathbf{r}_1$, $\mathbf{r}_2$, which satisfies $\mathbf{k} \cdot \mathbf{r}_1 = \mathbf{k} \cdot \mathbf{r}_2 = \mathbf{k} \cdot \mathbf{r}_0$, spans a plane orthogonal to $\mathbf{k}$. This plane is called a transverse plane. Since $\mathbf{r}_0$ is taken in the same direction as $\mathbf{k}$, the phase change is the largest in this direction.

In lossless and isotropic media, the wave number becomes,

$$k = \omega\sqrt{\varepsilon\mu} = \frac{\omega}{v} = \frac{2\pi}{\lambda} \tag{2.1.31}$$

where v is the velocity of electromagnetic wave in the medium, and λ is the wavelength. The naming of wave number comes from $\frac{2\pi}{\lambda}$, that is, how many λ exists in the range 2π. If the medium has relative dielectric constant ε_r, the wave number becomes

$$k = k_0\sqrt{\varepsilon_r}, \quad k_0 = \omega\sqrt{\varepsilon_0\mu_0} = \frac{2\pi}{\lambda_0} \tag{2.1.32}$$

The subscript 0 is added for the values in the free-space,

$$\text{phase velocity: } v = \frac{\omega}{k} = \frac{c_0}{\sqrt{\varepsilon_r}} = \frac{3\times10^8}{\sqrt{\varepsilon_r}} \tag{2.1.33}$$

$$\text{wavelength: } \lambda = \frac{\lambda_0}{\sqrt{\varepsilon_r}} \tag{2.1.34}$$

In the dielectric medium, the wavelength becomes shorter, and the velocity becomes slow because of $\varepsilon_r > 1$.

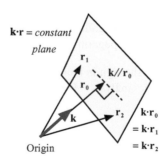

FIGURE 2.4 Constant phase plane. (From Yamaguchi, Y., *Radar Polarimetry from Basics to Applications: Radar Remote Sensing Using Polarimetric Information [in Japanese]*, IEICE, 2007.)

2.1.3 TRANSVERSE ELECTROMAGNETIC WAVE

Here, we examine the relation between the electric field $\mathbf{E}(\mathbf{r})$ and the magnetic field $\mathbf{H}(\mathbf{r})$ in the free-space. The constitutive parameters become $\varepsilon = \varepsilon_0$, $\mu = \mu_0$, and $\sigma = 0$. Maxwell's equation (2.1.1) becomes in the phasor notation

$$\nabla \times \mathbf{E}(\mathbf{r}) = -j\omega\mu_0\mathbf{H}(\mathbf{r}) \tag{2.1.35}$$

Substituting the electric field $\mathbf{E}_0(\mathbf{r})$ (2.1.27) propagating to the $+\mathbf{r}$ direction into (2.1.35), we can obtain the following:

$$\nabla \times \mathbf{E}_0(\mathbf{r}) = \begin{vmatrix} \mathbf{a}_x & \mathbf{a}_y & \mathbf{a}_z \\ \dfrac{\partial}{\partial x} & \dfrac{\partial}{\partial y} & \dfrac{\partial}{\partial z} \\ E_x & E_y & E_z \end{vmatrix} = \begin{vmatrix} \mathbf{a}_x & \mathbf{a}_y & \mathbf{a}_z \\ -jk_x & -jk_y & -jk_z \\ E_x & E_y & E_z \end{vmatrix} = -j\,\mathbf{k} \times \mathbf{E}_0(\mathbf{r}) = -j\omega\mu_0\mathbf{H}_0(\mathbf{r})$$

Therefore, equation (2.1.1) becomes

$$\mathbf{k} \times \mathbf{E}_0 = \omega\mu_0\mathbf{H}_0 \tag{2.1.36}$$

If we normalize $\mathbf{k}$ as $\hat{\mathbf{k}} = \dfrac{\mathbf{k}}{|\mathbf{k}|} = \dfrac{\mathbf{k}}{\omega\sqrt{\varepsilon_0\mu_0}}$: unit vector, equation (2.1.36) can be written as

$$\hat{\mathbf{k}} \times \mathbf{E}_0 = \frac{\omega\mu_0}{\omega\sqrt{\varepsilon_0\mu_0}}\mathbf{H}_0 = \eta_0\mathbf{H}_0 \tag{2.1.37}$$

where η_0 is the intrinsic impedance in the free-space,

$$\eta_0 = \sqrt{\frac{\mu_0}{\varepsilon_0}} = 120\pi\left[\Omega\right] \tag{2.1.38}$$

Similarly from equation (2.1.2), we obtain,

$$\eta_0\mathbf{H}_0 \times \hat{\mathbf{k}} = \mathbf{E}_0 \tag{2.1.39}$$

$$\left|\mathbf{E}_0\right| = \left|\eta_0\mathbf{H}_0\right| \tag{2.1.40}$$

In addition, $\nabla \cdot \mathbf{D} = 0$ and $\nabla \cdot \mathbf{B} = 0$ derive the following relations:

$$\hat{\mathbf{k}} \cdot \mathbf{E}_0 = 0 \tag{2.1.41}$$

$$\hat{\mathbf{k}} \cdot \mathbf{H}_0 = 0 \tag{2.1.42}$$

The vector relations among Equations (2.1.37) through (2.1.42) can be visualized as shown in Figure 2.5. $\mathbf{E}_0$ and $\eta_0\mathbf{H}_0$ have the same magnitude and are orthogonal to each other. They are also orthogonal to the propagation direction $\hat{\mathbf{k}}$. Since $\mathbf{E}_0$ and $\mathbf{H}_0$ reside in the same transverse plane perpendicular to the propagation direction, the wave is called the transverse electromagnetic (TEM) wave.

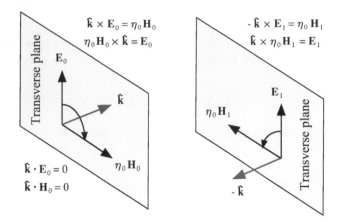

FIGURE 2.5   Vector relations of TEM wave propagation. (From Yamaguchi, Y., *Radar Polarimetry from Basics to Applications: Radar Remote Sensing Using Polarimetric Information [in Japanese]*, IEICE, 2007.)

For the electric field $\mathbf{E}_1$ (2.1.27) propagating in the $-\mathbf{r}(-\hat{\mathbf{k}})$ direction, we have similar vector relations,

$$-\hat{\mathbf{k}} \times \mathbf{E}_1 = \eta_0 \mathbf{H}_1, \quad \hat{\mathbf{k}} \times \eta_0 \mathbf{H}_1 = \mathbf{E}_1, \quad \hat{\mathbf{k}} \cdot \mathbf{E}_1 = 0, \quad \hat{\mathbf{k}} \cdot \mathbf{H}_1 = 0 \qquad (2.1.43)$$

In this case, the vector relations of equation (2.1.43) become the right one shown in Figure 2.5. $\mathbf{E}_1$ and $\eta_0 \mathbf{H}_1$ are in the same magnitude and are orthogonal to each other and perpendicular to the $-\hat{\mathbf{k}}$ direction. It is easy to understand these relations in a graphical way rather than in mathematical formulations.

To sum up the properties of TEM waves in both propagation directions, $\mathbf{E}$ and $\mathbf{H}$ are orthogonal to each other. The propagation direction is $\mathbf{E} \times \mathbf{H}$.

The propagation direction $\mathbf{k}$ can be taken arbitrarily. If we choose the propagation direction as the z-axis in the rectangular coordinate, $\mathbf{k}$ becomes $\mathbf{k} = k\mathbf{a}_z$. In this case, $\mathbf{E}$ and $\mathbf{H}$ are laid in the x-y plane as shown in Figure 2.6, and the x-y plane becomes transverse plane for the wave.

$$\mathbf{E}_0 \times \eta_0 \mathbf{H}_0 = \mathbf{a}_z, \qquad \mathbf{E}_1 \times \eta_0 \mathbf{H}_1 = -\mathbf{a}_z \qquad (2.1.44)$$

Note that the polarization of electromagnetic waves is defined for the electric field vector $\mathbf{E}$ only. Since the magnetic field vector $\mathbf{H}$ is orthogonal to $\mathbf{E}$, no similar definition is required in order to avoid confusion.

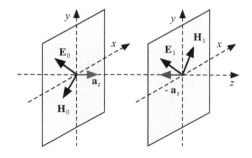

FIGURE 2.6   *E* and *H* in the transverse plane. (From Yamaguchi, Y., *Radar Polarimetry from Basics to Applications: Radar Remote Sensing Using Polarimetric Information [in Japanese]*, IEICE, 2007.)

2.1.4   TEM Wave Power

Next, we consider the power of a plane wave. According to Poynting theorem, the power is represented by the product of the instantaneous electric field **E** and the instantaneous magnetic field **H**, which is denoted as the Poynting vector **S** as

$$\mathbf{S}(\mathbf{r},t) = \mathbf{E}(\mathbf{r},t) \times \mathbf{H}(\mathbf{r},t) \tag{2.1.45}$$

S is a function of space **r** and time t, and therefore is not a phasor representation. Assuming the wave is propagating in the z-direction, and using Equations (2.1.30) and (2.1.45), we obtain the following form:

$$\mathbf{S}(z,t) = \mathbf{a}_z \frac{|\mathbf{E}_0|^2}{\eta_0}\cos^2(\omega t - kz) - \mathbf{a}_z \frac{|\mathbf{E}_1|^2}{\eta_0}\cos^2(\omega t + kz) \tag{2.1.46}$$

The time averaging of the instantaneous power through a constant z-plane yields the net flow of power,

$$\langle \mathbf{S}(z,t) \rangle = \frac{1}{T}\int_0^T \mathbf{S}(z,t)\,dt = \mathbf{a}_z \frac{|\mathbf{E}_0|^2}{2\eta_0} - \mathbf{a}_z \frac{|\mathbf{E}_1|^2}{2\eta_0} \tag{2.1.47}$$

From this equation, we can notice that the power is represented by $\frac{|\mathbf{E}|^2}{2\eta_0}$ and is carried out by $(\mathbf{E}_0,\mathbf{H}_0)$ and $(\mathbf{E}_1,\mathbf{H}_1)$ independently.

If we use phasor representation for power expression, it is convenient to employ a complex Poynting vector **P** defined as

$$\mathbf{P}(\mathbf{r}) = \mathbf{E}(\mathbf{r}) \times \mathbf{H}^*(\mathbf{r}) \qquad \text{or} \qquad \mathbf{P} = \mathbf{E} \times \mathbf{H}^* \tag{2.1.48}$$

where the superscript * denotes complex conjugation. Since $\mathbf{E}(\mathbf{r})$ and $\mathbf{H}(\mathbf{r})$ are complex valued, we can write them as

$$\mathbf{E}(\mathbf{r}) = \mathbf{E}_r + j\mathbf{E}_i, \mathbf{H}(\mathbf{r}) = \mathbf{H}_r + j\mathbf{H}_i \tag{2.1.49}$$

P can be expressed as

$$\mathbf{P} = \mathbf{E}_r \times \mathbf{H}_r + \mathbf{E}_i \times \mathbf{H}_i + j(\mathbf{E}_i \times \mathbf{H}_r - \mathbf{E}_r \times \mathbf{H}_i) \tag{2.1.50}$$

Since the instantaneous field vectors are given by

$$\mathbf{E}(\mathbf{r},t) = \mathrm{Re}\{\mathbf{E}(\mathbf{r})e^{j\omega t}\} = \mathbf{E}_r \cos \omega t - \mathbf{E}_i \sin \omega t$$

$$\mathbf{H}(\mathbf{r},t) = \mathrm{Re}\{\mathbf{H}(\mathbf{r})e^{j\omega t}\} = \mathbf{H}_r \cos \omega t - \mathbf{H}_i \sin \omega t \tag{2.1.51}$$

the instantaneous Poynting vector **S** can be expressed as

$$\mathbf{S}(\mathbf{r},t) = \mathbf{E}_r \times \mathbf{H}_r \cos^2\omega t + \mathbf{E}_i \times \mathbf{H}_i \sin^2\omega t - (\mathbf{E}_i \times \mathbf{H}_r + \mathbf{E}_r \times \mathbf{H}_i)\frac{\sin 2\omega t}{2} \tag{2.1.52}$$

Although the expression of **P** and **S** is different, the averaging over time yields the following relation:

$$\langle \mathbf{S}(\mathbf{r},t) \rangle = \frac{1}{T}\int_0^T \mathbf{S}(\mathbf{r},t)\,dt = \frac{1}{2}\left[\mathbf{E}_r \times \mathbf{H}_r + \mathbf{E}_i \times \mathbf{H}_i\right] = \frac{1}{2}\mathrm{Re}\left\{\mathbf{E}\times\mathbf{H}^*\right\} = \frac{1}{2}\,\mathrm{Re}\left\{\mathbf{P}\right\} \quad (2.1.53)$$

This equation indicates that the time averaging of **S** is equal to half of the real part of **P**. Therefore, without integral calculation, we can obtain the electromagnetic power (net flow) by the complex Poynting vector **P** using phasor notation.

For a plane wave propagating in the z-direction, the time average power flow is given by

$$\frac{1}{2}\mathrm{Re}\left\{\mathbf{P}\right\} = \frac{|\mathbf{E}|^2}{2\eta_0}\mathbf{a}_z \quad (2.1.54)$$

and is proportional to $|\mathbf{E}|^2$. If **E** has an x- and y-component, the power can be decomposed into the sum,

$$|\mathbf{E}|^2 = |E_x|^2 + |E_y|^2 \quad (2.1.55)$$

It is a simple vector relation; however, it is a very important from the radar point of view. Power is the most fundamental radar parameter for which we will see in the following chapter.

2.2 REPRESENTATION OF POLARIZATION

Polarization is the trace of the extremity of an electric field vector as a function of time at a fixed space as seen back from the propagation direction [2] as shown in Figure 2.1. The shape of the trace becomes elliptical in general form. Elliptical includes thin, round, and even oriented one. In the extreme case, the trance becomes a line or circle. Each shape corresponds to one polarization state. There are several ways to represent elliptical polarization, such as geometrical parameters by ellipticity angle, tilt angle, size, or polarization ratio, relative phase, Stokes parameters, and Poincaré sphere. In this section, we introduce some definitions for the representation of polarization and their relations.

2.2.1 MATHEMATICAL EXPRESSION

The electric field vector of a plane wave can be decomposed into two orthogonal components in the transverse plane. If we assume the wave is propagating in the $+z$ direction, there is no z-component. Then the instantaneous electric field vector can be written by

$$\varepsilon(z,t) = \begin{bmatrix} \varepsilon_x(z,t) \\ \varepsilon_y(z,t) \end{bmatrix} = \begin{bmatrix} |E_x|\cos(\omega t - kz + \phi_x) \\ |E_y|\cos(\omega t - kz + \phi_y) \end{bmatrix} \quad (2.2.1)$$

where, $|E_x|$ and $|E_y|$ are amplitudes, and ϕ_x and ϕ_y are the absolute phases. By the definition of polarization, we observe the time-varying field at $z = 0$ plane. Equation (2.2.1) reduces to $\varepsilon(0,t) = \varepsilon(t)$,

$$\varepsilon(t) = \begin{bmatrix} \varepsilon_\tau(t) \\ \varepsilon_y(t) \end{bmatrix} = \begin{bmatrix} |E_x|\cos(\omega t + \phi_x) \\ |E_y|\cos(\omega t + \phi_y) \end{bmatrix} \quad (2.2.2)$$

Expansion of the cosine function using the relative phase difference

$$\delta = \phi_y - \phi_x \quad (2.2.3)$$

leads to the next intermediate relation.

$$\left|E_y\right|\varepsilon_x(t)\cos\phi_y - \left|E_x\right|\varepsilon_y(t)\cos\phi_x = \left|E_x\right|\left|E_y\right|\sin\delta\,\sin\omega t$$

$$\left|E_y\right|\varepsilon_x(t)\sin\phi_y - \left|E_x\right|\varepsilon_y(t)\sin\phi_x = \left|E_x\right|\left|E_y\right|\sin\delta\,\cos\omega t \qquad (2.2.4)$$

By deleting time factor ωt, we reach the following equation.

$$\frac{\varepsilon_x^2(t)}{\left|E_x\right|^2} - \frac{2\varepsilon_x(t)\varepsilon_y(t)}{\left|E_x\right|\left|E_y\right|}\cos\delta + \frac{\varepsilon_y^2(t)}{\left|E_y\right|^2} = \sin^2\delta \qquad (2.2.5)$$

This is an equation of oriented ellipse. In general, the trace is an oriented (tilted) ellipse as shown in Figure 2.7.

If the relative phase difference δ has a specific value, the equation becomes familiar ones. For simplicity, we put $a=\left|E_x\right|, b=\left|E_y\right|, x=\varepsilon_x(t), y=\varepsilon_y(t)$, then the equation becomes

$$\frac{x^2}{a^2} - \frac{2xy}{ab}\cos\delta + \frac{y^2}{b^2} = \sin^2\delta \qquad (2.2.6)$$

If $\delta = \pm\dfrac{\pi}{2} \quad \Rightarrow \dfrac{x^2}{a^2} + \dfrac{y^2}{b^2} = 1$: normal ellipse

and if $a = b \quad \Rightarrow x^2 + y^2 = a^2$: circle

If $\delta = 0, \pi \quad \Rightarrow \dfrac{X^2}{a^2} + \dfrac{2xy}{ab} + \dfrac{y^2}{b^2} = \left(\dfrac{x}{a} \mp \dfrac{y}{b}\right)^2 = 0. \quad y = \pm\dfrac{b}{a}x$: straight line

The sense of rotation can be found by the time derivative of the phase. The phase angle ψ is taken as the angle spanned by the x-axis and $\varepsilon(t)$.

$$\tan\psi = \frac{\varepsilon_y(t)}{\varepsilon_x(t)} \qquad (2.2.7)$$

$$\frac{d\psi}{dt} = -\frac{\omega\left|E_x\right|\left|E_y\right|}{\left|\varepsilon_x(t)\right|^2}\sin\delta \qquad (2.2.8)$$

Using equation (2.2.8) and Figure 2.7, we can find

$$\text{For } 0 < \delta < \pi, \quad \frac{d\psi}{dt} < 0 : \text{Clockwise} \qquad (2.2.9a)$$

$$\text{For } -\pi < \delta < 0, \quad \frac{d\psi}{dt} > 0 : \text{Counterclockwise} \qquad (2.2.9b)$$

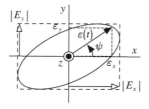

FIGURE 2.7   Instantaneous electric field.

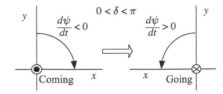

FIGURE 2.8 Rotation sense (left handed). (From Yamaguchi, Y., *Radar Polarimetry from Basics to Applications: Radar Remote Sensing Using Polarimetric Information [in Japanese]*, IEICE, 2007.)

The sense of rotation of a polarized wave is defined as seen from the backside. The above mathematical situation is opposite to the IEEE standard [3]. Figure 2.8 shows the case for $0 < \delta < \pi$. The left figure is for mathematical expression (2.2.9a) and looks like a right-handed rotation. However, the rotation sense is opposite in the right figure where the wave is seen from backside [3]. It is left-handed.

So, the sense of rotation can be checked by the relative phase $\delta = \phi_y - \phi_x$,

$$0 < \delta < \pi : \text{left-handed rotation sense}$$

$$-\pi < \delta < 0 : \text{right-handed rotation sense}$$

Figure 2.9 shows the general elliptical polarization using relative phase difference δ.

The general form of polarization can be written as $\begin{cases} \varepsilon_x(t) = |E_x| \cos(\omega t) \\ \varepsilon_y(t) = |E_y| \cos(\omega t + \delta) \end{cases}$

Specific polarization is characterized by the following:

If $|E_x| \neq 0, |E_y| = 0$, Horizontally polarized wave: $\begin{cases} \varepsilon_x(t) = |E_x| \cos(\omega t) \\ \varepsilon_y(t) = 0 \end{cases}$

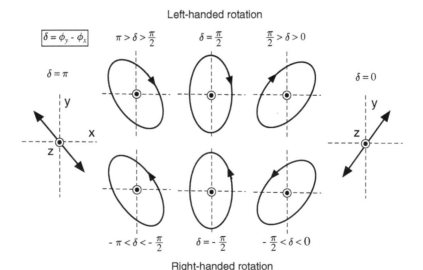

FIGURE 2.9 Elliptical polarization with a rotation sense. (From Yamaguchi, Y., *Radar Polarimetry from Basics to Applications: Radar Remote Sensing Using Polarimetric Information [in Japanese]*, IEICE, 2007.)

If $\left|E_x\right| = 0, \left|E_y\right| \neq 0$, Vertically polarized wave: $\begin{cases} \varepsilon_x(t) = 0 \\ \varepsilon_y(t) = \left|E_y\right| \cos(\omega t) \end{cases}$

For the case of $\left|E_x\right| \neq 0, \left|E_y\right| \neq 0$,

If $\delta = 0$, $\varepsilon_x(t)$ and $\varepsilon_y(t)$ are in-phase.

Linear polarization with positive angle orientation. $\begin{cases} \varepsilon_x(t) = \left|E_x\right| \cos \omega t \\ \varepsilon_y(t) = \left|E_y\right| \cos \omega t \end{cases}$

If $\delta = \pi$, $\varepsilon_x(t)$ and $\varepsilon_y(t)$ are out of phase.

Linear polarization with negative angle orientation. $\begin{cases} \varepsilon_x(t) = \left|E_x\right| \cos \omega t \\ \varepsilon_y(t) = -\left|E_y\right| \cos \omega t \end{cases}$

$0 < \delta < \pi$ indicates the left-handed elliptical polarization.

If $\delta = \dfrac{\pi}{2}$ and $\left|E_x\right| = \left|E_y\right|$, Left-handed circular polarization: $\begin{cases} \varepsilon_x(t) = \left|E_x\right| \cos \omega t \\ \varepsilon_y(t) = -\left|E_x\right| \sin \omega t \end{cases}$

$-\pi < \delta < 0$ corresponds to the right-handed elliptical polarization

If $\delta = -\dfrac{\pi}{2}$ and $\left|E_x\right| = \left|E_y\right|$, right-handed circular polarization: $\begin{cases} \varepsilon_x(t) = \left|E_x\right| \cos \omega t \\ \varepsilon_y(t) = \left|E_x\right| \sin \omega t \end{cases}$

Therefore, linear polarization and circular polarization are special cases of general elliptical polarization.

2.2.2 Representation by Geometrical Parameters

It is straightforward to use geometrical parameters (ε, τ, A) for representing a polarization ellipse. The geometrical parameters are ellipticity angle ε, tilt (orientation) angle τ, and size A. Polarization ellipse and the angles (ε, τ) are defined as shown in Figures 2.10 and 2.11.

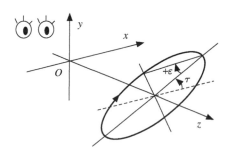

FIGURE 2.10 Coordinate and geometrical parameters (ε, τ, A). (From Yamaguchi, Y., *Radar Polarimetry from Basics to Applications: Radar Remote Sensing Using Polarimetric Information [in Japanese]*, IEICE, 2007.)

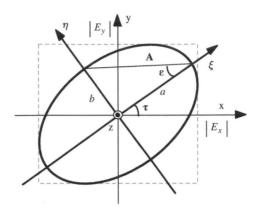

FIGURE 2.11 Geometrical parameters (ε, τ, A) of polarization ellipse. (From Yamaguchi, Y., *Radar Polarimetry from Basics to Applications: Radar Remote Sensing Using Polarimetric Information [in Japanese]*, IEICE, 2007.)

Ellipticity angle ε is defined as,

$$\varepsilon = \tan^{-1}\frac{b}{a}, \ \tan\varepsilon = \frac{b}{a} \quad \left(-\frac{\pi}{4} \le \varepsilon \le \frac{\pi}{4}\right) \tag{2.2.10}$$

where a is the major axis length of an ellipse, and b is the minor axis length. Ellipticity represents the degree of round shape and is reciprocal of the "axial ratio" in antenna engineering. If $a = b$, then $\varepsilon = \frac{\pi}{4}$, which corresponds to the complete circular shape. This becomes the circular polarization. If $b = 0$, then $\varepsilon = 0$, which represents the linear polarization. The sign of ε shows the sense of rotation, that is, $\varepsilon > 0$ for the left-handed rotation, and $\varepsilon < 0$ for the right-handed rotation.

Here, the symbol "ε" is used to denote the ellipticity angle. Since the first letter "e" is close to ε, we employed this notation referring to [4–8]. Other literatures use a different character to denote the ellipticity angle.

Tilt angle $\tau\left(-\frac{\pi}{2} < \tau < \frac{\pi}{2}\right)$ is spanned by the x-axis and the major axis of the ellipse as shown in Figure 2.6. We employed this symbol "τ" according to the first letter of tilt. Other literatures use different characters to denote tilt angle. It is also called the "orientation angle" [9–11].

The magnitude size of ellipse is given by

$$A = \sqrt{a^2 + b^2} \tag{2.2.11}$$

where A^2 corresponds to the power and has no direct relations to polarization. Therefore, the geometrical parameter mainly refers to ellipticity angle ε and tilt angle τ.

Using ellipticity and tilt angles, all polarization ellipses can be visualized as shown in Figure 2.12. This exhibition is easy to understand the polarization ellipse. If $\varepsilon = \pm\frac{\pi}{4}$, we have circular polarization for any value of τ. If $\varepsilon = 0$, we have linear polarizations with various orientations (tilt angles).

Now, let's check the relation between mathematical parameters and geometrical parameters.

New axes ξ and η are chosen as the major and minor axes. In this new $\xi - \eta$ coordinate system, the ellipse becomes without tilt. The electric field vector therefore can be written as

$$\begin{bmatrix} \varepsilon_\xi \\ \varepsilon_\eta \end{bmatrix} = \begin{bmatrix} a\,\cos\left(\omega t + \phi_\xi\right) \\ b\,\cos\left(\omega t + \phi_\eta\right) \end{bmatrix} \tag{2.2.12}$$

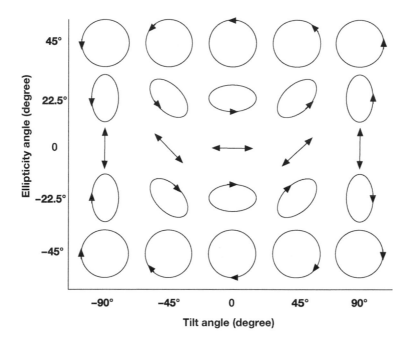

FIGURE 2.12 Polarization ellipse representation by geometrical parameters. (From Yamaguchi, Y., *Radar Polarimetry from Basics to Applications: Radar Remote Sensing Using Polarimetric Information [in Japanese]*, IEICE, 2007.)

Referring to equation (2.2.6), we have the relation $\delta = \phi_\eta - \phi_\xi = \frac{\pi}{2}$, so that equation (2.2.12) becomes

$$
\begin{bmatrix} \varepsilon_\xi \\ \varepsilon_\eta \end{bmatrix} = \begin{bmatrix} a \cos(\omega t + \phi_\xi) \\ -b \sin(\omega t + \phi_\xi) \end{bmatrix} \tag{2.2.13}
$$

where ϕ_ξ is the absolute phase, satisfying $a \cos \phi_\xi = \varepsilon_\varepsilon$.

Since the new axes $\xi - \eta$ can be obtained by rotation of the x-y axes by angle τ, the electric field can be written as

$$
\begin{bmatrix} \mathcal{E}_\xi \\ \varepsilon_\eta \end{bmatrix} = \begin{bmatrix} \cos\tau & \sin\tau \\ -\sin\tau & \cos\tau \end{bmatrix} \begin{bmatrix} \varepsilon_x \\ \varepsilon_y \end{bmatrix} \tag{2.2.14}
$$

Therefore, we obtain the next relation from equations (2.2.2), (2.2.13), and (2.2.14),

$$
\begin{bmatrix} a \cos(\omega t + \phi_\xi) \\ -b \sin(\omega t + \phi_\xi) \end{bmatrix} = \begin{bmatrix} \cos\tau & \sin\tau \\ -\sin\tau & \cos\tau \end{bmatrix} \begin{bmatrix} |E_x| \cos(\omega t + \phi_x) \\ |E_y| \cos(\omega t + \phi_y) \end{bmatrix} \tag{2.2.15}
$$

After arranging equation (2.2.15) using $\delta = \phi_y - \phi_x$, we obtain the following relations between mathematical expression and geometrical parameters.

$$
a^2 + b^2 = |E_x|^2 + |E_y|^2 \tag{2.2.16}
$$

$$
ab = |E_x||E_y| \sin \delta \tag{2.2.17}
$$

$$\tan 2\tau = \frac{2|E_x||E_y|}{|E_x|^2 - |E_y|^2} \cos \delta \qquad (2.2.18)$$

$$\sin 2\varepsilon = \frac{2|E_x||E_y|}{|E_x|^2 + |E_y|^2} \sin \delta \qquad (2.2.19)$$

In addition, equation (2.2.15) derives

$$\mathrm{Re}\left\{\begin{bmatrix} |E_x|e^{j\phi_x} \\ |E_y|e^{j\phi_y} \end{bmatrix}\right\} = \mathrm{Re}\left\{\begin{bmatrix} \cos\tau & -\sin\tau \\ \sin\tau & \cos\tau \end{bmatrix}\begin{bmatrix} a\,e^{j\phi_\xi} \\ -b\,e^{j\phi_\xi} \end{bmatrix}\right\} = A\begin{bmatrix} \cos\tau & -\sin\tau \\ \sin\tau & \cos\tau \end{bmatrix}\mathrm{Re}\left\{\begin{bmatrix} \cos\varepsilon \\ j\sin\varepsilon \end{bmatrix}e^{j\phi_\xi}\right\}$$

where the ellipticity angle ε is used, satisfying $\cos\varepsilon = \frac{a}{\sqrt{a^2+b^2}}$, $\sin\varepsilon = \frac{b}{\sqrt{a^2+b^2}}$.

From this equation, the electric field vector can be expressed using geometrical parameters as

$$\begin{bmatrix} |E_x|e^{j\phi_x} \\ |E_y|e^{j\phi_y} \end{bmatrix} = A\begin{bmatrix} \cos\tau & -\sin\tau \\ \sin\tau & \cos\tau \end{bmatrix}\begin{bmatrix} \cos\varepsilon \\ j\sin\varepsilon \end{bmatrix}e^{j\phi_\xi} \qquad (2.2.20)$$

We can assume that the vector **E** is obtained by the angular rotation by $-\tau$ of the Jones vector $\begin{bmatrix} \cos\varepsilon \\ j\sin\varepsilon \end{bmatrix}e^{j\phi_\xi}$. (See Figure 2.13).

In order to reduce unknown phases in equation (2.2.20), we modify it as

$$\begin{bmatrix} |E_x| \\ |E_y|e^{j\delta} \end{bmatrix} = A\begin{bmatrix} \cos\tau & -\sin\tau \\ \sin\tau & \cos\tau \end{bmatrix}\begin{bmatrix} \cos\varepsilon \\ j\sin\varepsilon \end{bmatrix}e^{j(\phi_\xi-\phi_x)} = A\begin{bmatrix} \cos\tau\cos\varepsilon & -j\sin\tau\sin\varepsilon \\ \sin\tau\cos\varepsilon & j\cos\tau\sin\varepsilon \end{bmatrix}e^{j\phi}$$

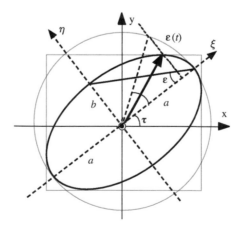

FIGURE 2.13 Electric field vector in the oriented elliptical coordinates. (From Yamaguchi, Y., *Radar Polarimetry from Basics to Applications: Radar Remote Sensing Using Polarimetric Information [in Japanese]*, IEICE, 2007.)

Since $|E_x|$ is real-valued, the phase should be zero so that

$$\mathrm{Arg}\left(\cos\tau\cos\varepsilon - j\sin\tau\sin\varepsilon\right) + \phi = 0$$

$$\therefore \phi = -\mathrm{Arg}\left(\cos\tau\cos\varepsilon - j\sin\tau\sin\varepsilon\right) = \tan^{-1}\left(\tan\tau\tan\varepsilon\right)$$

Therefore, a similar expression to (2.2.20) can be obtained, which relates to the geometrical parameters more precisely.

$$\begin{bmatrix}|E_x| \\ |E_y|e^{j\delta}\end{bmatrix} = A\begin{bmatrix}\cos\tau & -\sin\tau \\ \sin\tau & \cos\tau\end{bmatrix}\begin{bmatrix}\cos\varepsilon \\ j\sin\varepsilon\end{bmatrix}e^{j\tan^{-1}(\tan\tau\tan\varepsilon)} \tag{2.2.21}$$

2.2.3 JONES VECTOR REPRESENTATION

The electric field vector **E** in the *HV* polarization basis can be expressed as the two-dimensional column vector,

$$\mathbf{E}(HV) = \begin{bmatrix}E_H \\ E_V\end{bmatrix} = \begin{bmatrix}|E_H|e^{j\phi_H} \\ |E_V|e^{j\phi_V}\end{bmatrix} \tag{2.2.22}$$

This vector form is called the **Jones vector**. It should be noted in the Jones vector representation that the propagation direction is omitted.

The polarization ratio ρ in the *HV* polarization basis is defined as

$$\rho = \frac{E_V}{E_H} = \frac{|E_V|}{|E_H|}e^{j\delta} = |\rho|e^{j\delta} \tag{2.2.23}$$

where γ is the angle as shown in Figure 2.14.

$$|\rho| = \frac{|E_V|}{|E_H|} = \tan\gamma \tag{2.2.24}$$

The Jones vector of the electric field can be written as,

$$\mathbf{E}(HV) = \begin{bmatrix}E_H \\ E_V\end{bmatrix} = |E_H|e^{j\phi_H}\begin{bmatrix}1 \\ \rho\end{bmatrix} = |E_H|e^{j\phi_H}\frac{\sqrt{1+\dfrac{E_V E_V^*}{E_H E_H^*}}}{\sqrt{1+\dfrac{E_V E_V^*}{E_H E_H^*}}}\begin{bmatrix}1 \\ \rho\end{bmatrix} = \frac{|\mathbf{E}|e^{j\phi_H}}{\sqrt{1+\rho\rho^*}}\begin{bmatrix}1 \\ \rho\end{bmatrix} \tag{2.2.25}$$

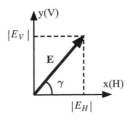

FIGURE 2.14 *E* and γ. (From Yamaguchi, Y., *Radar Polarimetry from Basics to Applications: Radar Remote Sensing Using Polarimetric Information [in Japanese]*, IEICE, 2007.)

$$|\mathbf{E}| = \sqrt{E_H E_H^* + E_V E_V^*} \tag{2.2.26}$$

This form is one of the expressions of the polarization state and is used for coherent analysis. If the magnitude is normalized as $|\mathbf{E}| = 1$, and if the absolute phase is set $\phi_H = 0$, the expression (2.2.25) becomes a rather simple form.

$$\mathbf{E}(HV) = \frac{1}{\sqrt{1 + \rho\rho^*}} \begin{bmatrix} 1 \\ \rho \end{bmatrix} \tag{2.2.27}$$

We can use ρ to specify the polarization state.

For example, for the case of a horizontally polarized wave, $\rho = 0$, so that

$$\mathbf{E}(HV) \Rightarrow \hat{\mathbf{H}} = \frac{1}{\sqrt{1 + 0 \cdot 0}} \begin{bmatrix} 1 \\ 0 \end{bmatrix} = \begin{bmatrix} 1 \\ 0 \end{bmatrix} \tag{2.2.28}$$

For a 45°-oriented linearly polarized wave, we have $\rho = 1$,

$$\mathbf{E}(HV) \Rightarrow \hat{\mathbf{X}} = \frac{1}{\sqrt{1 + 1 \cdot 1}} \begin{bmatrix} 1 \\ 1 \end{bmatrix} = \frac{1}{\sqrt{2}} \begin{bmatrix} 1 \\ 1 \end{bmatrix} \tag{2.2.29}$$

A vertically polarized wave corresponds to $\rho = \infty$, so that

$$\mathbf{E}(HV) \Rightarrow \hat{\mathbf{V}} = \frac{1}{\sqrt{1 + \infty \cdot \infty}} \begin{bmatrix} 0 \\ \infty \end{bmatrix} = \begin{bmatrix} 0 \\ 1 \end{bmatrix} \tag{2.2.30}$$

Left-handed circular (LHC) polarization is obtained from $\rho = j$,

$$\mathbf{E}(HV) \Rightarrow \hat{\mathbf{L}} = \frac{1}{\sqrt{1 + j \cdot (-j)}} \begin{bmatrix} 1 \\ j \end{bmatrix} = \frac{1}{\sqrt{2}} \begin{bmatrix} 1 \\ j \end{bmatrix} \tag{2.2.31}$$

Right-handed circular (RHC) polarization may be obtained from $\rho = -j$,

$$\mathbf{E}(HV) \Rightarrow \hat{\mathbf{R}} = \frac{1}{\sqrt{1 + (-j) \cdot j}} \begin{bmatrix} 1 \\ -j \end{bmatrix} = \frac{1}{\sqrt{2}} \begin{bmatrix} 1 \\ -j \end{bmatrix} \tag{2.2.32}$$

The notation $\hat{\mathbf{R}}$ (2.2.32) is employed in the IEEE definition [12]. However, this notation causes problems in vector transformation between the HV and the circular LR basis. The transformation matrix should be unitary. In order to avoid misleading in the polarization analyses, $\hat{\mathbf{R}}$ is chosen here as

$$\hat{\mathbf{R}} = \frac{j}{\sqrt{1 + (-j) \cdot j}} \begin{bmatrix} 1 \\ -j \end{bmatrix} = \frac{1}{\sqrt{2}} \begin{bmatrix} j \\ 1 \end{bmatrix} \tag{2.2.33}$$

This expression (2.2.33) satisfies the inner product and ensures the basis transformation because it can follow the unitary matrix condition. This expression also is the same as one in [11].

Some polarization states using the Jones vector are shown in Figure 2.15. The propagation direction is taken toward the reader.

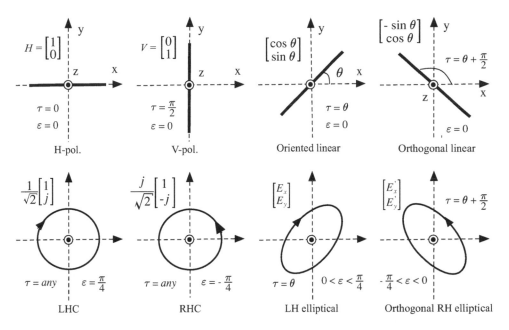

FIGURE 2.15 Polarization states by the Jones vector. (From Yamaguchi, Y., *Radar Polarimetry from Basics to Applications: Radar Remote Sensing Using Polarimetric Information [in Japanese]*, IEICE, 2007.)

2.2.4 STOKES VECTOR REPRESENTATION

Assume that we are observing wheat crops by radar in a windy environment. If the duration time of radar pulse is long (in the order of several milliseconds), the wheat crops can be considered as fluctuating objects within the observation time. The reflected wave to the radar is a sum of scattered waves from numerous scattering points in the crop field, and therefore the total phase changes randomly within the observation time. That is, the observed signal is fluctuating. On the other hand, if the duration time is short enough (in the order of nanoseconds), the wheat crops can be regarded as stationary. The magnitude of the reflected wave and its phase can be considered as constant. The observed signal is rather stationary, not fluctuating.

The extent of signal fluctuation depends on the observation time and the target's fluctuating speed. In general, a target can be considered as stationary if the observation time is short. In this case, the reflected signal to radar becomes a coherent wave. On the other hand, if the observation time is long, the reflected wave to radar becomes incoherent or fluctuating, which means the phase of each scattered wave is random.

Therefore, the reflected wave can be considered generally as a sum of a coherent and incoherent wave (See Figure 2.16). A coherent wave is also called "completely polarized wave." An incoherent wave with a random phase is called a "completely unpolarized wave." The mixture is called a "partially polarized wave."

In radar sensing, a coherent wave is transmitted. The scattered wave can be incoherent due to fluctuation of target or propagating medium. Coherent wave is decomposed into the sum of a coherent part and incoherent part. This phenomenon is called depolarization. However, depolarization is also used for the meaning of generation of cross-polarization. For example, if a *V*-pol component is generated by a target when *H*-pol is transmitted, it is sometimes called depolarization. In this case, it seems better to use "repolarization" [6], since both the transmitting and receiving waves are completely polarized.

We have been dealing with completely polarized waves, which have a constant value of $|E_H|, |E_V|, \delta_{HV}$ during observation time. But for a rapid changing phase or amplitude during the

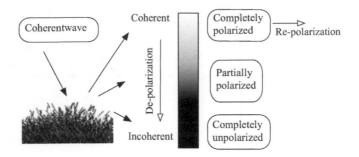

FIGURE 2.16 Coherent to incoherent wave. (From Yamaguchi, Y., *Radar Polarimetry from Basics to Applications: Radar Remote Sensing Using Polarimetric Information [in Japanese]*, IEICE, 2007.)

observation time, it is impossible to deal with the wave in the same way anymore. For dealing with completely polarized and partially polarized wave, we need to use the Stokes parameters. Stokes parameters were devised by Sir George Stokes in 1852. The advantages of Stokes parameters are all real-valued, and any polarization state can be determined by power measurement. Since phase measurement is quite difficult for high frequencies, such as a millimeter wave or optical wave, it is advantageous to derive all polarization information by the Stokes parameter by power measurement only.

2.2.4.1 Stokes Vector for Completely Polarized Wave

The Stokes vector is denoted here as **g**. The four components of the vector, g_0, g_1, g_2, g_3, are called Stokes parameters. The relations to the completely polarized wave are written as

$$\mathbf{g} = \begin{bmatrix} g_0 \\ g_1 \\ g_2 \\ g_3 \end{bmatrix} = \begin{bmatrix} |E_H|^2 + |E_V|^2 \\ |E_H|^2 - |E_V|^2 \\ 2\,\mathrm{Re}\{E_V E_H^*\} \\ 2\,\mathrm{Im}\{E_V E_H^*\} \end{bmatrix} = \begin{bmatrix} |E_H|^2 + |E_V|^2 \\ |E_H|^2 - |E_V|^2 \\ 2|E_H||E_V|\cos\delta \\ 2|E_H||E_V|\sin\delta \end{bmatrix} = A^2 \begin{bmatrix} 1 \\ \cos 2\tau \cos 2\varepsilon \\ \sin 2\tau \cos 2\varepsilon \\ \sin 2\varepsilon \end{bmatrix} \tag{2.2.34}$$

where $|E_H|, |E_V|$ are the amplitude, and $\delta = \phi_V - \phi_H$ is the relative phase of E_H and E_V. Geometrical parameters (A, ε, τ) correspond to size, ellipticity angle, and tilt angle, respectively. For a completely polarized wave, the next relation holds.

$$g_0^2 = g_1^2 + g_2^2 + g_3^2 \tag{2.2.35}$$

Stokes parameters correspond to the rectangular coordinate of Poincaré sphere as shown in Figure 2.19. The physical meaning of Stokes parameters can be derived from the following vector transformations.

The electric field vector components in the circular polarization basis can be transformed by those in the *HV* polarization basis as (Appendix A2.2)

$$E_L = \frac{1}{\sqrt{2}}\left(E_H - j\,E_V\right), \quad E_R = \frac{1}{\sqrt{2}}\left(-j\,E_H + E_V\right) \tag{2.2.36}$$

This leads to,

$$|E_L|^2 = E_L E_L^* = \frac{1}{2}\left(|E_H|^2 + |E_V|^2 + 2\,\mathrm{Im}\left\{E_V E_H^*\right\}\right)$$

$$|E_R|^2 = E_R E_R^* = \frac{1}{2}\left(|E_H|^2 + |E_V|^2 - 2\,\mathrm{Im}\left\{E_V E_H^*\right\}\right) \tag{2.2.37}$$

$$E_L E_R^* = \frac{1}{2}\left(j|E_H|^2 - j|E_V|^2 + 2\,\mathrm{Re}\left\{E_V E_H^*\right\}\right)$$

Therefore, Stokes parameters can be expressed by

$$\begin{cases} g_0 = |E_H|^2 + |E_V|^2 = |E_L|^2 + |E_R|^2 \\[2mm] g_1 = |E_H|^2 - |E_V|^2 = 2\,\mathrm{Im}\left\{E_L E_R^*\right\} \\[2mm] g_2 = 2\,\mathrm{Re}\left\{E_V E_H^*\right\} = 2\,\mathrm{Re}\left\{E_L E_R^*\right\} \\[2mm] g_3 = 2\,\mathrm{Im}\left\{E_V E_H^*\right\} = |E_L|^2 - |E_R|^2 \end{cases} \tag{2.2.38}$$

For a 45°-oriented linear polarization basis, we denote the basis $(45°-135°) = (X, Y)$. The electric field vector can be transformed to

$$E_X = \frac{1}{\sqrt{2}}\left(E_H + E_V\right),\ E_Y = \frac{1}{\sqrt{2}}\left(-E_H + E_V\right) \tag{2.2.39}$$

This leads to

$$|E_X|^2 = E_X E_X^* = \frac{1}{2}\left(|E_H|^2 + |E_V|^2 + 2\,\mathrm{Re}\left\{E_V E_H^*\right\}\right)$$

$$|E_Y|^2 = E_Y E_Y^* = \frac{1}{2}\left(|E_H|^2 + |E_V|^2 - 2\,\mathrm{Re}\left\{E_V E_H^*\right\}\right) \tag{2.2.40}$$

$$E_X E_Y^* = \frac{1}{2}\left(-|E_H|^2 + |E_V|^2 - j2\,\mathrm{Im}\left\{E_V E_H^*\right\}\right)$$

Therefore, Stokes parameters can be expressed by

$$\begin{cases} g_0 = |E_H|^2 + |E_V|^2 = |E_X|^2 + |E_Y|^2 \\[2mm] g_1 = |E_H|^2 - |E_V|^2 = -2\,\mathrm{Re}\left\{E_X E_Y^*\right\} \\[2mm] g_2 = 2\,\mathrm{Re}\left\{E_V E_H^*\right\} = |E_X|^2 - |E_Y|^2 \\[2mm] g_3 = 2\,\mathrm{Im}\left\{E_V E_H^*\right\} = -2\,\mathrm{Im}\left\{E_X E_Y^*\right\} \end{cases} \tag{2.2.41}$$

Using the expression of (2.2.38) and (2.2.41), the Stokes parameters can be expressed as listed in Table 2.1. From this table, we can understand the property of Stokes parameters from the physical point of view.

TABLE 2.1
Stokes Parameter Representation

Stokes Parameter	HV	45°/135° Linear	LR	Geometric Parameter
g_0	$\left\|E_H\right\|^2 + \left\|E_V\right\|^2$	$\left\|E_X\right\|^2 + \left\|E_Y\right\|^2$	$\left\|E_L\right\|^2 + \left\|E_R\right\|^2$	A^2
g_1	$\left\|E_H\right\|^2 - \left\|E_V\right\|^2$	$-2\,\mathrm{Re}\left\{E_X E_Y^*\right\}$	$2\,\mathrm{Im}\left\{E_L E_R^*\right\}$	$A^2 \cos 2\tau \cos 2\varepsilon$
g_2	$2\,\mathrm{Re}\left\{E_V E_H^*\right\}$	$\left\|E_X\right\|^2 - \left\|E_Y\right\|^2$	$2\,\mathrm{Re}\left\{E_L E_R^*\right\}$	$A^2 \sin 2\tau \cos 2\varepsilon$
g_3	$2\,\mathrm{Im}\left\{E_V E_H^*\right\}$	$-2\,\mathrm{Im}\left\{E_X E_Y^*\right\}$	$\left\|E_L\right\|^2 - \left\|E_R\right\|^2$	$A^2 \sin 2\varepsilon$

g_0 corresponds to total power, which is an invariant parameter with respect to polarization basis.
g_1 represents the power difference between horizontally and vertically polarized waves.
g_2 represents the power difference between the 45°- and 135°-oriented linearly polarized wave.
g_3 represents the power difference between left and right circularly polarized wave.

If there exist nonzero components in g_1, g_2, g_3, then a completely polarized wave also exists.

2.2.4.1.1 Polarization State Determination by Power Measurement

In addition, we find that Stokes parameters can be determined by power measurements. As seen in the table, the following equation is derived from channel powers.

$$\begin{cases} g_1 = \left|E_H\right|^2 - \left|E_V\right|^2 \\ g_2 = \left|E_X\right|^2 - \left|E_Y\right|^2 \\ g_3 = \left|E_L\right|^2 - \left|E_R\right|^2 \end{cases} \tag{2.2.42}$$

Using these relations and the scheme shown in Figure 2.17, all Stokes parameters can be determined.

Once Stokes parameters are determined, the polarization state (geometrical parameters) of the wave can be found from

$$\text{Total power} \quad A^2 = g_0 \tag{2.2.43}$$

$$\text{Orientation angle} \quad \sin 2\varepsilon = \frac{g_3}{g_0}, \quad \varepsilon = \frac{1}{2}\sin^{-1}\frac{g_3}{g_0} \tag{2.2.44}$$

$\left|E_H\right|^2 \quad \left|E_V\right|^2 \quad \left|E_X\right|^2 \quad \left|E_Y\right|^2 \quad \left|E_L\right|^2 \quad \left|E_R\right|^2 \qquad \text{Polarization state}$

FIGURE 2.17 Polarization state determination by power measurement. (From Yamaguchi, Y., *Radar Polarimetry from Basics to Applications: Radar Remote Sensing Using Polarimetric Information [in Japanese]*, IEICE, 2007.)

$$\text{Ellipticity angle } \tan 2\tau = \frac{g_2}{g_1}, \quad \tau = \frac{1}{2}\tan^{-1}\frac{g_2}{g_1} \tag{2.2.45}$$

$$\text{Axial ratio } \frac{a}{b} = \frac{1}{\tan \varepsilon} \tag{2.2.46}$$

Furthermore, the following information can be derived by Stokes parameters.
  Power polarization ratio in the *HV* basis:

$$|\rho_{HV}|^2 = \frac{|E_V|^2}{|E_H|^2} = \frac{g_0 - g_1}{g_0 + g_1} \tag{2.2.47}$$

Normalized power difference in the *HV* basis:

$$\frac{|E_H|^2 - |E_V|^2}{|E_H|^2 + |E_V|^2} = \frac{g_1}{g_0} \tag{2.2.48}$$

Power polarization ratio in the *LR* basis:

$$|\rho_{LR}|^2 = \frac{|E_R|^2}{|E_L|^2} = \frac{g_0 - g_3}{g_0 + g_3} \tag{2.2.49}$$

Normalized power difference in the *LR* basis:

$$\frac{|E_L|^2 - |E_R|^2}{|E_L|^2 + |E_R|^2} = \frac{g_3}{g_0} \tag{2.2.50}$$

2.2.4.2 Stokes Vector for Partially Polarized Wave

Ensemble averaging is used for dealing with incoherent waves or partially coherent waves. Assuming an ergodic hypothesis that time averaging can be replaced by space averaging, ensemble averaging is applied to the Stokes vector as

$$\mathbf{g} = \begin{bmatrix} \langle g_0 \rangle \\ \langle g_1 \rangle \\ \langle g_2 \rangle \\ \langle g_3 \rangle \end{bmatrix} = \begin{bmatrix} \langle |E_H|^2 \rangle + \langle |E_V|^2 \rangle \\ \langle |E_H|^2 \rangle - \langle |E_V|^2 \rangle \\ 2\langle \mathrm{Re}\{E_V E_H^*\} \rangle \\ 2\langle \mathrm{Im}\{E_V E_H^*\} \rangle \end{bmatrix} \tag{2.2.51}$$

where the symbol <> represents averaging. From this definition, the next inequality holds.

$$\langle g_0 \rangle^2 \geq \langle g_1 \rangle^2 + \langle g_2 \rangle^2 + \langle g_3 \rangle^2 \tag{2.2.52}$$

The degree of polarization (DoP) is defined to show how much of the coherent part is included in the total power.

2.2.4.2.1 Degree of Polarization

$$\mathrm{DoP} = \frac{\sqrt{\langle g_1 \rangle^2 + \langle g_2 \rangle^2 + \langle g_3 \rangle^2}}{\langle g_0 \rangle} \tag{2.2.53}$$

According to the value of DoP, the wave can be categorized as follows:

- DoP = 1 corresponds to "completely polarized wave." It shows a coherent wave. There exists the following relation in Stokes parameters:

$$\langle g_0 \rangle^2 = \langle g_1 \rangle^2 + \langle g_2 \rangle^2 + \langle g_3 \rangle^2 \tag{2.2.54}$$

- DoP = 0 corresponds to "completely unpolarized wave." The wave is incoherent. Stokes parameters become,

$$\langle g_0 \rangle^2 \neq 0, \quad \langle g_1 \rangle^2 = \langle g_2 \rangle^2 = \langle g_3 \rangle^2 = 0 \tag{2.2.55}$$

- $0 \leq \mathrm{DoP} \leq 1$ corresponds to a "partially polarized wave." Most of the observed wave falls in this category. A partially polarized wave is a sum of the coherent part and incoherent part. It can be written as

$$q = \langle g_1 \rangle^2 + \langle g_2 \rangle^2 + \langle g_3 \rangle^2 \tag{2.2.56}$$

$$\mathbf{g} = \begin{bmatrix} q \\ \langle g_1 \rangle \\ \langle g_2 \rangle \\ \langle g_3 \rangle \end{bmatrix} + \begin{bmatrix} \langle g_0 \rangle - q \\ 0 \\ 0 \\ 0 \end{bmatrix} \tag{2.2.57}$$

polarized + unpolarized

DoP is often used as a criterion for classification or decomposition of waves.

2.2.5 POLARIZATION PARAMETERS AND POINCARÉ SPHERE

As a visual representation of polarization, the Poincaré sphere is frequently used. The Poincaré sphere was invented by a French mathematician named A. Poincaré as shown in Figure 2.18. It can be considered a 3D expansion of Figure 2.12. A point on the sphere shows a specific polarization state. There is a one-to-one correspondence between a point on the sphere and a polarization state. If the sphere is mapped to the earth, the north pole corresponds to LHC polarization, and the south pole is RHC polarization, whereas the equator has oriented linear polarization. It is possible to specify the polarization state by a point on the sphere surface.

There are some parameters to specify a point on the sphere such as $(g_1, g_2, g_3), (\varepsilon, \tau)$, and (γ, δ) in Figure 2.19.

(g_1, g_2, g_3): Stokes parameters constitute the three rectangular axes of the sphere. They can be treated as the rectangular coordinates ($g_1 = x, g_2 = y, g_3 = z$).

(ε, τ): 2τ corresponds to the longitude on the equator measured from OH axis with range $(-\pi \leq 2\tau \leq \pi)$. $\tau = \frac{\pi}{4}$ corresponds to 45°-oriented linear polarization, while $\tau = \frac{\pi}{2}$ is V-polarization. On the other hand, 2ε is the angle between equator plane and OP axis with range $(-\frac{\pi}{2} \leq 2\varepsilon \leq \frac{\pi}{2})$. The north pole ($\varepsilon = \frac{\pi}{4}$) represents LHC, and the south pole ($\varepsilon = -\frac{\pi}{4}$) represents RHC. Therefore, we can assume that 2ε as the latitude, and 2τ as the longitude of the sphere.

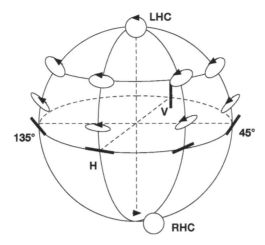

FIGURE 2.18 Poincaré sphere and polarization state (upper hemisphere shows left-handed rotation, and lower hemisphere is for right-handed rotation). (From Yamaguchi, Y., *Radar Polarimetry from Basics to Applications: Radar Remote Sensing Using Polarimetric Information [in Japanese]*, IEICE, 2007.)

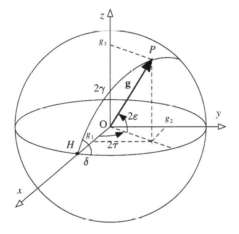

FIGURE 2.19 Poincaré sphere and polarization parameters.

(γ, δ): The arc 2γ starting from point H on the equator to point P on the sphere surface represents the distance HP on the largest circle. It finally reaches to the antipodal point V of the sphere with the range $(0 \leq 2\gamma \leq \pi)$. The phase difference $\delta = \phi_y - \phi_x$ represents the angle between the equator and HP $(-\pi \leq \delta \leq \pi)$.

These three parameters are well-organized to specify point P on the Poincaré sphere as shown in Figure 2.19. The same point P can be expressed by these three parameters:

(g_1, g_2, g_3) in the range $(-1 \leq g_1, g_2, g_3 \leq 1)$
$(2\varepsilon, 2\tau)$ in the range $(-\pi \leq 2\tau \leq \pi, -\frac{\pi}{2} \leq 2\varepsilon \leq \frac{\pi}{2})$
$(2\gamma, \delta)$ in the range $(0 \leq 2\gamma \leq \pi, -\pi \leq \delta \leq \pi)$

An interesting notice is that the polarization state of antipodal point Q is orthogonal to that of point P as shown in Figure 2.20. For example, LHC at the north pole is orthogonal to RHC at the

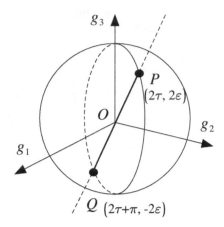

FIGURE 2.20 Orthogonal polarization states P and Q at the antipodal point.

south pole, and the opposite side of *H*-pol is *V*-pol (Figure 2.18). This property holds any points on the sphere. The straight line across the origin of the Poincaré sphere thus becomes the orthogonal polarization basis.

2.2.6 POLARIZATION VECTOR

Since the orthogonal polarization state can be obtained by the antipodal point on the Poincaré sphere, it is straightforward to check by the equation. Referring to (2.2.58),

$$\begin{bmatrix} |E_X| \\ |E_y|e^{j\delta} \end{bmatrix} = A \begin{bmatrix} \cos\tau & -\sin\tau \\ \sin\tau & \cos\tau \end{bmatrix} \begin{bmatrix} \cos\varepsilon \\ j\sin\varepsilon \end{bmatrix} e^{j\phi} \tag{2.2.58}$$

we can define the polarization vector **p**

$$\mathbf{p} = \mathbf{p}(\tau,\varepsilon) = \begin{bmatrix} \cos\tau & -\sin\tau \\ \sin\tau & \cos\tau \end{bmatrix} \begin{bmatrix} \cos\varepsilon \\ j\sin\varepsilon \end{bmatrix}. \tag{2.2.59}$$

The antipodal point on the Poincaré sphere has geometrical parameters as

$$\begin{cases} 2\tau \Rightarrow 2\tau + \pi & \therefore \tau \Rightarrow \tau + \dfrac{\pi}{2} \\[2mm] 2\varepsilon \Rightarrow -2\varepsilon & \therefore \varepsilon \Rightarrow -\varepsilon \end{cases} \tag{2.2.60}$$

Therefore, the orthogonal polarization vector $\mathbf{p}_\perp$ becomes,

$$\mathbf{p}_\perp = \mathbf{p}\left(\tau + \frac{\pi}{2}, -\varepsilon\right) = \begin{bmatrix} -\sin\tau & -\cos\tau \\ \cos\tau & -\sin\tau \end{bmatrix} \begin{bmatrix} \cos\varepsilon \\ -j\sin\varepsilon \end{bmatrix}. \tag{2.2.61}$$

This satisfies the inner product,

$$\mathbf{p} \cdot \mathbf{p}_\perp^* = 0 \tag{2.2.62}$$

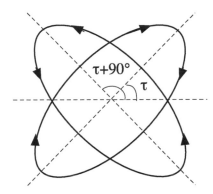

FIGURE 2.21 Orthogonal polarization. (From Yamaguchi, Y., *Radar Polarimetry from Basics to Applications: Radar Remote Sensing Using Polarimetric Information [in Japanese]*, IEICE, 2007.)

Therefore, Equations (2.2.59) and (2.2.61) apply to any orthogonal polarization pair (see Figure 2.21).

Typical polarization states can be represented as below.

$$\text{H-Pol} \quad \tau = 0, \varepsilon = 0 \quad \Rightarrow \quad \mathbf{p} = \mathbf{p}(0,0) = \begin{bmatrix} 1 & 0 \\ 0 & 1 \end{bmatrix} \begin{bmatrix} 1 \\ j\,0 \end{bmatrix} = \begin{bmatrix} 1 \\ 0 \end{bmatrix} \tag{2.2.63a}$$

$$\text{Orthogonal } (=\text{V-pol}) \quad \Rightarrow \mathbf{p}_\perp = \mathbf{p}\left(\frac{\pi}{2},0\right) = \begin{bmatrix} 0 & -1 \\ 0 & 0 \end{bmatrix} \begin{bmatrix} 1 \\ -j\,0 \end{bmatrix} = \begin{bmatrix} 0 \\ 1 \end{bmatrix} \tag{2.2.63b}$$

$$45° \text{ linear} \quad \tau = \frac{\pi}{4}, \varepsilon = 0 \Rightarrow \mathbf{p} = \mathbf{p}\left(\frac{\pi}{4},0\right) = \frac{1}{\sqrt{2}} \begin{bmatrix} 1 & -1 \\ 1 & 1 \end{bmatrix} \begin{bmatrix} 1 \\ j\,0 \end{bmatrix} = \frac{1}{\sqrt{2}} \begin{bmatrix} 1 \\ 1 \end{bmatrix} \tag{2.2.64a}$$

$$\text{Orthogonal } (=135° \text{ linear}) \quad \Rightarrow \mathbf{p}_\perp = \mathbf{p}\left(\frac{\pi}{4}+\frac{\pi}{2},0\right) = \frac{1}{\sqrt{2}} \begin{bmatrix} -1 & -1 \\ 1 & -1 \end{bmatrix} \begin{bmatrix} 1 \\ j\,0 \end{bmatrix} = \frac{1}{\sqrt{2}} \begin{bmatrix} -1 \\ 1 \end{bmatrix} \tag{2.2.64b}$$

$$\text{LHC} \quad \varepsilon = \frac{\pi}{4} \Rightarrow \mathbf{p} = \mathbf{p}\left(\tau,\frac{\pi}{4}\right) = \begin{bmatrix} \cos\tau & -\sin\tau \\ \sin\tau & \cos\tau \end{bmatrix} \frac{1}{\sqrt{2}} \begin{bmatrix} 1 \\ j \end{bmatrix} = \frac{e^{-j\tau}}{\sqrt{2}} \begin{bmatrix} 1 \\ j \end{bmatrix} \tag{2.2.65a}$$

$$\text{Orthogonal } (=\text{RHC}) \quad \Rightarrow \mathbf{p}_\perp = \mathbf{p}\left(\tau+\frac{\pi}{2}, -\frac{\pi}{4}\right) = \begin{bmatrix} -\sin\tau & -\cos\tau \\ \cos\tau & -\sin\tau \end{bmatrix} \frac{1}{\sqrt{2}} \begin{bmatrix} 1 \\ -j \end{bmatrix} = \frac{e^{j\tau}}{\sqrt{2}} \begin{bmatrix} j \\ 1 \end{bmatrix} \tag{2.2.65b}$$

The results are identical to those of the Jones vector representation.

In the expression of LHC and RHC, $e^{-j\tau}$ and $e^{j\tau}$ indicate a phase shift by the tilt angle. There is ambiguity in choosing the tilt angle. However, the polarization shape remains completely circular for any value of tilt angle. Therefore, we can set $\tau = 0$ without loss of generality so that polarization vectors of LHC and RHC become

$$\hat{\mathbf{L}} = \frac{1}{\sqrt{2}} \begin{bmatrix} 1 \\ j \end{bmatrix}, \quad \hat{\mathbf{R}} = \frac{1}{\sqrt{2}} \begin{bmatrix} j \\ 1 \end{bmatrix} \tag{2.2.66}$$

2.2.7   COVARIANCE MATRIX FOR PARTIALLY POLARIZED WAVE

Covariance matrix can be expressed as

$$\langle [J] \rangle = \langle \mathbf{E} \cdot \mathbf{E}^{*\mathrm{T}} \rangle = \begin{bmatrix} \left\langle |E_H|^2 \right\rangle & \left\langle E_H E_V^* \right\rangle \\ \left\langle E_V E_H^* \right\rangle & \left\langle |E_V|^2 \right\rangle \end{bmatrix} \tag{2.2.67}$$

where $\langle \cdots \rangle$ represents the time averaging or space averaging under the ergodic hypothesis. The sum of diagonal elements is the trace of the matrix and is equivalent to the total power.

$$\mathrm{Trace}\langle [J] \rangle = \left\langle |E_H|^2 \right\rangle + \left\langle |E_V|^2 \right\rangle = A^2 = \langle g_0 \rangle \tag{2.2.68}$$

The off-diagonal terms represent cross-correlation and are complex-valued.

Stokes parameters are defined as

$$\langle g_0 \rangle = \left\langle |E_H|^2 \right\rangle + \left\langle |E_V|^2 \right\rangle$$

$$\langle g_1 \rangle = \left\langle |E_H|^2 \right\rangle - \left\langle |E_V|^2 \right\rangle$$

$$\langle g_2 \rangle = \left\langle E_H E_V^* \right\rangle + \left\langle E_H^* E_V \right\rangle \tag{2.2.69}$$

$$\langle g_3 \rangle = j\langle E_H E_V^* \rangle - j\langle E_H^* E_V \rangle$$

Covariance matrix is expressed in terms of the Stokes parameters,

$$\langle [J] \rangle = \frac{1}{2} \begin{bmatrix} \langle g_0 \rangle + \langle g_1 \rangle & \langle g_2 \rangle - j\langle g_3 \rangle \\ \langle g_2 \rangle + j\langle g_3 \rangle & \langle g_0 \rangle - \langle g_1 \rangle \end{bmatrix} \tag{2.2.70}$$

Now, we expand the covariance matrix by eigenvalues and eigenvectors

$$\langle [J] \rangle = [U_2] \begin{bmatrix} \lambda_1 & 0 \\ 0 & \lambda_2 \end{bmatrix} [U_2]^{-1} = \lambda_1 \mathbf{u}_1 \mathbf{u}_1^T + \lambda_2 \mathbf{u}_2 \mathbf{u}_2^T \tag{2.2.71}$$

$$[U_2] = [\mathbf{u}_1 \mathbf{u}_2] : \mathrm{Unitary\ matrix}$$

The eigenvalues are given by the following equations.

$$\lambda_1 = \frac{1}{2}\left(\langle g_0 \rangle + \sqrt{\langle g_1 \rangle^2 + \langle g_2 \rangle^2 + \langle g_3 \rangle^2} \right) = \frac{\langle g_0 \rangle}{2}(1 + DoP) \tag{2.2.72}$$

$$\lambda_2 = \frac{1}{2}\left(\langle g_0 \rangle - \sqrt{\langle g_1 \rangle^2 + \langle g_2 \rangle^2 + \langle g_3 \rangle^2} \right) = \frac{\langle g_0 \rangle}{2}(1 - DoP) \tag{2.2.73}$$

From this expression, we understand that λ_1 corresponds to completely polarized wave, whereas λ_2 corresponds to completely unpolarized wave.

The DoP can be expressed as follows, and is equal to anisotropy.

$$\text{DoP} = \frac{\text{Polarized power}}{\text{Total power}} = \frac{\sqrt{\langle g_1 \rangle^2 + \langle g_2 \rangle^2 + \langle g_3 \rangle^2}}{\langle g_0 \rangle} = \frac{\lambda_1 - \lambda_2}{\lambda_1 + \lambda_2} = \text{Anisotropy} \qquad (2.2.74)$$

Entropy H, which represents statistical randomness, can be given by eigenvalues as

$$H = -p_1 \log_2 p_1 - p_2 \log_2 p_2, \quad p_i = \frac{\lambda_i}{\lambda_1 + \lambda_2} \qquad (2.2.75)$$

The wave can be classified according to DoP as shown in Figure 2.22.

- Completely polarized wave

    $\text{DoP} = 1$

    $\langle g_1 \rangle^2 + \langle g_2 \rangle^2 + \langle g_3 \rangle^2 = \langle g_0 \rangle^2$

    $\lambda_1 = \langle g_0 \rangle, \quad \lambda_2 = 0, \quad H = 0$

    $\text{Det} \langle [J] \rangle = 0$

It locates on the surface of the Poincaré sphere.

- Partially polarized wave

    $0 < \text{DoP} < 1$

    $\langle g_1 \rangle^2 + \langle g_2 \rangle^2 + \langle g_3 \rangle^2 < \langle g_0 \rangle^2$

    $\lambda_1 \neq \lambda_2, 0 < H < 1$

    $\text{Det} \langle [J] \rangle > 0$

Completely polarized part (•) locates inside the sphere

- Completely unpolarized wave

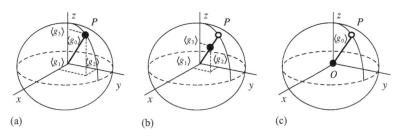

(a) (b) (c)

FIGURE 2.22 Polarized to unpolarized waves: (a) completely polarized wave, (b) partially polarized wave, and (c) completely unpolarized wave.

$$\text{DoP} = 0$$

$$\langle g_1 \rangle^2 + \langle g_2 \rangle^2 + \langle g_3 \rangle^2 = 0$$

$$\lambda_1 = \lambda_2 = \frac{\langle g_0 \rangle}{2}, H = 1$$

$$\langle E_H E_V^* \rangle = 0$$

2.3 RELATIONS AMONG POLARIZATION PARAMETERS AND SUMMARY

So far, we have seen various representations for wave polarization. The Poincaré sphere seems the most suitable for a visual representation of wave polarization. Any polarization state can be specified by a point on the surface. For polarimetric data analysis, we can use the most suitable one for the purpose because each representation has its own merit. The relations among polarization parameters are shown in Figure 2.23.

From equation (2.2.21), the polarization ratio is related to the geometrical parameters

$$\rho = \frac{E_H}{E_V} = \frac{\sin \tau \cos \varepsilon + j \cos \tau \sin \varepsilon}{\cos \tau \cos \varepsilon - j \sin \tau \sin \varepsilon} = \frac{\tan \tau + j \tan \varepsilon}{1 - j \tan \tau \tan \varepsilon} \qquad (2.3.1)$$

Therefore, geometrical parameters can be expressed in terms of the polarization ratio,

$$\tan 2\tau = \frac{2 \operatorname{Re}\{\rho\}}{1 - |\rho|^2}, \qquad \sin 2\varepsilon = \frac{2 \operatorname{Im}\{\rho\}}{1 + |\rho|^2} \qquad (2.3.2)$$

This relation is frequently used in the coherent polarimetric measurement and useful to determine the polarization state of the wave.

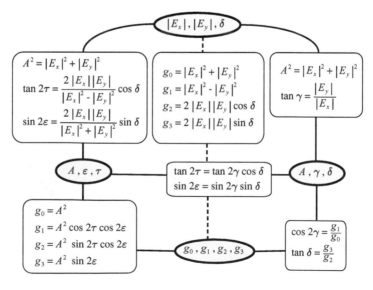

FIGURE 2.23 Mutual relations among polarization parameters.

APPENDIX

A2.1   FIRST DEFINITION

The definition is the starting point. If the definition is different, the next step and the final result becomes ambiguous. Polarization, unfortunately, also falls in this category. Confusion comes from this starting point. For example, wave representation is often used as e^{-jkr} in the engineering field, and e^{+ikr} in optics for propagating in the positive r direction. This is due to the first definition of e^{+ikr} in optics and $e^{+j\omega t}$ in engineering. The definition of polarization in engineering is based on the transverse plane seen from the backside of propagation, which is contrary to the definition in optics. Accordingly, the rotation sense of LHC and RHC polarization becomes totally opposite.

Famous confusion is cited [13,14] as,

...Circularly polarized waves have either a right-handed or left-handed sense, which is defined by convention. The TELSTAR satellite sent out circularly polarized microwaves. When it first passed over the Atlantic, the British station at Goonhilly and the French station at Pleumeur Bodou both tried to receive its signals. The French succeeded because their definition of polarization agreed with the American definition. The British station was set up to receive the wrong (orthogonal) polarisation because their definition of sense...was contrary to 'our' definition...

from J R Pierce, *Almost Everything about Waves,* **pp. 130–131, Cambridge MA, MIT Press, 1974 [15].**

Which is first, H or V, L or R in the polarization basis? These also make confusion in the development of polarimetric theory. In this book, H is firstly defined and then V is defined as orthogonal polarization. Similarly, LHC (L) is first defined, and then R is defined. If we take differently, the transformation matrix becomes a different form.

A2.2   VECTOR TRANSFORMATION USING POLARIZATION RATIO

Basis vector $\hat{\mathbf{E}}$ transformation from the arbitrary basis (A-B) to the basis (H-V) can be performed by

$$\hat{\mathbf{E}}(HV) = [T]\hat{\mathbf{E}}(AB) \tag{A2.1}$$

where

$$[T] = \frac{1}{\sqrt{1+\rho\rho^*}}\begin{bmatrix} 1 & -\rho^* \\ \rho & 1 \end{bmatrix}\begin{bmatrix} e^{-j\alpha} & 0 \\ 0 & e^{j\alpha} \end{bmatrix} \tag{A2.2}$$

is the basis transformation matrix, expressed in terms of the polarization ratio ρ and the geometrical parameters (ε,τ) as,

$$\rho = \frac{\tan\tau + j\tan\varepsilon}{1 - j\tan\tau\tan\varepsilon}, \quad \alpha = \tan^{-1}(\tan\tau\tan\varepsilon) \tag{A2.3}$$

On the other hand, the vector component $\mathbf{E}$ transformation from the basis (H-V) to the arbitrary basis (A-B) is carried out by,

$$\mathbf{E}(AB) = [U]\mathbf{E}(HV) = [T]^{-1}\mathbf{E}(HV) \tag{A2.4}$$

where

$$[U] = [T]^{-1} = \frac{1}{\sqrt{1+\rho\rho^*}} \begin{bmatrix} e^{-j\alpha} & 0 \\ 0 & e^{j\alpha} \end{bmatrix} \begin{bmatrix} 1 & \rho^* \\ -\rho & 1 \end{bmatrix} \tag{A2.5}$$

is the vector transformation matrix.

Therefore, the vector component (E_A, E_B) in the basis $(A\text{-}B)$ can be written by the vector component (E_H, E_V) in the basis $(H\text{-}V)$ as,

$$\begin{bmatrix} E_A \\ E_B \end{bmatrix} = \frac{1}{\sqrt{1+\rho\rho^*}} \begin{bmatrix} e^{-j\alpha} & 0 \\ 0 & e^{j\alpha} \end{bmatrix} \begin{bmatrix} 1 & \rho^* \\ -\rho & 1 \end{bmatrix} \begin{bmatrix} E_H \\ E_V \end{bmatrix} \tag{A2.6}$$

For example, the components of the circular polarization vector can be written as,

$$\rho = j, \alpha = 0 \Rightarrow [T] = \frac{1}{\sqrt{2}} \begin{bmatrix} 1 & j \\ j & 1 \end{bmatrix}, \ [U] = \frac{1}{\sqrt{2}} \begin{bmatrix} 1 & -j \\ -j & 1 \end{bmatrix}$$

$$\begin{bmatrix} E_L \\ E_R \end{bmatrix} = \frac{1}{2} \begin{bmatrix} 1 & -j \\ -j & 1 \end{bmatrix} \begin{bmatrix} E_H \\ E_V \end{bmatrix} = \frac{1}{\sqrt{2}} \begin{bmatrix} E_H - jE_V \\ -jE_H + E_V \end{bmatrix} \tag{A2.7}$$

It is confirmed that

$$\begin{bmatrix} E_H \\ E_V \end{bmatrix} = \frac{1}{\sqrt{2}} \begin{bmatrix} 1 \\ j \end{bmatrix} \text{ of LHC in the } HV \text{ basis makes } \begin{bmatrix} E_L \\ E_R \end{bmatrix} = \frac{1}{2} \begin{bmatrix} 1 & -j \\ -j & 1 \end{bmatrix} \begin{bmatrix} 1 \\ j \end{bmatrix} = \begin{bmatrix} 1 \\ 0 \end{bmatrix} \text{ in the } LR \text{ basis.}$$

$$\begin{bmatrix} E_H \\ E_V \end{bmatrix} = \frac{1}{\sqrt{2}} \begin{bmatrix} j \\ 1 \end{bmatrix} \text{ of RHC in the } HV \text{ basis makes } \begin{bmatrix} E_L \\ E_R \end{bmatrix} = \frac{1}{2} \begin{bmatrix} 1 & -j \\ -j & 1 \end{bmatrix} \begin{bmatrix} j \\ 1 \end{bmatrix} = \begin{bmatrix} 0 \\ 1 \end{bmatrix} \text{ in the } LR \text{ basis.}$$

For 45°- to 135°-oriented linear polarization, the $(X\text{-}Y)$ vector is,

$$\rho = 1, \alpha = 0 \Rightarrow [T] = \frac{1}{\sqrt{2}} \begin{bmatrix} 1 & -1 \\ 1 & 1 \end{bmatrix}, \ [U] = \frac{1}{\sqrt{2}} \begin{bmatrix} 1 & 1 \\ -1 & 1 \end{bmatrix}$$

$$\begin{bmatrix} E_X \\ E_Y \end{bmatrix} = \frac{1}{\sqrt{2}} \begin{bmatrix} 1 & 1 \\ -1 & 1 \end{bmatrix} \begin{bmatrix} E_H \\ E_V \end{bmatrix} = \frac{1}{\sqrt{2}} \begin{bmatrix} E_H + E_V \\ -E_H + E_V \end{bmatrix} \tag{A2.8}$$

We confirm that

$$\begin{bmatrix} E_H \\ E_V \end{bmatrix} = \frac{1}{\sqrt{2}} \begin{bmatrix} 1 \\ 1 \end{bmatrix} \text{ of 45° in the } HV \text{ basis makes } \begin{bmatrix} E_X \\ E_Y \end{bmatrix} = \frac{1}{2} \begin{bmatrix} 1 & 1 \\ -1 & 1 \end{bmatrix} \begin{bmatrix} 1 \\ 1 \end{bmatrix} = \begin{bmatrix} 1 \\ 0 \end{bmatrix} \text{ in the } XY \text{ basis.}$$

$$\begin{bmatrix} E_H \\ E_V \end{bmatrix} = \frac{1}{\sqrt{2}} \begin{bmatrix} -1 \\ 1 \end{bmatrix} \text{ of 135° in the } HV \text{ basis makes } \begin{bmatrix} E_X \\ E_Y \end{bmatrix} = \frac{1}{2} \begin{bmatrix} 1 & 1 \\ -1 & 1 \end{bmatrix} \begin{bmatrix} -1 \\ 1 \end{bmatrix} = \begin{bmatrix} 0 \\ 1 \end{bmatrix} \text{ in the } XY \text{ basis.}$$

This transformation (A2.1) through (A2.5) preserves the phase information contained in the wave because of unitary transform.

A2.3   AXIAL RATIO

In antenna engineering, "axial ratio" is used to evaluate circular polarization radiated from antenna. This value is defined as the ratio of the major axis over the minor axis of polarization ellipse.

$$\text{Axial Ratio} = \frac{a}{b} = 20 \log \frac{a}{b} \; [dB], \; 1 < AR < \infty$$

The value is connected to ellipticity, $\frac{a}{b} = \frac{1}{\tan \varepsilon}$

It is common to regard the radiated wave as a circular polarization if the value of AR is less than 3 dB. However, we find through calculation that

$$(AR = 0 \text{ dB}) \Rightarrow \tan \varepsilon = 1 \Rightarrow \varepsilon = 45° \quad \text{completely circular}$$

$$(AR = 3 \text{ dB}) \Rightarrow |\tan \varepsilon| = \frac{1}{\sqrt{2}} \Rightarrow \varepsilon = \pm 35.27° \quad \text{elliptical}$$

The range of AR less than 3 dB is mapped to the area around poles on the Poincaré sphere as shown in Figure A2.1. Too big of an area is assigned to the circular polarization, which may cause problems in polarimetric treatment.

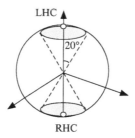

FIGURE A2.1   Area with AR less than 3 dB. (From Yamaguchi, Y., *Radar Polarimetry from Basics to Applications: Radar Remote Sensing Using Polarimetric Information [in Japanese]*, IEICE, 2007.)

REFERENCES

1. J. A. Stratton, *Electromagnetic Theory*, Chapter 5, McGraw-Hill, New York, 1941.
2. IEEE/ANSI, Standard No. 149-1979, *Test Procedures for Antennas*, IEEE Publications, New York, 1979.
3. IEEE Standard 145-1983, IEEE Standard Definition of terms for Antennas, 1983.
4. J. D. Kraus, *Electromagnetics*, 3rd ed., McGraw-Hill, New York, 1984.
5. C. A. Balanis, *Advanced Engineering Electromagnetics*, John Wiley & Sons, 1989.
6. H. Mott, *Remote Sensing with Polarimetric Radar*, John Wiley & Sons, New Jersey, 2007.
7. W. M. Boerner et al., eds., Direct and inverse methods in radar polarimetry, *Proceedings of the NATO-ARW*, September 18–24, 1988, 1987–91, NATO ASI Series C: Math & Phys. Sciences, vol. C-350, Parts 1&2, Kluwer Academic Publications, 1992.
8. W. L. Stutzman, *Polarization in Electromagnetic Systems*, p. 52, Artech House, Boston, 1993.
9. M. Born and E. Wolf, *Principles of Optics*, 6th ed., Pergamon Press, London, 1959.
10. E. Krogager, Aspects of polarimetric radar imaging, Doctoral Thesis, Technical University of Denmark, May 1993.

11. E. Pottier and F. Famil, Polarimetry from basics to applications, *Tutorials of IGARSS'04*, IEEE, 2004.
12. *IEEE Standard Dictionary of Electrical and Electronics Terms*, 3rd ed. IEEE, 1984.
13. Y. Yamaguchi, *Radar Polarimetry from Basics to Applications: Radar Remote Sensing using Polarimetric Information (in Japanese)*, IEICE, Tokyo, 2007.
14. J. S. Lee and E. Pottier, *Polarimetric Radar Imaging from Basics to Applications*, CRC Press, 2009.
15. J. R. Pierce, *Almost Everything about Waves*, MIT Press, Cambridge, MA, pp. 130–131, 1974.

3 Scattering Polarimetry

Radar is an instrument to determine the range distance of an object by time delay of electromagnetic waves and detect the object by the reflection amplitude. Radar is the abbreviation for "radio detection and ranging." Although invented during World War II, radar nowadays is a familiar name to our daily life through weather forecasts and automobile collision avoidance systems. Since there are so many books and literature on radar, it is impossible to cover all items. Some selected books are cited as references [1–15].

Monostatic radar refers to a radar system in which the position of the transmitting antenna is the same as that of the receiving antenna as shown in Figure 3.1a. On the other hand, bistatic radar has the transmitting and receiving antennas at different positions (3.1b). Monostatic radar is frequently used because of its maximum phase sensitivity. If a black-colored target moves ΔR to the position of gray color in the R direction of Figure 3.1, the phase difference of the electromagnetic wave becomes $\frac{4\pi\Delta R}{\lambda}$ for the monostatic radar. The phase difference by bistatic radar is always less than $\frac{4\pi\Delta R}{\lambda}$. The maximum phase difference is always obtained by the monostatic radar. Since this phase information plays the most important role in ranging, we deal with the monostatic radar system in this book.

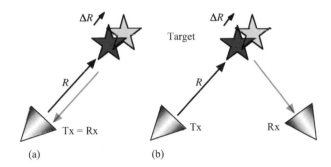

FIGURE 3.1 Locations of transmitting and receiving antenna for (a) monostatic and (b) bistatic radar. (From Yamaguchi, Y., *Radar Polarimetry from Basics to Applications: Radar Remote Sensing Using Polarimetric Information [in Japanese]*, IEICE, 2007.)

This chapter is devoted to the understanding of receiving power in polarimetric radar with an emphasis placed on the polarization state. The receiving power is dependent on the polarization states of the receiving antenna and incoming wave. For example, a horizontal wave antenna receives horizontally polarized waves, but it cannot receive vertically polarized waves. In this chapter, we see how the receiving power is expressed and changed by polarization. In the following, brief radar principle, radar equation, and power expression are reviewed. Then basic principles of radar polarimetry including reciprocity theorem, receiving antenna voltage, scattering matrix, and receiving power are introduced. Since the receiving power changes according to polarization states of the transmitter and receiver, it becomes possible to choose optimum polarization states for various applications. Using a specific polarization state, we can extend it, for example, to obtain characteristic polarization states of the target, polarimetric filtering for clutter reduction, detection and classification, identification, etc. Radar polarimetry is a general technology that makes use of fully polarimetric information contained in radar waves.

3.1 BASIC RADAR PRINCIPLE

Radar **ranging** is determined by the round trip time of pulsed wave as shown in Figure 3.2. If the distance between radar and target is R, the round-trip time of electromagnetic wave becomes

$$\tau = \frac{2R}{c} \tag{3.1.1}$$

where c is the speed of electromagnetic wave

$$c = 3 \times 10^8 \, \text{m/s} \tag{3.1.2}$$

The round-trip time τ is also called a time delay. The distance R is determined by measuring the time delay.

If a radar wave impinges on a target, scattering waves are generated and radiated in various directions. Scattering is a complicated phenomenon depending on target shape, material, wavelength, incidence angle, polarization, etc. Here we focus on radar power expressions in relation to receiving power in communication.

3.1.1 FRIIS TRANSMISSION EQUATION AND RADAR RANGE EQUATION

Two antennas are separated by distance r as shown in Figure 3.3. The distance r is assumed much larger than the wavelength. Let's consider the receiving power of antenna #2 when antenna #1 transmits a power P_t.

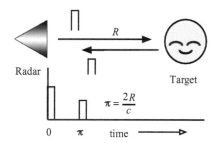

FIGURE 3.2 Radar ranging principle. (From Yamaguchi, Y., *Radar Polarimetry from Basics to Applications: Radar Remote Sensing Using Polarimetric Information [in Japanese]*, IEICE, 2007.)

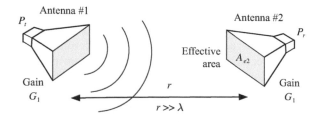

FIGURE 3.3 Receiving power of antenna #2 (communication).

When antenna #1 radiates the electromagnetic power P_t, the power density at r becomes,

$$\frac{P_t}{4\pi r^2} \tag{3.1.3}$$

If antenna #1 has gain G_1, which includes the directivity and the antenna efficiency, the power density is multiplied by G_1.

$$\frac{P_t G_1}{4\pi r^2} \tag{3.1.4}$$

Assuming antenna #2 has the effective area A_{e2}, the input power into antenna #2 can be expressed as

$$\frac{P_t G_1}{4\pi r^2} A_{e2} \tag{3.1.5}$$

Since the effective area and the antenna gain is related by

$$A_{e2} = \frac{\lambda^2 G_2}{4\pi} \tag{3.1.6}$$

the receiving power of antenna #2, P_r, can be written as

$$P_r = \frac{P_t G_1}{4\pi r^2} \frac{\lambda^2 G_2}{4\pi} = \left(\frac{\lambda}{4\pi r}\right)^2 G_1 G_2 P_t \tag{3.1.7}$$

This equation is known as the **Friis transmission equation** and it relates the power P_r (delivered to the receiver load) to the input power of the transmitting antenna P_t. The term $\left(\frac{\lambda}{4\pi r}\right)^2$ is called the free-space loss factor, and it takes into account the losses due to the spherical spreading of the energy by the antenna. This equation is the basis for communication systems. In this configuration, the receiving power decreases as r^{-2} with increasing r.

Next, we consider a radar measurement case as shown in Figure 3.4. Antenna #2 of Figure 3.3 is replaced by an object whose power reflection coefficient is σ.

The transmitted power is incident upon a target. Then it will be scattered in all directions with directivity by the target. The effective power reflection coefficient, scattering cross-section σ, is defined, which accounts for directional pattern. It serves to define a re-radiation source of scattering power. Using this parameter, the new power source can be expressed as

$$\frac{P_t G_1}{4\pi r^2} \sigma \tag{3.1.8}$$

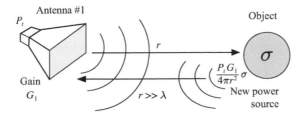

FIGURE 3.4 Receiving power of antenna #1 for monostatic radar sensing.

The power density at antenna #1 after traveling distance r from the target becomes,

$$\frac{1}{4\pi r^2} \frac{P_t G_1}{4\pi r^2} \sigma \tag{3.1.9}$$

Using the effective area A_{e1}, the receiving power of antenna #1 is given by

$$P_r = \frac{1}{4\pi r^2} \frac{P_t G_1}{4\pi r^2} \sigma A_{e1} = \frac{\lambda^2 G_1^2 P_t}{(4\pi)^3 r^4} \sigma \tag{3.1.10}$$

This is known as the **radar range equation**. It relates the receiving power P_r, radar cross-section σ, and transmitting power P_t. This equation establishes the basis of radar power relations. The receiving power decreases as r^{-4} with increasing r. This is a significant character in a radar-sensing scenario. Compared to the communication case, the radar coverage area is restricted by this significant decay of the power. Since this dependency r^{-4} is inevitable theoretically, some countermeasure techniques such as sensitivity time control [5] is employed.

3.1.2 MAXIMUM DETECTABLE RANGE

How far can we detect a target? It is an interesting question for radar applications. The far-most distance is called maximum detectable range, denoted as r_{max}, and is determined by the minimum power sensitivity S_{min} of the radar receiver. By equating $P_r = S_{min}$ in equation (3.1.10), the expression of maximum detectable range becomes,

$$r_{max} = \left[\frac{\lambda^2 G_1^2 P_t}{(4\pi)^3 S_{min}} \sigma \right]^{\frac{1}{4}} \tag{3.1.11}$$

where S_{min} is the minimum power that is almost equal to the noise level of the instrument.

$$S_{min} = kTB \tag{3.1.12}$$

k: Boltzmann's constant $\left(1.38 \times 10^{-23} \text{J/K}\right)$, T: temperature [K], B: bandwidth [Hz].

It can be understood that the detectable range does not increase so much if the transmitting power is doubled. Due to the expression of (3.1.10), it gives $2^{\frac{1}{4}} \approx 1.19$. So the detectable range increases 19% only compared to the original case. A 10-dB increase in the transmitting power results in 1.78 times the range expansion only, not 10 times!

In addition, if there is conductivity in the propagation medium, the maximum detectable range further decreases by the exponential function. The electric field in a lossy medium can be written as

$$\mathbf{E} = \mathbf{E}_0 \exp(-\alpha r) \tag{3.1.13}$$

where α [dB/m] is the attenuation constant. Therefore, the receiving power becomes

$$P_r = P_r^0 \exp(-2\alpha r) \tag{3.1.14}$$

Compared to the free-space power (3.1.10), a further attenuation factor $\exp(-2\alpha r)$ is multiplied to the receiving power P_r^0. For ground-penetrating radar (GPR) applications, this attenuation causes severe problems in synthetic aperture radar (SAR) imaging in the underground. Due to the attenuation and power dissipation, no signal from the deep target is obtained. So, it is really difficult to detect objects deep in the underground and in a highly conducting medium.

3.1.3 RADAR CROSS SECTION (RCS)

If we take a close look at the radar equation, all the information on the target is saved in σ. Other terms are related to radar systems and independent of target information. The same target should provide the same reflection coefficient regardless of radar system. In order to express the target itself, we need a criterion to define a quantity regardless of range and transmitting/receiving power. RCS σ is thus defined to express the target size and is written as

$$\sigma = \sigma(\theta,\varphi) = \lim_{r\to\infty} 4\pi r^2 \left| \frac{\mathbf{E}^s(\theta,\varphi)}{\mathbf{E}^i} \right|^2 \quad \left[m^2 \right] \tag{3.1.15}$$

where $\mathbf{E}^i$ is the electric field incident on the target, $\mathbf{E}^s(\theta,\varphi)$ is the scattered field, and (θ,φ) are the angular components of the spherical coordinate system.

$\sigma(\theta,\varphi)$ represents the directional pattern of scattering wave energy. The definition of (3.1.15) seems difficult; however, it is quite similar to the definition of antenna directivity. It represents the power ratio in a specific direction compared with the mean power in the omni-direction. The mean power is the input power divided by $4\pi r^2$.

The backscattering direction is defined as the direction toward the radar as shown in Figure 3.5. Among various scattering directions, the backscattering direction is particularly important because the monostatic radar operates in this scattering mode. Radar cross sections of various metallic objects are listed in Table 3.1. These objects are assumed to be much larger than the wavelength λ.

The sphere has the same spherical shape from all viewing directions. It is suitable for a calibration target because the setting and arrangement of the target are easy in the measurement. However, the value of RCS is not large enough compared to that of the square side trihedral corner reflector or of triangle side trihedral corner reflector. In a radar calibration task, a calibrator with small size and large RCS is desired. Since the trihedral corner reflector has the largest RCS and also has the retro-directive property, they are frequently used for radiometric calibration. The desirable size of trihedral corner reflector has been found to be more than 8 wavelengths by frequency difference and time domain (FDTD) analysis. An example of RCS values is depicted in Figure 3.6 for the case of L-band frequency. The square-sided trihedral has a 10-dB larger RCS than that of triangle-sided trihedral corner reflector. The dihedral corner reflector is also used for polarimetric calibration to adjust the amplitude and phase of the horizontal and the vertical polarized waves.

The RCS of the metallic sphere is πa^2, which is equal to the circle area of radius a and the projection of the sphere. If the sphere is not metallic, the RCS value becomes different as shown in

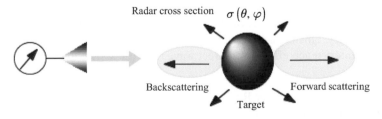

FIGURE 3.5 Backscattering and forward scattering. (From Yamaguchi, Y., *Radar Polarimetry from Basics to Applications: Radar Remote Sensing Using Polarimetric Information [in Japanese]*, IEICE, 2007.)

TABLE 3.1

Radar Cross Section (RCS)

Type	Shape	RCS (Max)
Sphere	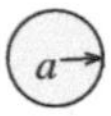	πa^2
Plate	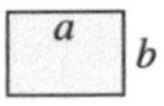	$\dfrac{4\pi\, a^2 b^2}{\lambda^2}$
Circular cylinder	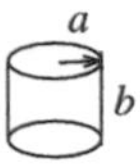	$\dfrac{2\pi\, ab^2}{\lambda}$
Dihedral	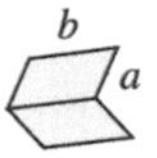	$\dfrac{8\pi\, a^2 b^2}{\lambda^2}$
Square side trihedral	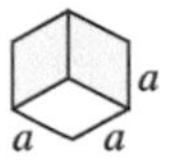	$\dfrac{12\pi\, a^4}{\lambda^2}$
Triangular side trihedral		$\dfrac{4\pi\, a^4}{3\lambda^2}$

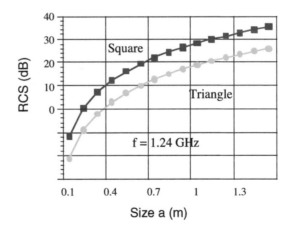

FIGURE 3.6 Radar cross section of trihedral corner reflector (f = 1.24 GHz). (From Yamaguchi, Y., *Radar Polarimetry from Basics to Applications: Radar Remote Sensing Using Polarimetric Information [in Japanese]*, IEICE, 2007.)

Figure 3.7. If the material is dielectric, the RCS becomes small and behaves in a random fashion as a function of frequency. This is due to the interaction of wave and materials. The oscillating ripples are caused by the interference effect by a direct wave from the sphere and multiple creeping waves on the surface.

The region of $\frac{2\pi a}{\lambda} \ll 1$ is called the Rayleigh scattering area where RCS is proportional to λ^{-4}. As seen in the figure, RCS changes drastically with the wavelength. Objects such as raindrops are the main target for this region. If two wavelength radars are employed, the receiving power difference will be a good indicator for target-size detection.

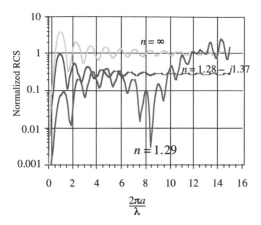

FIGURE 3.7 Normalized RCS due to the difference of material (*n*: refraction index). (From Yamaguchi, Y., *Radar Polarimetry from Basics to Applications: Radar Remote Sensing Using Polarimetric Information [in Japanese]*, IEICE, 2007.)

3.1.4 BACKSCATTERING FROM DISTRIBUTED TARGET

If the radar target is distributed and much larger than the radar footprint, the backscattering radar cross section becomes difficult to define. Just like a continuous cloud volume in the sky or extended terrain surface, there are several kinds of distributed targets. The receiving power of a radar system depends not only on target distribution but also on the radar footprint and resolution as shown in Figure 3.8. To define RCS more effectively, we use normalized RCS rather than absolute RCS. Since RCS has a dimension of $[\text{m}^2]$, we divide it by the target area and use the normalized $\sigma_0\,[\text{m}^2/\text{m}^2]$ as an averaged value in the footprint,

$$\sigma_0 = \left\langle \frac{\sigma_i}{\Delta A_i} \right\rangle \tag{3.1.16}$$

where ΔA_i represents the differential target area within the footprint. This normalized RCS is called backscattering coefficient or "sigma zero." It is a dimensionless value. If radar parameters in the footprint are constant, we have $\sigma = \sigma_0 \Delta A_i$.

Therefore, the radar equation for the distributed target can be written as

$$P_r = \iint_S \frac{\lambda^2 G_1^2 P_t}{(4\pi)^3 r^4} \sigma_0\, dS \tag{3.1.17}$$

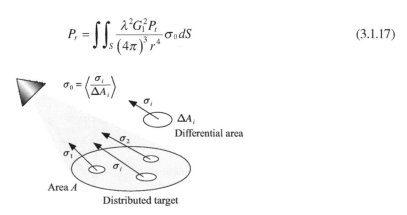

FIGURE 3.8 Normalized RCS from a distributed target. (From Yamaguchi, Y., *Radar Polarimetry from Basics to Applications: Radar Remote Sensing Using Polarimetric Information [in Japanese]*, IEICE, 2007.)

This equation is the most frequently used one in radar applications. The value of σ_0 is used for comparison of distributed target.

In addition to σ_0, the following normalized RCS is defined

$$\beta_0 = \frac{\sigma_0}{\sin\theta} \ , \ \gamma_0 = \frac{\sigma_0}{\cos\theta} \tag{3.1.18}$$

where θ is the incidence angle. The use of gamma zero is preferred for forest monitoring because the terrain modulation is eliminated [15].

3.1.5 Scattering by Polarization

So far, the radar power expression (3.1.15) has been derived for general purpose. However, one missing issue is polarimetric information. For example, if a vertically polarized wave impinges on a vertical grid as shown in Figure 3.9, it cannot go through the grid and is totally reflected. On the other hand, a horizontally polarized wave can go through the vertical grid without reflection. If a radar system operates in the vertical polarization, it obtains information from the grid. But a radar system of the horizontal polarization only cannot obtain any information from the grid.

Similarly, if an oriented dipole is placed in front of a vertical polarization radar as shown in Figure 3.10, an orthogonal component is created by the oriented dipole. Since the radar cannot obtain the orthogonal horizontal component, it loses the information about the oriented dipole. If a fully polarimetric radar is used, the orthogonal component can be obtained, and hence the same RCS can be obtained after the scattered wave reception. Therefore, it is necessary to formulate a radar equation taking into account the polarimetric information.

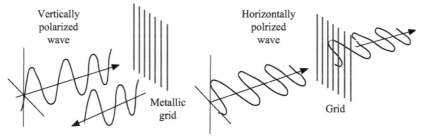

FIGURE 3.9 Reflection difference by polarization. (From Yamaguchi, Y., *Radar Polarimetry from Basics to Applications: Radar Remote Sensing Using Polarimetric Information [in Japanese]*, IEICE, 2007.)

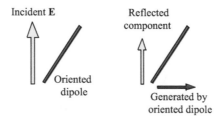

FIGURE 3.10 Reflected field by oriented dipole. (From Yamaguchi, Y., *Radar Polarimetry from Basics to Applications: Radar Remote Sensing Using Polarimetric Information [in Japanese]*, IEICE, 2007.)

In order to take into account the polarization directions, a new RCS σ_{pq} is introduced as

$$\sigma_{pq} = \sigma_{pq}(\theta,\varphi) = \lim_{r\to\infty} 4\pi r^2 \left| \frac{\mathbf{E}_p^s(\theta,\varphi)}{\mathbf{E}_q^i} \right|^2 \qquad (3.1.19)$$

where the subscript p refers to scattering wave polarization state and q indicates incident polarization state to the target. For example, σ_{HV} indicates the RCS when the vertically polarized wave is incident on the target ($q = V$), and the horizontally polarized scattering wave is received ($p = H$). Accordingly, we can obtain the following power reflection coefficients:

$$\begin{bmatrix} P_h^s \\ P_v^s \end{bmatrix} = K \begin{bmatrix} \sigma_{hh} & \sigma_{hv} \\ \sigma_{vh} & \sigma_{vv} \end{bmatrix} \begin{bmatrix} P_h^t \\ P_v^t \end{bmatrix}, \qquad K : \text{constant} \qquad (3.1.20)$$

By measuring the four RCS value, it becomes possible to examine the polarimetric dependency of the target to some extent. The power itself is a scalar value. Since phase information was difficult to measure accurately at the start-up time, it was impossible to obtain full information on scattering phenomena. With the advancement of technology, it is nowadays possible to obtain phase information very accurately.

The fundamental quantity of scattering phenomena is the electric field itself. It contains not only amplitude information but also phase information. Polarimetric phase information plays a key role in identifying objects. Amplitude only, or power only, does not yield sufficient information on the object. By considering the basic theory of electromagnetic wave scattering, radar equation and related issues were refined with the amplitude and phase of the electric field in 1950s. The theoretical work has been developed by several researchers. The phrase "radar polarimetry" was created during this period. In the 1980s, the Airborne AIRSAR system by NASA-JPL conducted experiments, verified the theoretical results, and demonstrated the importance of fully polarimetric information [16,17]. After the advent of new fully polarimetric SAR system, various airborne and spaceborne systems have been developed.

Here we start with the principles of electromagnetic waves with the emphasis placed on the vector nature of the electric field. In the following, the reciprocity theorem, antenna voltage, scattering matrix, and basic power equations of radar polarimetry are explained. The receiving power is expressed in terms of polarization parameters, and polarization signatures are presented to show how the power changes with polarization states.

3.2 RECIPROCITY THEOREM

As shown in Figure 3.11, let's consider the relation between two sources and the fields generated by them [2]. The medium is assumed to be isotropic and homogeneous, bounded by the volume V. The electric current and the magnetic current source pairs $(\mathbf{J}_1, \mathbf{M}_1)$ and $(\mathbf{J}_2, \mathbf{M}_2)$ produce

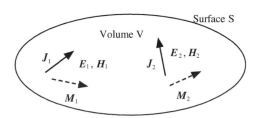

FIGURE 3.11 Source and field in isotropic and homogeneous medium. (From Yamaguchi, Y., *Radar Polarimetry from Basics to Applications: Radar Remote Sensing Using Polarimetric Information [in Japanese]*, IEICE, 2007.)

the electric field and the magnetic field pairs $(\mathbf{E}_1, \mathbf{H}_1)$ and $(\mathbf{E}_2, \mathbf{H}_2)$, respectively. These sources satisfy Maxwell's equation,

$$\nabla \times \mathbf{H}_1 = \mathbf{J}_1 + j\omega\varepsilon\,\mathbf{E}_1, \qquad \nabla \times \mathbf{E}_1 = -\mathbf{M}_1 - j\omega\mu\,\mathbf{H}_1$$
$$\nabla \times \mathbf{H}_2 = \mathbf{J}_2 + j\omega\varepsilon\,\mathbf{E}_2, \qquad \nabla \times \mathbf{E}_2 = -\mathbf{M}_2 - j\omega\mu\,\mathbf{H}_2 \tag{3.2.1}$$

The vector identity

$$\nabla \cdot (\mathbf{A} \times \mathbf{B}) = \mathbf{B} \cdot \nabla \times \mathbf{A} - \mathbf{A} \cdot \nabla \times \mathbf{B} \tag{3.2.2}$$

and (3.2.1) and (3.2.2) derive the following relation:

$$-\nabla \cdot (\mathbf{E}_1 \times \mathbf{H}_2 - \mathbf{E}_2 \times \mathbf{H}_1) = \mathbf{E}_1 \cdot \mathbf{J}_2 - \mathbf{E}_2 \cdot \mathbf{J}_1 + \mathbf{H}_2 \cdot \mathbf{M}_1 - \mathbf{H}_1 \cdot \mathbf{M}_2 \tag{3.2.3}$$

This equation can be transformed into an integral form,

$$-\iint (\mathbf{E}_1 \times \mathbf{H}_2 - \mathbf{E}_2 \times \mathbf{H}_1) \cdot d\mathbf{S} = \iiint (\mathbf{E}_1 \cdot \mathbf{J}_2 - \mathbf{E}_2 \cdot \mathbf{J}_1 + \mathbf{H}_2 \cdot \mathbf{M}_1 - \mathbf{H}_1 \cdot \mathbf{M}_2)\,dv \tag{3.2.4}$$

The equation (3.2.4) is called the **Lorentz reciprocity theorem**.

If there is no source inside the volume, the right-hand side of equation (3.2.4) vanishes,

$$\iint (\mathbf{E}_1 \times \mathbf{H}_2 - \mathbf{E}_2 \times \mathbf{H}_1) \cdot d\mathbf{S} = 0 \tag{3.2.5}$$

On the other hand, even if there are sources inside the volume, the left-hand side of equation (3.2.4) vanishes when the enclosing surface is taken at infinity. Then we obtain from the right-hand side of equation (3.2.4) as

$$\iiint (\mathbf{E}_1 \cdot \mathbf{J}_2 - \mathbf{H}_1 \cdot \mathbf{M}_2)\,dv = \iiint L(\mathbf{E}_2 \cdot \mathbf{J}_1 - \mathbf{H}_2 \cdot \mathbf{M}_1)\,dv \tag{3.2.6}$$

This relation of equation (3.2.6), called "reaction" by the source and the field, presents the connection between the two. We assign both sides of equation (3.2.6) as

$$\langle 1,2 \rangle = \iiint (\mathbf{E}_1 \cdot \mathbf{J}_2 - \mathbf{H}_1 \cdot \mathbf{M}_2)\,dv \tag{3.2.7}$$

$$\langle 2,1 \rangle = \iiint (\mathbf{E}_2 \cdot \mathbf{J}_1 - \mathbf{H}_2 \cdot \mathbf{M}_1)\,dv \tag{3.2.8}$$

Then the reciprocity theorem becomes a very simple form by this notation.

$$\langle 1,2 \rangle = \langle 2,1 \rangle \tag{3.2.9}$$

This means the reaction of field 1 caused by source 2 is the same as that of field 2 caused by source 1.

Now, this reciprocity theorem is applied to the voltage induced at the receiving antenna caused by the source current on the transmitting antenna. For simplicity, we assume there are

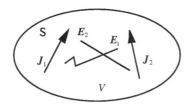

FIGURE 3.12 Reaction in source and field. (From Yamaguchi, Y., *Radar Polarimetry from Basics to Applications: Radar Remote Sensing Using Polarimetric Information [in Japanese]*, IEICE, 2007.)

no magnetic currents $\mathbf{M}_1 = \mathbf{M}_2 = 0$. Then the situation becomes as shown in Figure 3.12 where two current sources $\mathbf{J}_1, \mathbf{J}_2$ and corresponding electric fields $\mathbf{E}_1, \mathbf{E}_2$ exist in the infinite space.

The reaction (3.2.7) becomes $\qquad \langle 1, 2 \rangle = \iiint \mathbf{E}_1 \cdot \mathbf{J}_2 dv$

The volume integral is replaced to the line integral along $\mathbf{J}_2$. $\iiint \mathbf{E}_1 \cdot \mathbf{J}_2 dv = \int \mathbf{E}_1 \cdot I_2 d\mathbf{L}$

From this equation, we can derive $\int \mathbf{E}_1 \cdot I_2 \, d\mathbf{L} = I_2 \int \mathbf{E}_1 \cdot d\mathbf{L} = -I_2 V_{2,1}$

Therefore, $\qquad\qquad\qquad\qquad\qquad \langle 1, 2 \rangle = -I_2 V_{2,1} \qquad\qquad\qquad\qquad (3.2.10)$

where $V_{2,1}$ is the voltage across source 2 due to the field $\left(\mathbf{E}_1 \right)$ produced by source 1 $\left(\mathbf{J}_1 \right)$.
Similarly, we have

$$\langle 2, 1 \rangle = \iiint \mathbf{E}_2 \cdot \mathbf{J}_1 dv = \int \mathbf{E}_2 \cdot I_1 d\mathbf{L} = I_1 \int \mathbf{E}_2 \cdot d\mathbf{L} = -I_1 V_{1,2} \qquad (3.2.11)$$

$V_{1,2}$ is the voltage across source 1 due to the fields $\left(\mathbf{E}_2 \right)$ by source 2 $\left(\mathbf{J}_2 \right)$ is $V_{1,2}$. Therefore, the following equality holds from the reciprocity theorem.

$$\langle 1, 2 \rangle = \langle 2, 1 \rangle \ \Rightarrow \ I_2 V_{2,1} = I_1 V_{1,2} \qquad\qquad (3.2.12)$$

If we let $I_1 = I_2$, then we have the relation $V_{1,2} = V_{2,1}$.
This means the antenna voltages become the same if we use the same magnitude current excitation in each antenna.

3.3   RECEIVING VOLTAGE

We consider here the antenna voltage expression using the effective length as shown in Figure 3.13 [7]. If we flow the same current I on the two antennas, the radiation field from each antenna becomes

$$\mathbf{E}^{\mathrm{t}} = \frac{j \eta_0 I}{2\lambda} \frac{e^{-jkr}}{r} \mathbf{h}, \qquad \mathbf{E}^{\mathrm{i}} = \frac{j \eta_0 I}{2\lambda} \frac{e^{-jkr}}{r} \mathbf{L} \qquad\qquad (3.3.1)$$

where $\mathbf{h}$ is a vector form of the effective length of antenna #1, and $\mathbf{L}$ is that of antenna #2. If a small dipole antenna with length L is used, the effective length is $\mathbf{L} = L \sin \theta \, \mathbf{a}_\theta$. The vector effective length represents the electric field vector seen from the antenna.

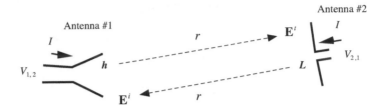

FIGURE 3.13 Radiation fields of two antennas and vector effective lengths h and L.

The radiated field of equation (3.3.1) becomes the incident field to each other. Then the open voltage $V_{2,1}$ across antenna #2 due to $\mathbf{E}^t$ can be written as

$$V_{2,1} = -\int_{\#2} \mathbf{E}^t \cdot d\mathbf{l} = \mathbf{E}^t \cdot \mathbf{L} = \frac{j\eta_0 I}{2\lambda} \frac{e^{-jkr}}{r} \mathbf{h} \cdot \mathbf{L} \tag{3.3.2}$$

This should be equal to the voltage of antenna #1 from the reciprocity theorem, so that

$$V_{1,2} = -\int_{\#1} \mathbf{E}^i \cdot d\mathbf{l} = \mathbf{E}^i \cdot \mathbf{h} = \frac{j\eta_0 I}{2\lambda} \frac{e^{-jkr}}{r} \mathbf{L} \cdot \mathbf{h} \tag{3.3.3}$$

Therefore, the voltage across antenna #1 can be expressed by the vector effective length $\mathbf{h}$ and the incoming wave $\mathbf{E}^i$

$$V_{1,2} = \mathbf{E}^j \cdot \mathbf{h} = \mathbf{h} \cdot \mathbf{E}^j \tag{3.3.4}$$

Note that the origin of $\mathbf{h}$ and $\mathbf{E}^i$ is different; however, the power of equation (3.3.6) associated with this voltage remains the same. $\mathbf{E}^i$ is seen from antenna #1 and treated as Jones vector representation (see Section 3.6.1).

As can be seen from equation (3.3.4), the antenna voltage is determined by the effective length $\mathbf{h}$ and the incident field $\mathbf{E}^i$. The effective length $\mathbf{h}$ is a vector representing the polarization state of radiating field from the antenna. The origin of $\mathbf{h}$ and the incident field $\mathbf{E}^i$ is taken at antenna #1 of the same coordinate.

Note that the voltage across antenna (3.3.4) yields a complex scalar value. If $\mathbf{h}$ and $\mathbf{E}^i$ are orthogonal to each other, then the voltage is zero, and if they are in complex conjugate relation, the voltage becomes maximum. The form of equation (3.3.4) looks like the "inner product," but it is not correct since it does not apply the complex conjugate operation. It seems better to rewrite as

$$V = \mathbf{h}^T \mathbf{E}^i = h_\theta E_\theta + h_\varphi E_\varphi \tag{3.3.5}$$

to avoid confusion. The superscript T in equation (3.3.5) denotes transpose.

Once the antenna voltage is given, the receiving power P is derived from the circuit theory as

$$P = \frac{1}{8R_a} VV^* = \frac{1}{8R_a} |V|^2 \tag{3.3.6}$$

where R_a is the matching load connected to the receiving antenna.

3.4 SCATTERING MATRIX

Polarimetric radar transmits a horizontally polarized wave to the target and receives horizontally and vertically polarized waves simultaneously. Then the radar switches the transmitting antenna from horizontal to vertical and transmits the vertically polarized wave. It receives horizontally and vertically polarized wave simultaneously again. By this procedure, the radar acquires 2×2 polarimetric scattering information. This 2×2 matrix is called Sinclair scattering matrix [9].

Figure 3.14 shows three coordinate systems of the radar and target. The transmitter (Tx) has (x_1, y_1, z_1), the scatterer coordinate is (x_2, y_2, z_2), and the receiver (Rx) has (x_3, y_3, z_3). The coordinate system with a forward scattering Alignment (FSA) is suitable for describing forward scattering in which the z-axis is the propagation direction. Bistatic scattering alignment (BSA) has a transmitter and receiver at different locations. Monostatic scattering alignment (MSA) deals with a backscattering measurement where Tx and Rx are co-located. MSA is a special case of bistatic configuration where the locations of Tx and Rx coincide. In the monostatic case, the coordinate of Tx and that of Rx is the same. It is convenient to use a single coordinate system for measurement.

3.4.1 DEFINITION OF SCATTERING MATRIX

The Jones vector expression of the transmitting field Tx is

$$\mathbf{E}^t = \begin{bmatrix} E_{x1} \\ E_{y1} \end{bmatrix} \tag{3.4.1}$$

This becomes the incident field to target after propagating distance r,

$$\mathbf{E}^i = \frac{e^{-jkr}}{\sqrt{4\pi r}} \mathbf{E}^t \tag{3.4.2}$$

The incident field is scattered on the target. The backscattering field is translated to

$$\mathbf{E}^s = \begin{bmatrix} E_{x2}^s \\ E_{y2}^s \end{bmatrix} = \begin{bmatrix} A_{x2x1} A_{x2y1} \\ A_{y2x1} A_{y2y1} \end{bmatrix} \begin{bmatrix} E_{x1} \\ E_{y1} \end{bmatrix} \left(\text{expressed in FSA} \right) \tag{3.4.3}$$

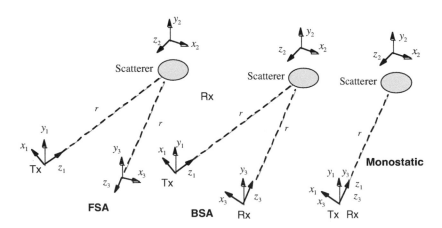

FIGURE 3.14 The coordinates of radar system and target. FSA, BSA, and MSA. (From Yamaguchi, Y., *Radar Polarimetry from Basics to Applications: Radar Remote Sensing Using Polarimetric Information [in Japanese]*, IEICE, 2007.)

This goes back to radar. The propagation direction is opposite to the incoming wave. If this wave is expressed in the receiver coordinate (x_3, y_3, z_3), it becomes,

$$\mathbf{E}^s = \begin{bmatrix} E_{x3}^s \\ E_{y3}^s \end{bmatrix} = \begin{bmatrix} A_{x3x1} A_{x3y1} \\ A_{y3x1} A_{y3y1} \end{bmatrix} \begin{bmatrix} E_{x1} \\ E_{y1} \end{bmatrix} \quad \text{(BSA)} \tag{3.4.4}$$

Since the coordinate system is the same, we can take $x_1 = x_3 = x$, $y_1 = y_3 = y$, $z_1 = z_3 = z$. We define this 2×2 matrix

$$[S] = \begin{bmatrix} S_{xx} & S_{xy} \\ S_{yx} & S_{yy} \end{bmatrix} \tag{3.4.5}$$

as the **Sinclair scattering matrix** [8] defined in the radar coordinates. It can be simply written as

$$\mathbf{E}^s = [S] \mathbf{E}^i \tag{3.4.6}$$

The 2×2 matrix $[A]$ in FSA (3.4.3) is called the Jones matrix and is different from the scattering matrix.

The elements of the scattering matrix are complex-valued and independent of each other. There are four independent elements in the bistatic case, whereas there are three elements in the monostatic scattering due to the condition $S_{xy} = S_{yx}$. The reason $S_{xy} = S_{yx}$ for the monostatic case is derived from the reciprocity theorem.

Let's denote the superscript t as transmission, and r as reception. Then antenna voltage can be simply expressed as

$$V = \mathbf{h}_r^T \mathbf{E}^s = \begin{bmatrix} h_x^r & h_y^r \end{bmatrix} \begin{bmatrix} S_{xx} & S_{xy} \\ S_{yx} & S_{yy} \end{bmatrix} \begin{bmatrix} E_x^t \\ E_y^t \end{bmatrix} \tag{3.4.7}$$

If the transmitting antenna and the receiving antenna are changed, the voltage expression becomes,

$$V = \begin{bmatrix} E_x^t & E_y^t \end{bmatrix} \begin{bmatrix} S_{xx} & S_{xy} \\ S_{yx} & S_{yy} \end{bmatrix} \begin{bmatrix} h_x^r \\ h_y^r \end{bmatrix} \tag{3.4.8}$$

These two should be identical. Therefore, we have

$$S_{xy} = S_{yx} \tag{3.4.9}$$

This condition is often used as $S_{HV} = S_{VH}$ in the scattering matrix of a monostatic radar in the HV polarization basis. For anisotropic media with the Faraday rotation, this assumption does not hold.

Note that the subscript of the scattering matrix element Spq has the following meaning. The first p represents receiver polarization, whereas the second q denotes transmitter polarization as defined in the scattering matrix formulation (3.4.4) through (3.4.8). So, if the subscript HV appears, it means V-pol transmission and H-pol reception. This is often confused in our intuitive convention as H-transmission and V-reception.

Figure 3.15 shows some examples of the scattering mechanism and scattering matrix. The coordinate system is taken as the x-axis to the horizontal polarization (H) and the y-axis to the vertical polarization (V). The direction of electric field vector is described in the scattering matrix as 1 for

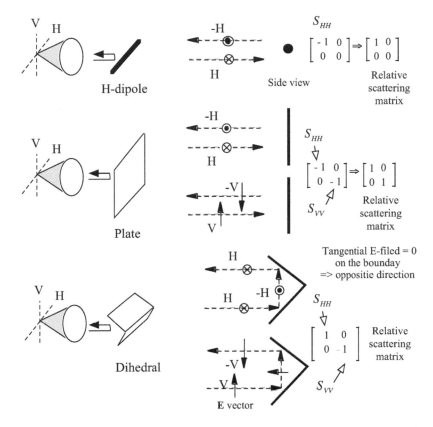

FIGURE 3.15 Scattering mechanism and scattering matrix. (From Yamaguchi, Y., *Radar Polarimetry from Basics to Applications: Radar Remote Sensing Using Polarimetric Information [in Japanese]*, IEICE, 2007.)

the same direction and −1 for the opposite direction. Due to the boundary condition, that is, the tangential electric field must be zero on the metallic surface, the direction of the scattered field can be understood for each target.

The scattering matrix is often expressed in a relative matrix form such that the *HH* component becomes a real number. There are five independent parameters in the relative scattering matrix. Not only the magnitude information, but the relations among the components are important for target recognition.

3.5 FUNDAMENTAL EQUATION OF RADAR POLARIMETRY

The fundamental principle of radar polarimetry is given by the following equations. The origin of the coordinate is located in the radar. Letting $\mathbf{E}^t$ as the transmitting polarization state, the scattered wave from the target is given as

$$\mathbf{E}^s = \begin{bmatrix} S \end{bmatrix} \mathbf{E}^t \tag{3.5.1}$$

where [S] represents a 2 × 2 complex scattering matrix. The polarization state of the scattered wave can be recovered by $\mathbf{E}^s$ using [S]. Therefore, [S] can be regarded as a polarization transformer from $\mathbf{E}^t$ to $\mathbf{E}^s$. The voltage induced on the receiving antenna is given by

$$V = \mathbf{h}^T \mathbf{E}^s \tag{3.5.2}$$

where $\mathbf{h}$ is the vector effective length when the receiving antenna acts as a transmitter. The receiving power P becomes

$$P = |V|^2 = |\mathbf{h}^T \mathbf{E}^s|^2 \tag{3.5.3}$$

Equations (3.5.1) through (3.5.3) establish the fundamental principles of radar polarimetry.

3.6   RECEIVING POWER EXPRESSION

Now we consider the expressions of receiving power of polarimetric radar. The receiving power is dependent on the incident field $\mathbf{E}^s$ to the receiver and the effective length $\mathbf{h}$. Since the scattered field $\mathbf{E}^s$ is dependent of $\mathbf{E}^t$, we have two independent parameters $\mathbf{E}^t$ and $\mathbf{h}$. Next, we express the power using the polarization ratio, geometrical parameters, and Stokes vector.

3.6.1   RECEIVING POWER BY POLARIZATION RATIO

The transmitting field with unit magnitude can be written in the Jones vector as

$$\mathbf{E}^t = \begin{bmatrix} E_x \\ E_y \end{bmatrix} = \frac{1}{\sqrt{1+\rho\rho^*}} \begin{bmatrix} 1 \\ \rho \end{bmatrix} \tag{3.6.1}$$

This field becomes the incidence to target, and the scattered field is multiplied by the scattering matrix,

$$\mathbf{E}^s = \begin{bmatrix} E_x^s \\ E_y^s \end{bmatrix} = [S]\mathbf{E}^t = \frac{[S]}{\sqrt{1+\rho\rho^*}} \begin{bmatrix} 1 \\ \rho \end{bmatrix} \tag{3.6.2}$$

where $\mathbf{E}^t$ and $\mathbf{E}^s$ are defined in the same radar coordinate system. If we transmit the left-handed circular (LHC) wave $(\rho = j)$ toward a flat plate, the scattered wave will return to the radar in the same rotation sense as shown in Figure 3.16. The rotation sense may be anticipated from the boundary condition of the tangential electric field ($E = 0$) and the time process.

The Jones vector expression becomes

$$\mathbf{E}^t = \frac{1}{\sqrt{1+jj^*}} \begin{bmatrix} 1 \\ j \end{bmatrix} = \frac{1}{\sqrt{2}} \begin{bmatrix} 1 \\ j \end{bmatrix} \qquad \text{LHC} \tag{3.6.3a}$$

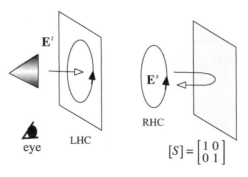

FIGURE 3.16   Polarization state of transmission and reception.

$$\mathbf{E}^s = [S]\mathbf{E}^t = \frac{1}{\sqrt{2}}\begin{bmatrix} 1 & 0 \\ 0 & 1 \end{bmatrix}\begin{bmatrix} 1 \\ j \end{bmatrix} = \frac{1}{\sqrt{2}}\begin{bmatrix} 1 \\ j \end{bmatrix} \tag{3.6.3b}$$

It should be noted that $\mathbf{E}^s$ has the same rotation sense as that of $\mathbf{E}^t$ when seen from the radar; however, it is a right-handed circular (RHC) because of the opposite propagation direction. We keep $\mathbf{E}^s$ just as the Jones vector as seen from the radar.

It is known experimentally that the LHC antenna cannot receive this $\mathbf{E}^s$. The situation is the same for the raindrop measurement using the same circular antennas, because the scattering matrix of raindrops is the same as that of the plate. In order to confirm the validity of the power expressions of (3.5.2) through (3.5.3), we take the receiver polarization as $\mathbf{h}$ when it acts as a transmitter

$$\mathbf{h} = \frac{1}{\sqrt{1 + \rho\rho^*}}\begin{bmatrix} 1 \\ \rho \end{bmatrix} \tag{3.6.4}$$

If the LHC antenna ($\rho = j$) is used for reception, the voltage and power expression is written as follows:

$$V = \mathbf{h}^T\mathbf{E}^s = \frac{1}{2}\begin{bmatrix} 1 & j \end{bmatrix}\begin{bmatrix} 1 \\ j \end{bmatrix} = 0 \qquad \therefore P = 0, \qquad \text{Min.} \tag{3.6.5a}$$

For the RHC antenna, we have

$$V = \mathbf{h}^T\mathbf{E}^s = \frac{1}{2}\begin{bmatrix} j & 1 \end{bmatrix}\begin{bmatrix} 1 \\ j \end{bmatrix} = j \qquad \therefore P = |j|^2 = 1, \qquad \text{Max.} \tag{3.6.5b}$$

These results are in a complete agreement with experimental results.

For H-pol transmission and reception,

$$V = \mathbf{h}^T[S]\mathbf{E}^t = \begin{bmatrix} 1 & 0 \end{bmatrix}\begin{bmatrix} 1 & 0 \\ 0 & 1 \end{bmatrix}\begin{bmatrix} 1 \\ 0 \end{bmatrix} = 1 \qquad \therefore P = 1 \qquad \text{Max.} \tag{3.6.6a}$$

For H-pol transmission and V-pol reception,

$$V = \mathbf{h}^T[S]\mathbf{E}^t = \begin{bmatrix} 0 & 1 \end{bmatrix}\begin{bmatrix} 1 & 0 \\ 0 & 1 \end{bmatrix}\begin{bmatrix} 1 \\ 0 \end{bmatrix} = 1 \qquad \therefore P = 0 \qquad \text{Min.} \tag{3.6.6b}$$

For H-pol transmission and RHC reception,

$$V = \frac{1}{\sqrt{2}}\begin{bmatrix} j & 1 \end{bmatrix}\begin{bmatrix} 1 & 0 \\ 0 & 1 \end{bmatrix}\begin{bmatrix} 1 \\ 0 \end{bmatrix} = \frac{j}{\sqrt{2}} \qquad \therefore P = \frac{1}{2} \tag{3.6.6c}$$

All the results agree with the experimental ones. Therefore, the validity of the power expressions of Equations (3.5.2) and (3.5.3) is confirmed. The Jones vector form is effective for power expression.

3.6.2 RECEIVING POWER OF THE POLARIMETRIC RADAR CHANNEL USING POLARIZATION RATIO AND GEOMETRICAL PARAMETERS

There are two independent parameters in polarimetric radar: one in transmitting polarization state $\mathbf{E}^t$, and another one is receiving polarization state $\mathbf{h}$. There is an infinite number of polarization states (see polarization map in Figure 2.12) for transmission and reception. Among them, we take three hypothetical combinations as shown in Figure 3.17. They are the co-polarization channel, cross-polarization channel, and matched polarization channel.

1. Co-polarization Channel, $\mathbf{h} = \mathbf{E}^t$

    The polarization state of the receiving antenna $\mathbf{h}$ is the same as that of transmitting antenna. $\left(\mathbf{h} = \mathbf{E}^t\right)$. This polarimetric channel is called co-pol channel. The receiving power P_c in the co-pol channel can be written in terms of ρ, or geometrical parameters as follows:

    $$P_c = \left|V_c\right|^2 = \left|\mathbf{E}^{tT}\left[S\right]\mathbf{E}^t\right|^2 = \left|\frac{\left[1\ \rho\right]\left[S\right]\begin{bmatrix}1\\\rho\end{bmatrix}}{1+\rho\rho^*}\right|^2 \qquad (3.6.7)$$

    $$V_c = \left[\cos\varepsilon\ j\sin\varepsilon\right]\begin{bmatrix}\cos\tau & \sin\tau\\-\sin\tau & \cos\tau\end{bmatrix}\begin{bmatrix}S_{xx} & S_{xy}\\S_{yx} & S_{yy}\end{bmatrix}\begin{bmatrix}\cos\tau & -\sin\tau\\\sin\tau & \cos\tau\end{bmatrix}\begin{bmatrix}\cos\varepsilon\\j\sin\varepsilon\end{bmatrix} \qquad (3.6.8)$$

2. Cross(X)-polarization channel, $\mathbf{h} = \mathbf{E}_\perp^t$

    The polarization state of the receiving antenna $\mathbf{h}$ is orthogonal to that of the transmitting antenna. This channel is called cross-pol or X-pol channel. The symbol $\pm$ denotes orthogonal.

    $$P_x = \left|V_x\right|^2 = \left|\mathbf{E}_\perp^{t\ T}\left[S\right]\mathbf{E}^t\right|^2 = \left|\frac{\left[\rho^*\ -1\right]\left[S\right]\begin{bmatrix}1\\\rho\end{bmatrix}}{1+\rho\rho^*}\right|^2 \qquad (3.6.9)$$

    $$V_x = \left[\cos\varepsilon\ -j\sin\varepsilon\right]\begin{bmatrix}-\sin\tau & \cos\tau\\\cos\tau & -\sin\tau\end{bmatrix}\left[S\right]\begin{bmatrix}\cos\tau & -\sin\tau\\\sin\tau & \cos\tau\end{bmatrix}\begin{bmatrix}\cos\varepsilon\\j\sin\varepsilon\end{bmatrix} \qquad (3.6.10)$$

3. Matched-polarization channel, $\mathbf{h} = \left(\mathbf{E}^s\right)^*$

    The polarization state of the receiving antenna is always matched to the scattered wave, so that $\mathbf{h} = \left(\mathbf{E}^s\right)^*$ holds for any polarization transmission. It is also a hypothetical channel

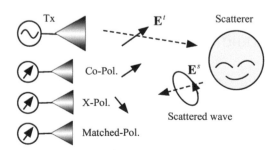

FIGURE 3.17 Three hypothetical polarization channels.

$$P_m = |V_m|^2 = \left| \left([S]\mathbf{E}^t \right)^{*T} [S]\mathbf{E}^t \right|^2 = \left| \frac{\begin{bmatrix} 1 & \rho^* \end{bmatrix} [S]^{*T} [S] \begin{bmatrix} 1 \\ \rho \end{bmatrix}}{1 + \rho\rho^*} \right|^2 \tag{3.6.11}$$

$$V_m = \begin{bmatrix} \cos\varepsilon - j\sin\varepsilon \end{bmatrix} \begin{bmatrix} \cos\tau & \sin\tau \\ -\sin\tau & \cos\tau \end{bmatrix} [S]^{*T} [S] \begin{bmatrix} \cos\tau & -\sin\tau \\ \sin\tau & \cos\tau \end{bmatrix} \begin{bmatrix} \cos\varepsilon \\ j\sin\varepsilon \end{bmatrix} \tag{3.6.12}$$

3.6.3 Receiving Power by the Stokes Vector

So far, we have been dealing with coherent scattering matrix, which has complex-valued elements. This coherent scattering matrix cannot be averaged directly for statistical analysis. On the other hand, the Stokes vector and Mueller matrix representation are convenient for the statistical analyses, such as addition or averaging, because they have the real-valued second-order elements. Therefore, the power expression based on Stokes vector formulation should be derived. In this section, the final expressions are given below. The details should be referred to [7–9].

Starting from the fundamental scattering equation (3.5.1), the power expression using the Stokes vector is given as

$$P = \frac{1}{2}\mathbf{g}_t^T [K]\mathbf{g}_t = \frac{1}{2} = \mathbf{g}_r^T \begin{bmatrix} 1 & 0 & 0 & 0 \\ 0 & 1 & 0 & 0 \\ 0 & 0 & 1 & 0 \\ 0 & 0 & 0 & -1 \end{bmatrix} [M]\mathbf{g}_t \tag{3.6.13}$$

where $\mathbf{g}_t$ and $\mathbf{g}_r$ are Stokes vectors (2.2.40) of transmitting and receiving, respectively. $[K]$ is the Kennaugh matrix and is related to the Mueller matrix $[M]$ in the following relation:

$$[K] = \begin{bmatrix} 1 & 0 & 0 & 0 \\ 0 & 1 & 0 & 0 \\ 0 & 0 & 1 & 0 \\ 0 & 0 & 0 & -1 \end{bmatrix} [M] \tag{3.6.14}$$

The Kennaugh matrix is used for radar (backscattering case), whereas the Mueller matrix is used in optics for forward scattering. The elements of a 4×4 Mueller matrix are given as follows:

$$m_{00} = \frac{1}{2}\left(|S_{xx}|^2 + |S_{xy}|^2 + |S_{yx}|^2 + |S_{yy}|^2 \right) \tag{3.6.15a}$$

$$m_{01} = \frac{1}{2}\left(|S_{xx}|^2 - |S_{xy}|^2 + |S_{yx}|^2 - |S_{yy}|^2 \right) \tag{3.6.15b}$$

$$m_{02} = \mathrm{Re}\left\{ S_{xx}S_{xy}^* + S_{yx}S_{yy}^* \right\} \tag{3.6.15c}$$

$$m_{03} = \mathrm{Im}\left\{ S_{xx}S_{xy}^* + S_{yx}S_{yy}^* \right\} \tag{3.6.15d}$$

$$m_{10} = \frac{1}{2}\left(|S_{xx}|^2 + |S_{xy}|^2 - |S_{yx}|^2 - |S_{yy}|^2 \right) \tag{3.6.15e}$$

$$m_{11} = \frac{1}{2}\left(\left|S_{xx}\right|^2 - \left|S_{xy}\right|^2 - \left|S_{yx}\right|^2 + \left|S_{yy}\right|^2 \right) \tag{3.6.15f}$$

$$m_{12} = \mathrm{Re}\left\{ S_{xx}S_{xy}^* - S_{yx}S_{yy}^* \right\} \tag{3.6.15g}$$

$$m_{13} = \mathrm{Im}\left\{ S_{xx}S_{xy}^* - S_{yx}S_{yy}^* \right\} \tag{3.6.15h}$$

$$m_{20} = \mathrm{Re}\left\{ S_{xx}S_{yx}^* + S_{xy}S_{yy}^* \right\} \tag{3.6.15i}$$

$$m_{21} = \mathrm{Re}\left\{ S_{xx}S_{yx}^* - S_{xy}S_{yy}^* \right\} \tag{3.6.15j}$$

$$m_{22} = \mathrm{Re}\left\{ S_{xx}S_{yy}^* + S_{xy}S_{yx}^* \right\} \tag{3.6.15k}$$

$$m_{23} = \mathrm{Re}\left\{ S_{xx}S_{yy}^* + S_{yx}S_{xy}^* \right\} \tag{3.6.15l}$$

$$m_{30} = -\mathrm{Im}\left\{ S_{xx}S_{yx}^* + S_{xy}S_{yy}^* \right\} \tag{3.6.15m}$$

$$m_{31} = -\mathrm{Im}\left\{ S_{xx}S_{yx}^* - S_{xy}S_{yy}^* \right\} \tag{3.6.15n}$$

$$m_{32} = -\mathrm{Im}\left\{ S_{xx}S_{yy}^* - S_{yx}S_{xy}^* \right\} \tag{3.6.15o}$$

$$m_{33} = \mathrm{Re}\left\{ S_{xx}S_{yy}^* - S_{xy}S_{yx}^* \right\} \tag{3.6.15p}$$

For a monostatic backscattering case where the relation $S_{yx} = S_{xy}$ holds, we have the relations,

$$m_{01} = m_{10}, \, m_{02} = m_{20}, \, m_{03} = -m_{30}, \, m_{12} = m_{21}, \, m_{13} = -m_{31}, \, m_{23} = -m_{32} \tag{3.6.16}$$

The Kennaugh matrix becomes a symmetric matrix.

$$[K] = \begin{bmatrix} k_{00} & k_{01} & k_{02} & k_{03} \\ k_{10} & k_{11} & k_{12} & k_{13} \\ k_{20} & k_{21} & k_{22} & k_{23} \\ k_{30} & k_{31} & k_{32} & k_{33} \end{bmatrix} = \begin{bmatrix} m_{00} & m_{01} & m_{02} & m_{03} \\ m_{01} & m_{11} & m_{12} & m_{13} \\ m_{02} & m_{12} & m_{22} & m_{23} \\ m_{03} & m_{13} & m_{23} & -m_{33} \end{bmatrix} \tag{3.6.17}$$

Among 10 elements, there is a relation such that

$$k_{00} = k_{11} + k_{22} + k_{33} \tag{3.6.18}$$

Therefore, nine elements are independent of each other in the Kennaugh and Mueller matrices. Using the Stokes vector, polarimetric radar channel power can be derived as follows:

1. *Co-polarization channel,* $\mathbf{g}_r = \mathbf{g}_t$

   Since the polarization state is the same in Tx and Rx, we put $\mathbf{g}_r = \mathbf{g}_t$ in (3.6.13). This yields

$$P_c = \frac{1}{2}\mathbf{g}_t^T \begin{bmatrix} 1 & 0 & 0 & 0 \\ 0 & 1 & 0 & 0 \\ 0 & 0 & 1 & 0 \\ 0 & 0 & 0 & -1 \end{bmatrix} [M]\mathbf{g}_t = \frac{1}{2}\mathbf{g}_t^T [K]_c \mathbf{g}_t \qquad (3.6.19)$$

where

$$[K]_c = \begin{bmatrix} 1 & 0 & 0 & 0 \\ 0 & 1 & 0 & 0 \\ 0 & 0 & 1 & 0 \\ 0 & 0 & 0 & -1 \end{bmatrix} [M] \qquad (3.6.20)$$

2. *Cross(X)-polarization channel,* $\mathbf{g}_r = \mathbf{g}_{t\perp}$

   Stokes vectors of the transmitting antenna and receiving antenna are orthogonal in the cross(X)-pol channel. The orthogonal Stokes vector can be easily obtained by viewing the graphical location of the Poincaré sphere. It locates on the antipodal point where the polarization state is orthogonal. If we let $\mathbf{g}_t$ as

$$\mathbf{g}_t = \left(1, g_1, g_2, g_3\right)^T \qquad (3.6.21)$$

then the antipodal point location is

$$\mathbf{g}_r = \left(1, -g_1, -g_2, -g_3\right)^T \qquad (3.6.22)$$

This form is substituted into equation (3.6.13). After rearranging the equation, we get

$$P_x = \frac{1}{2}\mathbf{g}_t^T \begin{bmatrix} 1 & 0 & 0 & 0 \\ 0 & -1 & 0 & 0 \\ 0 & 0 & -1 & 0 \\ 0 & 0 & 0 & 1 \end{bmatrix} [M]\mathbf{g}_t = \frac{1}{2}\mathbf{g}_t^T [K]_x \mathbf{g}_t \qquad (3.6.23)$$

where

$$[K]_x = \begin{bmatrix} 1 & 0 & 0 & 0 \\ 0 & -1 & 0 & 0 \\ 0 & 0 & -1 & 0 \\ 0 & 0 & 0 & 1 \end{bmatrix} [M] \qquad (3.6.24)$$

3. *Matched-Polarization Channel,* $\mathbf{g}_0^s$

   The matched-pol channel receives the total energy of the scattered wave. In Stokes vector notation, it receives the first element g_0^s. This can be written as

$$P_m = \left| \mathbf{E}_s \right|^2 = g_0^s = m_{00} + m_{01}x_1 + m_{02}x_2 + m_{03}x_3 \qquad (3.6.25)$$

Therefore, the power can be written in terms of the Stokes vector as

$$P_m = \mathbf{g}_t^T \begin{bmatrix} 1 & 0 & 0 & 0 \\ 0 & 0 & 0 & 0 \\ 0 & 0 & 0 & 0 \\ 0 & 0 & 0 & 0 \end{bmatrix} [M]\mathbf{g}_t = \mathbf{g}_t^T [K]_m \mathbf{g}_t \qquad (3.6.26)$$

$$[K]_m = \begin{bmatrix} 1 & 0 & 0 & 0 \\ 0 & 0 & 0 & 0 \\ 0 & 0 & 0 & 0 \\ 0 & 0 & 0 & 0 \end{bmatrix} [M] \qquad (3.6.27)$$

Since the relation

$$[K]_m = \frac{1}{2}[K]_c + \frac{1}{2}[K]_x \qquad (3.6.28)$$

derives

$$P_m = P_c + P_x, \qquad (3.6.29)$$

it is understood that the matched-pol channel power is the sum of the co-pol and cross(X)-pol channel powers.

3.7 POLARIZATION SIGNATURE

Once the scattering matrix or Kennaugh matrix is obtained, it is possible to calculate polarimetric channel powers as a function of geometrical parameters of polarization. By the combination of tilt angle τ and ellipticity angle ε, any polarization state can be realized (2.2.40), (2.2.64). Therefore, if the tilt (orientation) angle and ellipticity angle are taken as variables to calculate co-pol channel power (3.6.19) and/or X-pol channel power (3.6.23), it is possible to make 3D figures in which the height corresponds to the power. This 3D power pattern is called "polarization signature" [16] as shown in Figure 3.18.

Figure 3.18 displays some examples of polarization signature for the canonical target. The polarization signature exhibits a characteristic power pattern for each target. The 2D axes are taken in this figure so that H-pol locates the center of the bottom plane.

If we take a close look at an example of trihedral corner reflector, plate, and sphere, the power pattern looks like a mountain range. The mountain peak of the co-pol channel is obtained when $\varepsilon = 0$. $\varepsilon = 0$ means the linear polarization with arbitrary tile angle τ. Since the plate target reflects any linear polarized wave to the same polarization wave, the co-pol channel receives the maximum power at $\varepsilon = 0$. If the RHC polarized $(\varepsilon = -45°)$ wave is transmitted and received in the same RHC channel, the power becomes zero. This is because the reflected wave by plate turns out to be LHC $(\varepsilon = 45°)$. Since RHC and LHC are orthogonal, the power becomes null for the co-pol channel. Therefore, the power pattern looks like a mountain range.

For the case of the X-pol channel, the polarization state of the transmitter and receiver is orthogonal. So the power pattern is reversed that of the co-pol channel.

Trihedral corner reflector

Plate, sphere

$$[S] = \begin{bmatrix} 1 & 0 \\ 0 & 1 \end{bmatrix}$$

$$[K] = \begin{bmatrix} 1 & 0 & 0 & 0 \\ 0 & 1 & 0 & 0 \\ 0 & 0 & 1 & 0 \\ 0 & 0 & 0 & -1 \end{bmatrix}$$

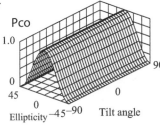
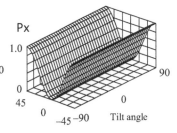

Dihedral corner reflector

$$[S] = \begin{bmatrix} 1 & 0 \\ 0 & -1 \end{bmatrix}$$

$$[K] = \begin{bmatrix} 1 & 0 & 0 & 0 \\ 0 & 0 & 0 & 0 \\ 0 & 0 & 0 & 0 \\ 0 & 0 & 0 & 1 \end{bmatrix}$$

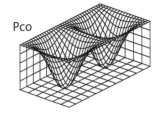
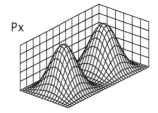

Horizontal dipole

$$[S] = \begin{bmatrix} 1 & 0 \\ 0 & 0 \end{bmatrix}$$

$$[K] = \frac{1}{4}\begin{bmatrix} 2 & 2 & 0 & 0 \\ 2 & 1 & 0 & 0 \\ 0 & 0 & 1 & 0 \\ 0 & 0 & 0 & 0 \end{bmatrix}$$

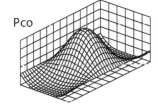
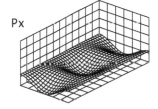

Vertical dipole

$$[S] = \begin{bmatrix} 0 & 0 \\ 0 & 1 \end{bmatrix}$$

$$[K] = \frac{1}{4}\begin{bmatrix} 2 & -2 & 0 & 0 \\ -2 & 1 & 0 & 0 \\ 0 & 0 & 1 & 0 \\ 0 & 0 & 0 & 0 \end{bmatrix}$$

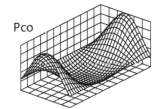
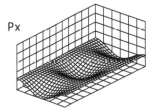

Left helix

$$[S] = \frac{1}{2}\begin{bmatrix} 1 & j \\ j & -1 \end{bmatrix}$$

$$[K] = \frac{1}{2}\begin{bmatrix} 1 & 0 & 0 & -1 \\ 0 & 0 & 0 & 0 \\ 0 & 0 & 0 & 0 \\ -1 & 0 & 0 & 1 \end{bmatrix}$$

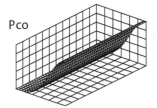
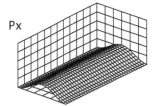

FIGURE 3.18 Examples of polarization signature. (From Yamaguchi, Y., *Radar Polarimetry from Basics to Applications: Radar Remote Sensing Using Polarimetric Information [in Japanese]*, IEICE, 2007.) *(Continued)*

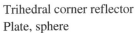

Right helix

$$[S] = \frac{1}{2}\begin{bmatrix} 1 & -j \\ -j & -1 \end{bmatrix}$$

$$[K] = \frac{1}{2}\begin{bmatrix} 1 & 0 & 0 & 1 \\ 0 & 0 & 0 & 0 \\ 0 & 0 & 0 & 0 \\ 1 & 0 & 0 & 1 \end{bmatrix}$$

$\frac{\lambda}{8}$ Orthogonal dipoles

$$[S] = \begin{bmatrix} 1 & 0 \\ 0 & -j \end{bmatrix}$$

$$[K] = \frac{1}{2}\begin{bmatrix} 2 & 0 & 0 & 0 \\ 0 & 1 & 0 & 0 \\ 0 & 0 & 1 & 2 \\ 0 & 0 & 2 & 0 \end{bmatrix}$$

$\frac{3\lambda}{8}$ Orthogonal dipoles

$$[S] = \begin{bmatrix} 1 & 0 \\ 0 & j \end{bmatrix}$$

$$[K] = \frac{1}{2}\begin{bmatrix} 2 & 0 & 0 & 0 \\ 0 & 1 & 0 & 0 \\ 0 & 0 & 1 & -2 \\ 0 & 0 & -2 & 0 \end{bmatrix}$$

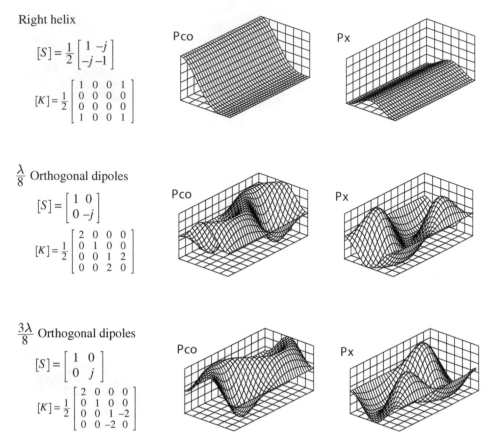

FIGURE 3.18 (Continued)   Examples of polarization signature. (From Yamaguchi, Y., *Radar Polarimetry from Basics to Applications: Radar Remote Sensing Using Polarimetric Information [in Japanese]*, IEICE, 2007.)

For the *H*-dipole case, the maximum power can be obtained at *H*-polarized wave transmission and reception. The peak power point occurs at the *H*-polarization point. The *V*-polarization combination cannot obtain any power for *H*-dipole scattering in the co-pol channel. Therefore, the co-pol power pattern looks like a mountain with its summit at the center. As the tilt angle of dipole changes from horizontal to vertical, the mountain peaks move along the tilt angle axis up to *V*-polarization.

Dihedral scattering has a characteristic power pattern with two depressions in the co-pol channel, whereas it has a special pattern with two mountain peaks in the X-pol channel.

The advantage of the polarization signature is its 3D shape pattern, which serves intuitive recognition of the target type. If accustomed to 3D shapes for various targets, we can recognize a target by shape. Furthermore, we can evaluate a calibration accuracy by the 3D shape distortion. The disadvantage may be a misjudgment of a 3D shape if the range of variable changed (e.g., *V*-pol is taken at the bottom center). A 3D representation is not suitable for precise value retrieval. When an accurate power value or peak-point retrieval is required, a gray scale 2D map is used.

3.8   CHARACTERISTIC POLARIZATION STATE

The polarization signature shows receiving power changes with polarization. We can recognize in the 3D power pattern that there exist several stationary points such as mountain peaks, bottoms, and saddle points in the co-pol channel and X-pol channel as shown in Figure 3.19. These

Scattering Polarimetry

75

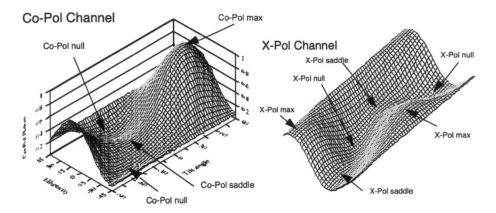

FIGURE 3.19 Polarization signature and characteristic polarization states that give stationary points of receiving power. (From Yamaguchi, Y., *Radar Polarimetry from Basics to Applications: Radar Remote Sensing Using Polarimetric Information [in Japanese]*, IEICE, 2007.)

stationary points (corresponding to ε, τ) are inherent to the target and are called characteristic polarization states [18–20].

Once the characteristic polarization states are obtained, we can use them to a special polarimetric filtering. For example, we can recalculate the receiving power with the maximum polarization state to derive the maximum power of the target, or with the minimum polarization state to null out the target. Furthermore, we can choose the maximum contrast polarization state to distinguish two targets. In this way, we can use them as a special polarimetric filtering to the receiving power.

In this section, characteristic polarization states are introduced, and some applications to polarimetric filtering are shown with experimental results.

3.8.1 The Number of Characteristic Polarization States and Their Relations

If we count the number of stationary points in Figure 3.19, arrows show that:

1 Peak point	Co-pol maximum (=X-Pol null)
1 Saddle point	Co-pol saddle (=X-Pol null)
2 Zero points	Co-pol nulls
2 Peak points	X-pol maximums
2 Saddle points	X-pol saddles
2 Zero points	X-pol nulls

There are 10 points in total. The names of these 10 points indicate the power situation. Co-pol maximum means that the receiving power of the co-pol channel becomes maximum. Nulls mean zero powers. Among them, the co-pol maximum and co-pol saddle locate the same positions as X-pol nulls, and two points are overlapped. Therefore, there exist eight stationary points at most for each target in principle. There may be further overlapping points for the specific target as shown in Figure 3.19.

These stationary points are inherent to the target. They are called characteristic polarization states [19] and are mapped on the Poincaré sphere as shown in Figure 3.20.

An interesting and attractive feature of characteristic polarization states on the Poincaré sphere is the shape of co-pol channel. Since the co-pol max, saddle, and nulls form the shape of a fork, it has been called the "polarization fork". The polarization fork is derived by the co-pol channel, but from Figure 3.20, we can see the axis of co-pol max to the saddle and the axis of X-pol nulls

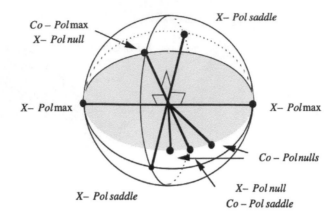

FIGURE 3.20 Polarization fork by characteristic polarization states on the Poincaré sphere. (From Yamaguchi, Y., *Radar Polarimetry from Basics to Applications: Radar Remote Sensing Using Polarimetric Information [in Japanese]*, IEICE, 2007.)

are the same. So, except for X-pol saddles, all characteristic polarization states locate on the great circle (on the gray area in Figure 3.20).

If we check these locations carefully, we recognize the importance of co-pol nulls. Once two co-pol nulls are obtained, then the co-pol saddle is decided, which locates in between. The co-pol max resides on the antipodal point of the saddle. Co-pol max and saddle axis are the same as that of X-pol nulls. So the co-pol max and saddle are the same as X-pol nulls. The X-pol max axis on the great plane (gray) is orthogonal to the X-pol null axis, and the X-pol saddle axis is orthogonal to the great circle. From these orthogonalities, we can determine X-pol maxes and X-pol saddles. Therefore, once co-pol nulls are found, it is possible to determine other characteristic polarization states by the graphical method [20].

3.8.2 RECEIVING POWER BY CHARACTERISTIC POLARIZATION STATE

In order to derive characteristic polarization states, we need to find stationary points of power equations.

$$\text{Co-pol channel power,} \quad P_c = \left| \frac{\begin{bmatrix} 1 & \rho \end{bmatrix}}{1 + \rho \rho^*} \begin{bmatrix} S_{HH} & S_{HV} \\ S_{VH} & S_{VV} \end{bmatrix} \begin{bmatrix} 1 \\ \rho \end{bmatrix} \right|^2 \tag{3.8.1}$$

$$\text{X-pol channel power,} \quad P_x = \left| \frac{\begin{bmatrix} \rho^* & -1 \end{bmatrix}}{1 + \rho \rho^*} \begin{bmatrix} S_{HH} & S_{HV} \\ S_{VH} & S_{VV} \end{bmatrix} \begin{bmatrix} 1 \\ \rho \end{bmatrix} \right|^2 \tag{3.8.2}$$

This problem can be solved more easily by diagonalizing the scattering matrix as

$$[S] \Rightarrow \begin{bmatrix} \lambda_1 & 0 \\ 0 & \lambda_2 \end{bmatrix} = \begin{bmatrix} |\lambda_1| e^{j\phi_1} & 0 \\ 0 & |\lambda_2| e^{j\phi_2} \end{bmatrix} \tag{3.8.3}$$

where λ_1 and λ_2 are eigenvalues with $|\lambda_1| > |\lambda_2|$. Then the Equations (3.8.1) and (3.8.2) can be translated into

TABLE 3.2

Characteristic Polarization States and Channel Powers

Polarization States	Power	Pol. Ratio	Pol. Ratio in the *HV* Basis
Co-pol max	$\|\lambda_1\|^2$	0	$\dfrac{-B \pm \sqrt{B^2 + 4\|A\|^2}}{2A}$
Co-pol saddle	$\|\lambda_2\|^2$	∞	$A = S_{HH}^* S_{HV} + S_{HV}^* S_{VV}$ $B = \|S_{HH}\|^2 - \|S_{VV}\|^2$
2 Co-pol nulls	0	$\pm j \left\|\dfrac{\lambda_1}{\lambda_2}\right\|^{1/2} e^{j2v}$	$\dfrac{-S_{HV} \pm \sqrt{S_{HV}^2 - S_{HH}S_{VV}}}{S_{VV}}$
2 X-pol Maxes	$\dfrac{1}{4}(\|\lambda_1\| + \|\lambda_2\|)^2$	$\pm j\, e^{j2v}$	$\|E_L\|^2 - \|E_R\|^2$
2 X-pol Saddles	$\dfrac{1}{4}(\|\lambda_1\| - \|\lambda_2\|)^2$	$\pm e^{j2v}$	$v = \dfrac{1}{4}(\phi_1 - \phi_2)$
2 X-pol nulls	0	$0, \infty$	$\dfrac{-B \pm \sqrt{B^2 + 4\|A\|^2}}{2A}$

$$P_c = \left\|\frac{[1\ \rho]}{1+\rho\rho^*}\begin{bmatrix}\lambda_1 & 0\\ 0 & \lambda_1\end{bmatrix}\begin{bmatrix}1\\ \rho\end{bmatrix}\right\|^2, \qquad P_x = \left\|\frac{[\rho^*\ -1]}{1+\rho\rho^*}\begin{bmatrix}\lambda_1 & 0\\ 0 & \lambda_1\end{bmatrix}\begin{bmatrix}1\\ \rho\end{bmatrix}\right\|^2 \qquad (3.8.4)$$

By Solving
$$\frac{\partial P_c}{\partial \rho} = 0, \quad \frac{\partial P_x}{\partial \rho} = 0, \qquad (3.8.5)$$

we can derive eight characteristic polarization states [19,21].

Table 3.2 lists characteristic polarization states and corresponding powers. In addition to the polarization ratio in the eigen basis, the expression in the *HV* polarization basis is also shown together. The most interesting polarization states are co-pol max, which gives the maximum power, and co-pol min, which gives the minimum power.

3.9 POLARIMETRIC FILTERING BY CHARACTERISTIC POLARIZATION STATES

In this section, we develop polarimetric filtering by using a specific polarization. Sometimes we would like to enhance a target from other object or eliminate clutter in a scene. In such a case, we may think of the polarization basis. Which polarization basis is suitable? *HV* or *LR* basis? Since the characteristic polarization state P on the Poincaré sphere and the antipodal point Q in Figure 3.21 forms a polarization basis, we can choose any basis for specific application. The polarization state of Q is orthogonal to that of P.

As seen in the previous section, there are several characteristic polarization states. The characteristic polarization states serve the purpose. The most frequently used ones are the co-pol max and co-pol nulls.

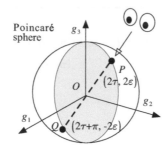

FIGURE 3.21   Selection of polarization state and polarization basis PQ.

Co-Pol Max, X-Pol Null
Co-pol max is the polarization state that maximizes the receiving power. Since co-pol max is equal to X-pol null, the polarization state of co-pol max can be derived from $P_x = 0$ in (3.8.4). This leads to

$$P_x = \left| \frac{-S_{HV} + S_{HV}\rho\rho^* + S_{HH}\rho^* - S_{VV}\rho}{1+\rho\rho^*} \right|^2 = 0 \qquad (3.9.1)$$

The roots can be written as $\rho_{cm1,2} = \rho_{xn1,2}$

$$\rho_{cm1,2} = \rho_{xn1,2} = \frac{-B \pm \sqrt{B^2 + 4|A|^2}}{2A} \qquad (3.9.2)$$

where

$$A = S_{HH}^* S_{HV} + S_{HV}^* S_{VV}, B = |S_{HH}|^2 - |S_{VV}|^2 \qquad (3.9.3)$$

Co-Pol Null
Co-pol null is the polarization state that forces the channel power to be zero, that is, the elimination of receiving power. $P_c = 0$ in equation (3.8.4) leads to the following equation:

$$P_c = \left| \frac{S_{HH} + 2S_{HV}\rho + S_{VV}\rho^2}{1+\rho\rho^*} \right|^2 = 0 \qquad (3.9.4)$$

The roots ρ_{cn1}, ρ_{cn2} are given as

$$\rho_{cn1}, \rho_{cn2} = \frac{-S_{HV} \pm \sqrt{S_{HV}^2 - S_{HH}S_{VV}}}{S_{VV}} \qquad (3.9.5)$$

These roots are listed in Table 3.1.
Once the polarization ratio is obtained, we can use it to recalculate powers (3.8.1) and (3.8.2) again in each pixel. This yields a specific polarimetric filtering. In order to find the polarization state itself, the geometrical parameters of the ellipse can be derived by the following relation:

$$\tan 2\tau = \frac{2\,\text{Re}\{\rho\}}{1-|\rho|^2} \quad \sin 2\varepsilon = \frac{2\,\text{Im}\{\rho\}}{1+|\rho|^2} \tag{3.9.6}$$

3.9.1 Two Orthogonal Wire Image

Polarimetric SAR imaging was conducted in an echoic chamber to confirm the polarimetric filtering [22]. Two orthogonal wire targets were used to show the polarimetric response as shown in Figure 3.22. The two antennas were scanned over these targets at $\lambda/2$ incremental intervals. The measurement specifications are:

Frequency	8.2–9.2 GHz
Target range	70 cm
Scan interval	1.5 cm
Scanning points	64 × 64
Polarization	*VV, HH, VH*

Two wire targets are metallic linear dipoles of length 50 cm and diameter 6 mm. They can be regarded as linear wire objects. Target 1 is −30°-oriented, and target 2 is 60°-oriented with respect to the X-direction, so that they are orthogonal. The X-direction is taken for *H*-polarization, while the Y-direction is for *V*-polarization. After a 2D scan of antennas, fully polarimetric and synthetic aperture radar data were calculated.

Figure 3.23 shows 2D SAR images of each polarimetric channel (*HH, VV,* and *HV*). Target 1 is clearly seen in the *HH*-channel, whereas target 2 is clear in the *VV*-channel. The X-pol *HV* channel has smaller magnitude images compared to those of co-pol channels. These images are in agreement of physical scattering phenomena. The total power image is also depicted. Since total power is roll-invariant parameter, it shows two wires clearly regardless of target orientation.

To examine the data in more detail, we picked up the scattering matrix around the center point of each target.

For target 1 oriented at −30°,
$$[S]_1 = \begin{bmatrix} -0.7656 - j0.0902 & 0.3529 - j0.1807 \\ 0.3529 - j0.1807 & -0.2056 + j0.2405 \end{bmatrix}$$

For target 2 oriented at +60°
$$[S]_2 = \begin{bmatrix} -0.0640 + j0.4191 & -0.2378 + j0.2607 \\ -0.2378 + j0.2607 & -0.0961 + j0.8162 \end{bmatrix}$$

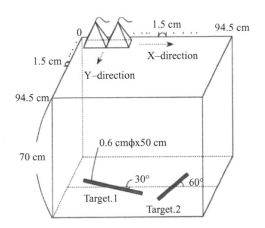

FIGURE 3.22 Target alignment. (From Yamaguchi, Y., *Radar Polarimetry from Basics to Applications: Radar Remote Sensing Using Polarimetric Information [in Japanese]*, IEICE, 2007.)

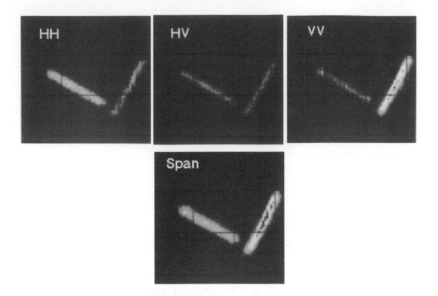

FIGURE 3.23   2D SAR image of two dipoles. (From Yamaguchi, Y., *Radar Polarimetry from Basics to Applications: Radar Remote Sensing Using Polarimetric Information [in Japanese]*, IEICE, 2007.)

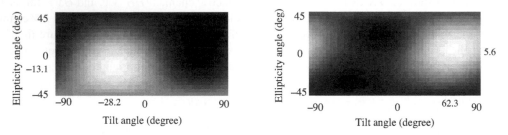

FIGURE 3.24   Polarization signature of targets 1 and 2 [36]. (From Yamaguchi, Y., *Radar Polarimetry from Basics to Applications: Radar Remote Sensing Using Polarimetric Information [in Japanese]*, IEICE, 2007.)

Polarization signatures based on these scattering matrices become as shown in Figure 3.24.

The brightest point in the co-pol signature corresponds to the co-pol max. The geometrical parameters for target 1 are found to be ($\tau = -28.2°$, $\varepsilon = -13.1°$), and for target 2, ($\tau = 62.3°$, $\varepsilon = 5.6°$) in Figure 3.24. We can understand that the measured tilt angles are very close to actual target orientation angles and that the ellipticity angles are also close to 0°, which means linear polarization.

Since two targets are placed orthogonal, we can expect the orthogonality of their co-pol maxes. The polarization ratio of each target can be calculated as

$$\rho_1 = 1.822 + j0.435, \quad \rho_2 = -0.498 - j0.294$$

The orthogonal condition for the polarization ratio is $\rho_1 p_2^* = -1$. We have

$$\rho_1 \rho_2^* = -1.034 + j0.316$$

in this experimental case. This is very close to theoretical orthogonal condition.

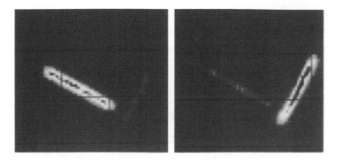

FIGURE 3.25 Co-pol max image of targets 1 and 2 [36]. (From Yamaguchi, Y., *Radar Polarimetry from Basics to Applications: Radar Remote Sensing Using Polarimetric Information [in Japanese]*, IEICE, 2007.)

Figure 3.25 shows co-pol max images for both targets. If the polarimetric filtering is adjusted to the co-pol max for target 1, we have an enhanced image of target 1, and the orthogonal target 2 is suppressed. That is, target 1 is maximized while target 2 is minimized, simultaneously, in this measurement. On the other hand, if the filtering is taken as co-pol max for target 2, then target 2 is enhanced and the orthogonal target 1 is suppressed. This figure shows the property of polarimetric filtering using a characteristic polarization state very well.

3.9.2 Contrast Enhancement by Polarization

Suppose we have two targets in a PolSAR image as shown in Figure 3.26. We sometimes would like to enhance target A against target B. This is related to target enhancement. There are two methods to achieve enhancement, that is, one is the maximization of target A, and another is the elimination or minimizing undesired target B. The latter is effective for clutter suppression or detection of underground objects in GPR. In this section, we consider target enhancement by polarimetric filtering [22–25].

At first, we define power contrast by the next equation and call this value as polarimetric contrast enhancement factor

$$C = \frac{\text{Desired power}}{\text{Undesired power}} \tag{3.9.7}$$

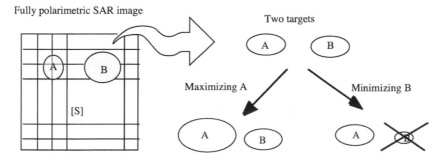

FIGURE 3.26 Method of contrast enhancement of A against B. (From Yamaguchi, Y., *Radar Polarimetry from Basics to Applications: Radar Remote Sensing Using Polarimetric Information [in Japanese]*, IEICE, 2007.)

where the desired power is the polarimetric power from our target, and undesired power is that caused by an unwanted target such as clutter or noise. If the subscript 1 and 2 are used for desired and undesired, respectively, we can define three contrast factors for three polarization channels.

Co-pol channel:

$$C^c = \left| \frac{\mathbf{E}_t^T [S]_1 \mathbf{E}_t}{\mathbf{E}_t^T [S]_2 \mathbf{E}_t} \right|^2 = \frac{\mathbf{g}_t^T [K]_1^c \mathbf{g}_t}{\mathbf{g}_t^T [K]_2^c \mathbf{g}_t} \tag{3.9.8}$$

X-pol channel:

$$C^x = \left| \frac{\mathbf{E}_{t\perp}^T [S]_1 \mathbf{E}_t}{\mathbf{E}_{t\perp}^T [S]_2 \mathbf{E}_t} \right|^2 = \frac{\mathbf{g}_t^T [K]_1^x \mathbf{g}_t}{\mathbf{g}_t^T [K]_2^x \mathbf{g}_t} \tag{3.9.9}$$

Matched-pol channel:

$$C^m = \left| \frac{\mathbf{E}_t^{*T} [G]_1 \mathbf{E}_t}{\mathbf{E}_t^{*T} [G]_2 \mathbf{E}_t} \right|^2 = \frac{\mathbf{g}_t^T [K]_1^m \mathbf{g}}{\mathbf{g}_t^T [K]_2^m \mathbf{g}} \tag{3.9.10}$$

To optimize contrast is to choose the polarization state that maximizes C. If the imaging is performed using the optimal polarization state, we can obtain the maximum contrast image. For a completely polarized wave (coherent) case, the corresponding polarization state is the null polarization state, which eliminates the undesired target. It is just like using an optical polarization filter to suppress reflection light from water surface in order to see objects inside water by camera.

Figure 3.27 shows various polarimetric SAR images over the Bonanza River area acquired with the NASA JPL AIRSAR system. Polarimetric channel images by typical polarization combination (Tx = H: horizontal, V: vertical, and L: LHC, and Rx = co-pol, X-pol, and matched-pol) are compared.

The image brightness and contrast are sensitive to polarization combination in the co-pol and in the X-pol channels (left two rows). The HH-image is the brightest, and the 45°–45° image has low contrast in the co-pol channel (left row). The HV image is dark, and the 45°–135° image has a very high contrast in the X-pol channel image (middle row). The matched-pol channel images are not sensitive to polarization change.

To evaluate images quantitatively, the averaged power of polarization images were calculated and listed in Table 3.3. For comparison of each image, the average power of HH is selected as the basis value, and other images were normalized by it. As can be seen in Table 3.3, the matched-pol power has the biggest value, while the X-pol channel power is the smallest (Table 3.3).

For target classification or identification, the polarimetric contrast enhancement factor will serve the purpose. In the polarimetric filtering, we choose A: riverside and B: forest near Bonanza River. To see the polarimetric characteristics, Kennaugh matrices were retrieved by selecting 40 pixels belonging to each area,

$$\text{Riverside} \quad [K]_1 = \begin{bmatrix} 2.5903 & 0.3716 & 0.0391 & 0.0060 \\ 0.3716 & 2.0150 & 0.0426 & -0.0274 \\ 0.0391 & 0.0426 & -0.9294 & -0.1669 \\ 0.0060 & -0.0274 & -0.1669 & 1.5047 \end{bmatrix}$$

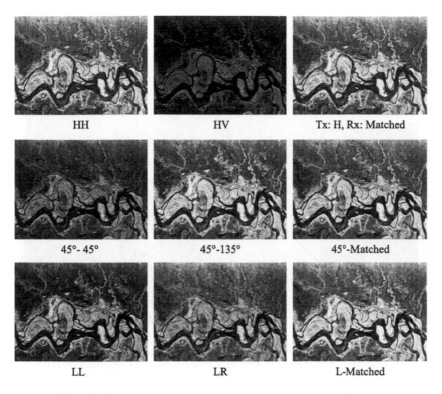

FIGURE 3.27 Polarimetric channel power images. (From Yamaguchi, Y., *Radar Polarimetry from Basics to Applications: Radar Remote Sensing Using Polarimetric Information [in Japanese]*, IEICE, 2007.)

TABLE 3.3
Mean Power in Three Polarimetric Radar Channels

Polarization State	Co-Pol	X-Pol	M-Pol
Linear horizontal (H)	1.00	0.14	1.14
Linear vertical (V)	0.73	0.14	0.88
45° linear	0.54	0.48	1.01
135° linear	0.53	0.48	1.01
Left-handed circular	0.63	0.39	1.02
Right-handed circular	0.61	0.39	1.00
Average	0.67	0.34	1.01

$$\text{Forest area} \quad \left[K\right]_2 = \begin{bmatrix} 1.2749 & 0.3539 & -0.0614 & -0.0298 \\ 0.3539 & 1.0870 & -0.0007 & 0.0010 \\ -0.0614 & -0.0007 & 0.3154 & 0.7949 \\ -0.0298 & 0.0010 & 0.7949 & -0.1276 \end{bmatrix}$$

Based on these matrices and Equations (3.9.7) through (3.9.9), a maximum contrast polarization state can be derived. The contrast signature is a signature of the riverside divided by the forest as shown in Figure 3.28. The maximum point (polarization state) of the signature yields the maximum contrast images of co-pol and X-pol channels as shown in Figures 3.28 and 3.29, respectively. Note that co-pol max of the river side is not identical to the polarization state of max contrast.

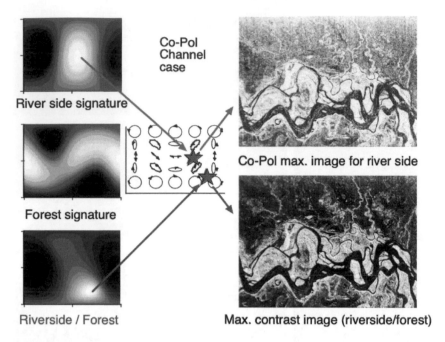

FIGURE 3.28   Contrast enhanced image in the co-pol channel. (From Yamaguchi, Y., *Radar Polarimetry from Basics to Applications: Radar Remote Sensing Using Polarimetric Information [in Japanese]*, IEICE, 2007.)

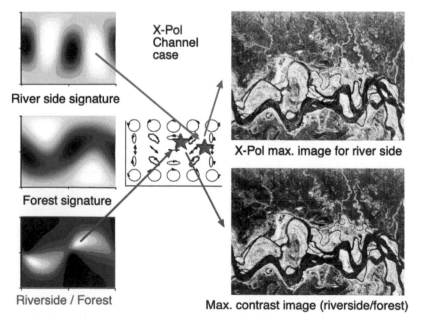

FIGURE 3.29   Contrast enhanced image in the X-pol channel. (From Yamaguchi, Y., *Radar Polarimetry from Basics to Applications: Radar Remote Sensing Using Polarimetric Information [in Japanese]*, IEICE, 2007.)

Therefore, the characteristic polarization state is not necessarily identical with that of maximum contrast. The difference can be confirmed in the final contrast max image and in the co-pol max image of the river side.

This polarimetric contrast enhancement scheme has been further investigated by J. Yang and successfully extended to classify/identify a narrow road in a forest [30]. In addition to the Kennaugh matrix and Stokes vector formulation, there is a method of optimization using the covariance matrix [26–30].

3.9.3 ELIMINATION BY POLARIZATION—SUPPRESSION OF CLUTTER

3.9.3.1 Applications to Ground-Penetrating Radar

GPR tries to detect underground objects such as historical remains, gas pipes, electric cables, or even mines. Due to the reflection from the surface, together with severe attenuation inside an inhomogeneous underground medium, mapping of objects over a wide area is a challenging task for GPR. GPR or snow radar scans antennae on a surface or along a line to make a B-scan image, which shows depth information along a scan line.

We developed a fully polarimetric frequency modulated and continuous wave (FMCW) synthetic aperture radar to test detection performance. By scanning rectangular horn antennas operating at 350–1000 MHz, B-scan images in the *HV*-polarization basis were obtained [24]. The target was a small metallic plate of 50 cm long buried at 40 cm depth. The detection images are shown in Figure 3.30.

As can be seen in this figure, we have very strong surface clutter. This strong clutter sometimes masks the echo from the target. In order to suppress the surface clutter, co-pol null imaging of the clutter was calculated. It is shown in Figure 3.31 together with the co-pol max image, which maximizes the target.

It is seen in the co-pol max image that the echo is enhanced; however, the surface clutter is also big. In GPR applications, the information of the desired target is unknown from the outset. Once we obtain the information, we can manipulate it to certain signal processing. But we cannot know the information in advance. In that sense, the co-pol max is not necessarily the best polarization state. What we know in advance is the surface information. By suppressing or eliminating the surface echo, we can see objects inside the medium more clearly. The reduced amplitude can be compensated by

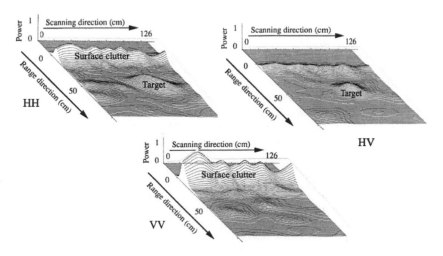

FIGURE 3.30 Detection result in the *HV*-pol basis. (From Yamaguchi, Y., *Radar Polarimetry from Basics to Applications: Radar Remote Sensing Using Polarimetric Information [in Japanese]*, IEICE, 2007.)

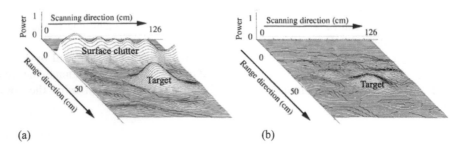

FIGURE 3.31 Polarimetric filtering images: (a) co-pol max image of target and (b) co-pol null image of the surface. (From Yamaguchi, Y., *Radar Polarimetry from Basics to Applications: Radar Remote Sensing Using Polarimetric Information [in Japanese]*, IEICE, 2007.)

amplification. Therefore, the suppression of clutter by the null polarization state serves, that is, the full utilization of polarimetric information increases the detection performance considerably. This advantageous method comes from the utilization of fully polarimetric information.

3.10 SUMMARY

In this chapter, the radar principle is reviewed, and the importance of polarimetric information is explained. The basic principles of radar polarimetry, the receiving antenna voltage starting from the reciprocity theorem, Sinclair scattering matrix, and the receiving power are defined. An important point is how the receiving power is changed by polarization and the scattering matrix. Since the receiving power changes according to polarization states of the transmitter and receiver, it becomes possible to choose the optimum polarization state for various applications. Using a specific polarization state, we can extend it, for example, to obtain characteristic polarization states of target, polarimetric filtering for clutter reduction, and target contrast enhancement. Contrast enhancement by polarization state is a very promising technique for detection and classification, identification of targets in a severely cluttered environment. This superb technique can be accomplished by polarimetric radar only.

REFERENCES

1. E. Yamashita, *Introduction to Electromagnetic Waves*, in Japanese, Sangyo-Tosho, Tokyo, 1980.
2. C. A. Balanis, *Antenna Theory: Analysis and Design*, 2nd ed., Ch. 2, Wiley, Hoboken, NJ, 1982.
3. F. T. Ulaby, R. K. Moore, and A. K. Fung, *Microwave Remote Sensing: Active and Passive*, vol. I, Artech House, Boston, 1986.
4. J. P. Fitch, *Synthetic Aperture Radar*, Springler-Verlag, New York, 1988.
5. M. I. Skolnik, ed., *Radar Handbook*, 2nd ed., McGraw-Hill, New York, 1990.
6. D. L. Mensa, *High Resolution Radar Cross-Section Imaging*, Artech House, Boston, 1991.
7 H. Mott, *Antennas for Radar and Communications: A Polarimetric Approach*, John Wiley & Sons, New York, 1992.
8. W.-M. Boerner, et al. (eds), Direct and inverse methods in radar polarimetry, *Proceedings of the NATO-ARW*, September 18–24, 1988, 1987–1991, NATO ASI Series C: Math & Phys. Sciences, vol. C-350, Parts 1&2, Kluwer Academic Publication, the Netherlands, 1992.
9. F. M. Henderson and A. J. Lewis, *Principles & Applications of Imaging Radar, Manual of Remote Sensing*, 3rd ed., vol. 2, ch. 5, pp. 271–357, John Wiley & Sons, New York, 1998.
10. G. W. Stimson, *Introduction to Airborne Radar*, Scitech Publishing, Mendham, NJ, 1998.
11. B. R. Mahafza, *Introduction to Radar Analysis*, CRC Press, 1998.
12. K. Ouchi, *Fundamentals of Synthetic Aperture Radar for Remote Sensing*, Tokyo Denki University Press, Tokyo, 2003 and 2009.

13. I. G. Cumming and F. H. Wong, *Digital Processing of Synthetic Aperture Radar Data*, Artech House, Boston, 2005.

14. J. S. Lee and E. Pottier, *Polarimetric Radar Imaging from Basics to Applications*, CRC Press, 2009.

15. M. Shimada, *Imaging from Spaceborne SARs, Calibration, and Applications*, CRC Press, 2019.

16. J. J. van Zyl, H. A. Zebker, and C. Elachi, "Imaging radar polarization signatures: Theory and observation," *Radio Sci.*, vol. 22, no. 4, pp. 529–543, 1987.

17. D. L. Evans, T. G. Farr, J. J. van Zyl, and H. A. Zebker, "Radar polarimetry: Analysis tools and applications," *IEEE Trans. Geosci. Remote Sens.*, vol. 26, no. 6, pp. 774–789, 1988.

18. A. P. Agrawal and W.-M. Boerner, "Redevelopment of Kennaugh target characteristic polarization state theory using the polarization transformation ratio for the coherent case," *IEEE Trans. Geosci. Remote Sens.*, vol. GE-27, pp. 2–14, 1989.

19. W.-M. Boerner, W. L. Yan, A.-Q. Xi, and Y. Yamaguchi, "On the basic principles of radar polarimetry: The target characteristic polarization state theory of Kennaugh, Huynen's polarization fork concept, and its extension to the partially polarized case," *Proc. IEEE*, vol. 79, no. 10, pp. 1538–1550, 1991.

20. J. Yang, Y. Yamaguchi, H. Yamada, and M. Sengoku, "Simple method for obtaining characteristic polarization states," *Electron. Lett.*, vol. 34, no. 5, pp. 441–443, 1998.

21. Y. Yamaguchi, *Radar Polarimetry from Basics to Applications: Radar Remote Sensing Using Polarimetric Information (in Japanese)*, IEICE, Tokyo, 2007.

22. Y. Yamaguchi, T. Nishikawa, M. Sengoku, and W.-M. Boerner, "Two-dimensional and full polarimetric imaging by a synthetic aperture FM-CW radar," *IEEE Trans. Geosci. Remote Sens.*, vol. 33, no. 2, pp. 421–427, 1995.

23. Y. Yamaguchi, Y. Takayanagi, W.-M. Boerner, H. J. Eom, and M. Sengoku, "Polarimetric enhancement in radar channel imagery," *IEICE Trans. Commun.*, vol. E78-B, no. 12, pp. 1571–1579, 1995.

24. T. Moriyama, Y. Yamaguchi, H. Yamada, and M. Sengoku, "Reduction of surface clutter by a polarimetric FM-CW radar in underground target detection," *IEICE Trans. Commun.*, vol. E78-B, no. 4, pp. 625–629, 1995.

25. J. Yang, Y. Yamaguchi, H. Yamada, and S. Lin, "The formulae of the characteristic polarization states in the Co-Pol channel and the optimal polarization state for contrast enhancement," *IEICE Trans. Commun.*, vol. E80-B, no. 10, pp. 1570–1575, 1997.

26. J. A. Kong, A. A. Swartz, H. A. Yueh, L. M. Novak, and R. T. Shin, "Identification of terrain cover using the optimal polarimetric classifier," *J. Electromagnet. Wave*, vol. 2, no. 2, pp. 171–194, 1988.

27. A. A. Swartz, H. A. Yueh, J. A. Kong, and R. T. Shin, "Optimal polarization for achieving maximum contrast in radar images," *J. Geophys. Res.* vol. 93, no. B12, pp. 15252–15260, 1988.

28. J. A. Kong, ed., *Polarimetric Remote Sensing*, PIER-3, Elsevier, New York, 1990.

29. J. Yang, Y. Yamaguchi, H. Yamada, et al., "The optimal problem for contrast enhancement in polarimetric radar remote sensing," *IEICE Trans. Commun.*, vol. E82-B, no. 1, pp. 174–183, 1999.

30. J. Yang, Y. Yamaguchi, W.-M. Boerner, and S. Lin, "Numerical methods for solving the optimal problem of contrast enhancement," *IEEE Trans. Geosci. Remote Sens.*, vol. 38, no. 2, pp. 965–971, 2000.

4 Polarization Matrices

This chapter describes polarization matrices such as the scattering matrix, covariance matrix, coherency matrix, and Kennaugh matrix, in addition to their relations. Starting from the scattering matrix, polarization matrices in various forms can be created as shown in Figure 4.1. When we deal with PolSAR observation data, we often take ensemble average to retrieve reliable information from the data set. In this case, we use the ensemble term as $\langle S_{HH}S_{VV}^* \rangle$ rather than the scattering matrix itself. There are 3×3 and 4×4 polarization matrices suitable for ensemble averaging data [1–4]. Among them, the covariance matrix has a specific characteristic such that the element is related to actual radar channel power or correlation, while the coherency matrix has advantages in physical interpretations of the scattering mechanism and in mathematical operations. The Kennaugh matrix consists of real-valued elements and is convenient for generating a polarization signature. Even if the form of the polarization matrix is different, nine parameters exist inside the matrix. In this chapter, we deal with representative polarization matrices shown in Figure 4.1 and derive theoretical average matrices to find key parameters of PolSAR data.

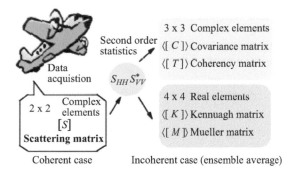

FIGURE 4.1 Various polarization matrices. (From Yamaguchi, Y., *Radar Polarimetry from Basics to Applications: Radar Remote Sensing Using Polarimetric Information (in Japanese)*, IEICE, 2007.)

4.1 SCATTERING MATRIX DATA

Data arrangement or data save in the scattering matrix by the airborne or spaceborne PolSAR system is shown in Figure 4.2, where H stands for the horizontal polarization and V for the vertical polarization. Pulse radiated from transmitting antenna spreads spherically. The main beam direction is set to the slant range direction. The range resolution becomes $\frac{c}{2B}$, where B is the bandwidth of pulse signal, and c is the speed of light. The ground-range resolution becomes $\frac{c}{2B}\frac{1}{\sin\theta}$ according to the incidence angle θ. Therefore, the ground-range resolution of the raw data becomes coarse for small θ, whereas it becomes fine for large θ. It is anticipated that the SAR image becomes coarse in the near range and becomes finer in the far range. It is impossible to image just underneath ($\theta = 0$) the radar system. This is why the SAR system illuminates toward the side-looking direction and is called the side-looking system. This point is completely different from the optical system.

Scattering matrix data are saved in pixels corresponding to range r_1, r_2, r_3 as shown in Figure 4.2. Assume that pixel size is the same as the range resolution for the sake of simplicity. Let's focus on

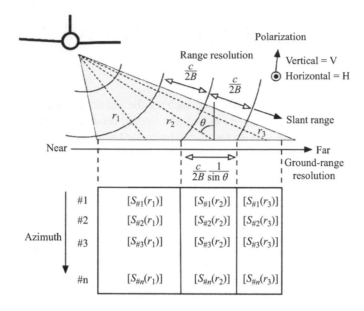

FIGURE 4.2   Scattering matrix data arrangement.

the phase of the scattering matrix in the azimuth and range directions. In the same range r_1, scattering matrices aligned in the azimuth direction

$$\left[S_{\#1}\left(r_1\right) \right], \left[S_{\#2}\left(r_1\right) \right], \left[S_{\#3}\left(r_1\right) \right], \ldots, \left[S_{\#n}\left(r_1\right) \right]$$

have the same phase due to the propagation path $2r_1$. However, scattering matrices aligned in the range direction $r_1, r_2, r_3, \ldots$

$$\left[S_{\#1}\left(r_1\right) \right], \left[S_{\#1}\left(r_2\right) \right], \left[S_{\#1}\left(r_3\right) \right], \ldots$$

have different propagation paths, and hence the propagation phase changes pixel by pixel. This situation is depicted on the complex plane as shown in Figure 4.3.

For example, if we have a scattering matrix $\left[S_{\#1}(r_1) \right]$ at r_1, the scattering matrix element *HH* and *VV* can be represented as $S_{r1}^{HH} = \left| S_{r1}^{HH} \right| \angle \phi_{r1}^{HH}$ and $S_{r1}^{VV} = \left| S_{r1}^{VV} \right| \angle \phi_{r1}^{VV}$, respectively.

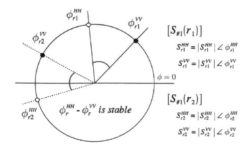

FIGURE 4.3   Phase in the complex plane. (From Yamaguchi, Y., *Radar Polarimetry from Basics to Applications: Radar Remote Sensing Using Polarimetric Information (in Japanese)*, IEICE, 2007.)

The phases ϕ_{r1}^{HH} and ϕ_{r1}^{VV} locate at ○ and ● in Figure 4.3. In a similar way, these elements become $S_{r2}^{HH} = \left|S_{r2}^{HH}\right| \angle \phi_{r2}^{HH}$ and $S_{r2}^{VV} = \left|S_{r2}^{VV}\right| \angle \phi_{r2}^{VV}$ for the scattering matrix $\left[S_{\#1}\left(r_2\right)\right]$ at r_2.

The corresponding phases locate different positions due to the propagation path difference.

From this simple fact, it becomes clear that we cannot add scattering matrix elements in the range direction directly, even if the scattering matrices are from the same object. This means "no direct sum of scattering matrix in the range direction."

On the other hand, the phase difference $\phi_{r2}^{HH} - \phi_{r2}^{VV}$ is close to $\phi_{r1}^{HH} - \phi_{r1}^{VV}$, and provides us with a stable value regardless of range. This phase difference is inherent polarimetric information of the object and plays a very important role in object classification. Therefore, data save is usually performed in the relative scattering matrix form as equation (4.1.1) by extracting $e^{j\phi^{HH}}$

$$\left[S\left(HV\right)\right] = \begin{bmatrix} S_{HH} & S_{HV} \\ S_{VH} & S_{VV} \end{bmatrix} = e^{j\phi^{HH}}\left[S\right]_{\text{relative}}$$

$$\left[S\right]_{\text{relative}} = \begin{bmatrix} \left|S_{HH}\right| & \left|S_{HV}\right|\angle\left(\phi^{HV} - \phi^{HH}\right) \\ \left|S_{VH}\right|\angle\left(\phi^{VH} - \phi^{HH}\right) & \left|S_{VV}\right|\angle\left(\phi^{VV} - \phi^{HH}\right) \end{bmatrix} \qquad (4.1.1)$$

The advantage of relative scattering matrix is keeping polarimetric characteristics in the range direction. As can be seen in equation (4.1.1), the number of polarimetric information becomes 5 for monostatic radar measurement. As far as polarimetric data utilization, such as classification and detection, it is desirable to use the $S_{VV}S_{HH}^*$ form rather than S_{HH} or S_{VV} alone, since the propagation phase is already removed in the form of $S_{VV}S_{HH}^*$, which is directly related to object information.

In this chapter, polarization matrices derived from scattering matrix are described. Sometimes people ask a question, "Which polarization is the best?" If we understand the polarization basis transformation, the answer is quite simple. It is just unitary transformation! Starting from the scattering matrix, the equivalent scattering vector, 3 × 3 covariance matrix in various polarization basis, coherency matrix, etc., are derived as shown in Figure 4.4. When we deal with ensemble averaging, we need theoretical ensemble values. The theoretical averaging can be performed through angular integration using certain probability functions. The averaging derives four important parameters of polarimetric information. Using these parameters, we can further go into efficient analyses of PolSAR data.

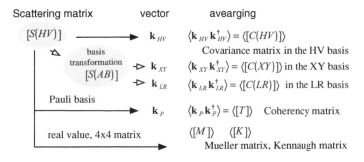

FIGURE 4.4   Creation of various polarization matrices.

4.2   SCATTERING MATRIX TRANSFORMATION

In this section, the scattering matrix in the arbitrary polarization basis is derived. We are familiar with the *HV* polarization basis, and we often use it for polarimetric measurement. But if we look at the object in the circular polarization basis, what happens? How about in the 45°-oriented linear polarization basis? Is there any advantage in other polarization bases? These questions might be quite natural for beginners. This section provides the basis for transformation of polarization matrices [4]. The reason why *HV* polarization basis is preferred is such that (1) it is easy to design antenna in the linear polarization, (2) polarization purity can be attained in wide band, (3) it is easy to understand scattering phenomena, and (4) it is transformable to other polarization bases. Therefore, once the polarimetric measurement is conducted in the *HV* polarization basis, it can be transformed to any polarization basis.

The voltage equation in the *HV* linear polarization basis is written as

$$V(HV) = \mathbf{E}_r(HV)^t \big[S(HV) \big] \mathbf{E}_t(HV) \tag{4.2.1}$$

where the superscript "*t*" denotes transpose. The voltage value should be the same as in the new polarization basis (*AB*),

$$V(AB) = \mathbf{E}_r(AB)^t \big[S(AB) \big] \mathbf{E}_t(AB) \tag{4.2.2}$$

There exists the following relation between **E**(*AB*) and **E**(*HV*) through the transformation matrix $\big[T\big]$ (Appendix A2.3)

$$\mathbf{E}(HV) = \big[T\big] \mathbf{E}(AB) \tag{4.2.3}$$

$$V(HV) = \mathbf{E}(AB)^t \big[T\big]^t \big[S(HV) \big]\big[T\big] \mathbf{E}(AB) = V(AB) \tag{4.2.4}$$

Therefore, the scattering matrix in the new polarization basis (*AB*) becomes

$$\big[S(AB) \big] = \big[T\big]^t \big[S(HV) \big]\big[T\big] = \begin{bmatrix} S_{AA} & S_{AB} \\ S_{BA} & S_{BB} \end{bmatrix} \tag{4.2.5}$$

More specifically, the scattering matrix can be written in the new basis as

$$\begin{bmatrix} S_{AA} & S_{AB} \\ S_{BA} & S_{BB} \end{bmatrix} = \frac{1}{1+\rho\rho^*} \begin{bmatrix} e^{j\alpha} & 0 \\ 0 & e^{-j\alpha} \end{bmatrix} \begin{bmatrix} 1 & \rho \\ -\rho^* & 1 \end{bmatrix} \begin{bmatrix} S_{HH} & S_{HV} \\ S_{VH} & S_{VV} \end{bmatrix} \begin{bmatrix} 1 & -\rho^* \\ \rho & 1 \end{bmatrix} \begin{bmatrix} e^{j\alpha} & 0 \\ 0 & e^{-j\alpha} \end{bmatrix} \tag{4.2.6}$$

where

$$\rho = \frac{\tan \tau + j \tan \varepsilon}{1 - j \tan \tau \tan \varepsilon}, \quad \alpha = \tan^{-1}\big(\tan \tau \tan \varepsilon \big)$$

For the backscattering case $\big(S_{HV} = S_{VH} \big)$, the element is written as follows:

$$\begin{cases} S_{AA} = \dfrac{e^{j2\alpha}}{1+\rho\rho^*}\left(S_{HH} + 2\rho S_{HV} + \rho^2 S_{VV}\right) \\[3mm] S_{BB} = \dfrac{e^{-j2\alpha}}{1+\rho\rho^*}\left(\rho^{*2} S_{HH} - 2\rho^* S_{HV} + S_{VV}\right) \\[3mm] S_{AB} = \dfrac{e^{-j2\alpha}}{1+\rho\rho^*}\left[\rho S_{VV} - \rho^* S_{HH} + \left(1-\rho\rho^*\right)S_{HV}\right] \end{cases} \qquad (4.2.7)$$

Therefore, we can derive the scattering matrix in any polarization basis using this transformation. For example, the scattering matrix in the circular polarization basis can be obtained by setting $\rho = j$ and $\alpha = 0$,

$$\left[S(LR)\right] = \begin{bmatrix} S_{LL} & S_{LR} \\ S_{RL} & S_{RR} \end{bmatrix} = \frac{1}{2}\begin{bmatrix} 1 & j \\ j & 1 \end{bmatrix}\begin{bmatrix} S_{HH} & S_{HV} \\ S_{VH} & S_{VV} \end{bmatrix}\begin{bmatrix} 1 & j \\ j & 1 \end{bmatrix} \quad (S_{HV} = S_{VH}) \qquad (4.2.8)$$

$$\begin{cases} S_{LL} = \dfrac{1}{2}\left(S_{HH} - S_{VV} + j2S_{HV}\right) \\[3mm] S_{RR} = \dfrac{1}{2}\left(S_{VV} - S_{HH} + j2S_{HV}\right) \\[3mm] S_{LR} = S_{RL} = \dfrac{j}{2}\left(S_{HH} + S_{VV}\right) \end{cases}$$

A linear basis rotation by angle θ yields the scattering matrix, by setting $\rho = \tan\theta$ and $\alpha = 0$

$$\left[S(\theta)\right] = \begin{bmatrix} S_{hh} & S_{hv} \\ S_{vh} & S_{vv} \end{bmatrix} = \begin{bmatrix} \cos\theta & \sin\theta \\ -\sin\theta & \cos\theta \end{bmatrix}\begin{bmatrix} S_{HH} & S_{HV} \\ S_{VH} & S_{VV} \end{bmatrix}\begin{bmatrix} \cos\theta & -\sin\theta \\ \sin\theta & \cos\theta \end{bmatrix} \qquad (4.2.9)$$

$$\begin{cases} S_{hh} = S_{HH}\cos^2\theta + S_{VV}\sin^2\theta + S_{HV}\sin 2\theta \\[2mm] S_{vv} = S_{HH}\sin^2\theta + S_{VV}\cos^2\theta - S_{HV}\sin 2\theta \\[2mm] S_{hv} = S_{vh} = S_{HV}\cos 2\theta + \dfrac{S_{VV} - S_{HH}}{2}\sin 2\theta \end{cases}$$

The XY polarization basis with rotation angle $\theta = 45°$ corresponds to the situation of $\rho = 1$, and the scattering matrix becomes

$$\left[S(XY)\right] = \begin{bmatrix} S_{XX} & S_{XY} \\ S_{YX} & S_{YY} \end{bmatrix} = \frac{1}{2}\begin{bmatrix} 1 & 1 \\ -1 & 1 \end{bmatrix}\begin{bmatrix} S_{HH} & S_{HV} \\ S_{VH} & S_{VV} \end{bmatrix}\begin{bmatrix} 1 & -1 \\ 1 & 1 \end{bmatrix} \qquad (4.2.10)$$

$$\begin{cases} S_{XX} = \dfrac{1}{2}\left(S_{HH} + S_{VV} + 2S_{HV}\right) \\[2mm] S_{YY} = \dfrac{1}{2}\left(S_{HH} + S_{VV} - 2S_{HV}\right) \\[2mm] S_{XY} = S_{YX} = \dfrac{1}{2}\left(S_{VV} - S_{HH}\right) \end{cases}$$

Scattering matrix in the eigen-basis

Eigen-basis is a special polarization basis such that the scattering matrix is in diagonal form. Therefore, we choose the following polarization ratio so as to force the off-diagonal terms to zero. The polarization ratio, which satisfies $S_{AB} = 0$ (4.2.7), is given by

$$\rho_{1,2} = \frac{B \pm \sqrt{B^2 + 4|A|^2}}{2A} \tag{4.2.11}$$

where

$$A = S_{HH}^* S_{HV} + S_{HV}^* S_{VV}, \ B = |S_{HH}|^2 - |S_{VV}|^2 \tag{4.2.12}$$

From equation (4.2.7), $\left[S(AB)\right]$ is diagonalized and becomes an eigenvalue matrix.

$$\left[S(AB)\right] \Rightarrow \begin{vmatrix} S_{AA} & 0 \\ 0 & S_{BB} \end{vmatrix} = \begin{vmatrix} \lambda_1 & 0 \\ 0 & \lambda_2 \end{vmatrix} = \left[S\right]_{\text{diag}} \tag{4.2.13}$$

$$\begin{cases} \lambda_1 = S_{AA}(\rho_1) = \dfrac{1}{1+\rho_1\rho_1^*}\left(S_{HH} + 2\rho_1 S_{HV} + \rho_1^2 S_{VV}\right) \\[3mm] \lambda_2 = S_{BB}(\rho_1) = \dfrac{1}{1+\rho_1\rho_1^*}\left(\rho_1^{*2} S_{HH} - 2\rho_1^* S_{HV} + S_{VV}\right) \end{cases} \tag{4.2.14}$$

4.3   SCATTERING VECTOR

Vector notation is sometimes convenient for mathematical operations. The scattering vector is a vector form equivalent to the scattering matrix. The scattering vector $\mathbf{k}_L$ is defined using a lexicographic order of scattering matrix elements. Other scattering vectors $\mathbf{k}_{HV}, \mathbf{k}_{XY}, \mathbf{k}_{LR}$ are also defined at different polarization bases. For the backscattering case, the scattering vector becomes as follows:

- **Scattering vector in the linear polarization basis (HV)**

$$\mathbf{k}_{HV} = \begin{bmatrix} S_{HH} \\ \sqrt{2}\,S_{HV} \\ S_{VV} \end{bmatrix} = \mathbf{k}_L \tag{4.3.1}$$

- **Scattering vector in the rotated *hv* basis**

$$
\mathbf{k}_{hv} = \begin{bmatrix} S_{hh} \\ \sqrt{2}\, S_{hv} \\ S_{vv} \end{bmatrix} = [U_\theta]\mathbf{k}_L, \qquad [U_\theta] = \begin{bmatrix} \cos^2\theta & \dfrac{\sin 2\theta}{\sqrt{2}} & \sin^2\theta \\[2mm] -\dfrac{\sin 2\theta}{\sqrt{2}} & \cos 2\theta & \dfrac{\sin 2\theta}{\sqrt{2}} \\[2mm] \sin^2\theta & -\dfrac{\sin 2\theta}{\sqrt{2}} & \cos^2\theta \end{bmatrix} \tag{4.3.2}
$$

- **Scattering vector in the *XY* (45° rotation) basis**

$$
\mathbf{k}_{XY} = \begin{bmatrix} S_{XX} \\ \sqrt{2}\, S_{XY} \\ S_{YY} \end{bmatrix} = [U_{XY}]\mathbf{k}_L, \qquad [U_{XY}] = \frac{1}{2}\begin{bmatrix} 1 & \sqrt{2} & 1 \\ -\sqrt{2} & 0 & \sqrt{2} \\ 1 & -\sqrt{2} & 1 \end{bmatrix} \tag{4.3.3}
$$

- **Scattering vector in the circular polarization basis (*LR*)**

$$
\mathbf{k}_{LR} = \begin{bmatrix} S_{LL} \\ \sqrt{2}\, S_{LR} \\ S_{RR} \end{bmatrix} = [U_C]\mathbf{k}_L, \qquad [U_C] = \frac{1}{2}\begin{bmatrix} 1 & j\sqrt{2} & -1 \\ j\sqrt{2} & 0 & j\sqrt{2} \\ -1 & j\sqrt{2} & 1 \end{bmatrix} \tag{4.3.4}
$$

These scattering vectors are equivalent mathematically because of unitary transformation of $\mathbf{k}_L$. The factor $\sqrt{2}$ is to ensure the equivalence of the vector norm and the span of the matrix.

$$
\|\mathbf{k}_L\|^2 = |S_{HH}|^2 + 2|S_{HV}|^2 + |S_{VV}|^2 = \mathrm{Span}[S] \tag{4.3.5}
$$

$\mathbf{k}_{HV}, \mathbf{k}_{XY}, \mathbf{k}_{LR}$ will be the basis vector for covariance matrix in the corresponding polarization basis.

- **Pauli scattering vector**

Pauli scattering vector $\mathbf{k}_P$ is defined as an orthogonal basis [2].

$$
\mathbf{k}_P = \left[\mathrm{Trace}\big([S]\sigma_0\big),\, \mathrm{Trace}\big([S]\sigma_1\big),\, \mathrm{Trace}\big([S]\sigma_2\big) \right]^t \tag{4.3.6}
$$

$\sigma_0, \sigma_1, \sigma_2$: Pauli basis matrix

$$
\sigma_0 = \frac{1}{\sqrt{2}}\begin{bmatrix} 1 & 0 \\ 0 & 1 \end{bmatrix},\ \sigma_1 = \frac{1}{\sqrt{2}}\begin{bmatrix} 1 & 0 \\ 0 & -1 \end{bmatrix},\ \sigma_2 = \frac{1}{\sqrt{2}}\begin{bmatrix} 0 & 1 \\ 1 & 0 \end{bmatrix} \tag{4.3.7}
$$

$\mathbf{k}_P$ is also called the Pauli vector or coherency vector.

$$
\mathbf{k}_P = \frac{1}{\sqrt{2}}\begin{bmatrix} S_{HH} + S_{VV} \\ S_{HH} - S_{VV} \\ 2S_{HV} \end{bmatrix} \tag{4.3.8}
$$

The vector norm is equal to the span of the scattering matrix. The characteristic feature is its physical meaning; $S_{HH} + S_{VV}$ corresponds to odd-bounce scattering, and $S_{HH} - S_{VV}$ represents even-bounce scattering. There is a unitary transformation relation between $\mathbf{k}_P$ and $\mathbf{k}_L$.

$$\mathbf{k}_P = \frac{1}{\sqrt{2}} \begin{bmatrix} S_{HH} + S_{VV} \\ S_{HH} - S_{VV} \\ 2S_{HV} \end{bmatrix} = \frac{1}{\sqrt{2}} \begin{bmatrix} 1 & 0 & 1 \\ 1 & 0 & -1 \\ 0 & \sqrt{2} & 0 \end{bmatrix} \begin{bmatrix} S_{HH} \\ \sqrt{2}\, S_{HV} \\ S_{VV} \end{bmatrix} = \begin{bmatrix} U_P \end{bmatrix} \mathbf{k}_L \qquad (4.3.9)$$

$$\begin{bmatrix} U_P \end{bmatrix} = \frac{1}{\sqrt{2}} \begin{bmatrix} 1 & 0 & 1 \\ 1 & 0 & -1 \\ 0 & \sqrt{2} & 0 \end{bmatrix} : \text{ unitary transformation matrix}$$

If we normalize to define a unit vector that represents scattering mechanism,

$$\mathbf{w}_P = \frac{\mathbf{k}_P}{|\mathbf{k}_P|} \qquad (4.3.10)$$

it is possible to further create scattering mechanism vector as,

$$\mathbf{w}_0 = \begin{bmatrix} 1 \\ 0 \\ 0 \end{bmatrix} \text{ odd bounce, } \mathbf{w}_1 = \begin{bmatrix} 0 \\ 1 \\ 0 \end{bmatrix} \text{ even bounce, } \mathbf{w}_2 = \begin{bmatrix} 0 \\ 0 \\ 1 \end{bmatrix} \text{ cross polarization}$$

Table 4.1 shows the scattering matrix and vector of canonical objects in various polarization bases.

4.4   ENSEMBLE AVERAGE OF POLARIZATION MATRIX

It is common to express or save polarimetric data in the form of $S_{VV}S_{HH}^{*}$. This form can be used as the second-order statistics and can be used for mathematical addition, subtraction, and multiplication operations directly.

The scattering vector in the previous section is used to create a 3×3 covariance matrix with complex elements, coherency matrix, and 4×4 Kennaugh matrix with real numbers. Their expressions become as follows: the superscript $\mathbf{k}^{*}$ denotes complex conjugation, $\mathbf{k}^{\dagger}$ denotes complex conjugate and transpose, and $< >$ represents ensemble averaging.

- **Ensemble Average Covariance Matrix**
  HV polarization basis

$$\begin{bmatrix} C(HV) \end{bmatrix} = \mathbf{k}_{HV}\mathbf{k}_{HV}^{\dagger} = \begin{bmatrix} |S_{HH}|^2 & \sqrt{2}\, S_{HH}S_{HV}^{*} & S_{HH}S_{VV}^{*} \\ \sqrt{2}\, S_{HV}S_{HH}^{*} & 2|S_{HV}|^2 & \sqrt{2}\, S_{HV}S_{VV}^{*} \\ S_{VV}S_{HH}^{*} & \sqrt{2}\, S_{VV}S_{HV}^{*} & |S_{VV}|^2 \end{bmatrix}$$

TABLE 4.1

Scattering Matrix and Vector of Canonical Objects in Various Polarization Basis

	HV		XY (±45°)		LR (Circular)		Pauli
	$[S(HV)]$	$\mathbf{k}_{HV}$	$[S(XY)]$	$\mathbf{k}_{XY}$	$[S(LR)]$	$\mathbf{k}_{LR}$	$\mathbf{k}_P$
Sphere, Plate	$\begin{bmatrix}1 & 0\\0 & 1\end{bmatrix}$	$\begin{bmatrix}1\\0\\1\end{bmatrix}$	$\begin{bmatrix}1 & 0\\0 & 1\end{bmatrix}$	$\begin{bmatrix}1\\0\\1\end{bmatrix}$	$\begin{bmatrix}0 & j\\j & 0\end{bmatrix}$	$\begin{bmatrix}0\\j\sqrt{2}\\0\end{bmatrix}$	$\sqrt{2}\begin{bmatrix}1\\0\\0\end{bmatrix}$
Dihedral	$\begin{bmatrix}1 & 0\\0 & -1\end{bmatrix}$	$\begin{bmatrix}1\\0\\-1\end{bmatrix}$	$\begin{bmatrix}0 & -1\\-1 & 0\end{bmatrix}$	$\begin{bmatrix}0\\-\sqrt{2}\\0\end{bmatrix}$	$\begin{bmatrix}1 & 0\\0 & -1\end{bmatrix}$	$\begin{bmatrix}1\\0\\-1\end{bmatrix}$	$\sqrt{2}\begin{bmatrix}0\\1\\0\end{bmatrix}$
H-dipole	$\begin{bmatrix}1 & 0\\0 & 0\end{bmatrix}$	$\begin{bmatrix}1\\0\\0\end{bmatrix}$	$\frac{1}{2}\begin{bmatrix}1 & -1\\-1 & 1\end{bmatrix}$	$\frac{1}{2}\begin{bmatrix}1\\-\sqrt{2}\\1\end{bmatrix}$	$\frac{1}{2}\begin{bmatrix}1 & j\\j & -1\end{bmatrix}$	$\frac{1}{2}\begin{bmatrix}1\\j\sqrt{2}\\-1\end{bmatrix}$	$\frac{1}{\sqrt{2}}\begin{bmatrix}1\\1\\0\end{bmatrix}$
V-dipole	$\begin{bmatrix}0 & 0\\0 & 1\end{bmatrix}$	$\begin{bmatrix}0\\0\\1\end{bmatrix}$	$\frac{1}{2}\begin{bmatrix}1 & 1\\1 & 1\end{bmatrix}$	$\frac{1}{2}\begin{bmatrix}1\\\sqrt{2}\\1\end{bmatrix}$	$\frac{1}{2}\begin{bmatrix}-1 & j\\j & 1\end{bmatrix}$	$\frac{1}{2}\begin{bmatrix}-1\\j\sqrt{2}\\1\end{bmatrix}$	$\frac{1}{\sqrt{2}}\begin{bmatrix}1\\-1\\0\end{bmatrix}$
Left helix	$\frac{1}{2}\begin{bmatrix}1 & j\\j & -1\end{bmatrix}$	$\frac{1}{2}\begin{bmatrix}1\\j\sqrt{2}\\-1\end{bmatrix}$	$\frac{1}{2}\begin{bmatrix}j & -1\\-1 & -j\end{bmatrix}$	$\frac{1}{2}\begin{bmatrix}j\\-\sqrt{2}\\-j\end{bmatrix}$	$\begin{bmatrix}0 & 0\\0 & -1\end{bmatrix}$	$\begin{bmatrix}0\\0\\-1\end{bmatrix}$	$\frac{1}{\sqrt{2}}\begin{bmatrix}0\\1\\j\end{bmatrix}$
Right helix	$\frac{1}{2}\begin{bmatrix}1 & -j\\-j & -1\end{bmatrix}$	$\frac{1}{2}\begin{bmatrix}1\\-j\sqrt{2}\\-1\end{bmatrix}$	$\frac{1}{2}\begin{bmatrix}-j & -1\\-1 & j\end{bmatrix}$	$\frac{1}{2}\begin{bmatrix}-j\\\sqrt{2}\\j\end{bmatrix}$	$\begin{bmatrix}1 & 0\\0 & 0\end{bmatrix}$	$\begin{bmatrix}1\\0\\0\end{bmatrix}$	$\frac{1}{\sqrt{2}}\begin{bmatrix}0\\1\\-j\end{bmatrix}$

$$\left\langle\left[C(HV)\right]\right\rangle = \left\langle \mathbf{k}_{HV}\mathbf{k}_{HV}^{\dagger}\right\rangle = \frac{1}{n}\sum^{n}\mathbf{k}_{HV}\mathbf{k}_{HV}^{\dagger} = \begin{bmatrix} \left\langle |S_{HH}|^2\right\rangle & \sqrt{2}\left\langle S_{HH}S_{HV}^*\right\rangle & \left\langle S_{HH}S_{VV}^*\right\rangle \\ \sqrt{2}\left\langle S_{HV}S_{HH}^*\right\rangle & 2\left\langle |S_{HV}|^2\right\rangle & \sqrt{2}\left\langle S_{HV}S_{VV}^*\right\rangle \\ \left\langle S_{VV}S_{HH}^*\right\rangle & \sqrt{2}\left\langle S_{VV}S_{HV}^*\right\rangle & \left\langle |S_{VV}|^2\right\rangle \end{bmatrix}$$

(4.4.1)

$$\text{Circular polarization basis}: \left\langle\left[C(LR)\right]\right\rangle = \left\langle \mathbf{k}_{LR}\mathbf{k}_{LR}^{\dagger}\right\rangle = \frac{1}{n}\sum^{n}\mathbf{k}_{LR}\mathbf{k}_{LR}^{\dagger}$$

(4.4.2)

$$\text{45(degree)-oriented linear basis}: \left\langle\left[C(XY)\right]\right\rangle = \left\langle \mathbf{k}_{XY}\mathbf{k}_{XY}^{\dagger}\right\rangle = \frac{1}{n}\sum^{n}\mathbf{k}_{XY}\mathbf{k}_{XY}^{\dagger}$$

(4.4.3)

- **Ensemble Average Coherency Matrix**

$$\langle [T] \rangle = \frac{1}{n} \sum^{n} \mathbf{k}_P \mathbf{k}_P^{\dagger}$$

$$= \frac{1}{2} \begin{bmatrix} \langle |S_{HH} + S_{VV}|^2 \rangle & \langle (S_{HH} + S_{VV})(S_{HH} - S_{VV})^* \rangle & \langle 2S_{HV}^* (S_{HH} + S_{VV}) \rangle \\ \langle (S_{HH} - S_{VV})(S_{HH} + S_{VV})^* \rangle & \langle |S_{HH} - S_{VV}|^2 \rangle & \langle 2S_{HV}^* (S_{HH} - S_{VV}) \rangle \\ \langle 2S_{HV} (S_{HH} + S_{VV})^* \rangle & \langle 2S_{HV} (S_{HH} - S_{VV})^* \rangle & \langle 4|S_{HV}|^2 \rangle \end{bmatrix} \quad (4.4.4)$$

Covariance and coherency matrices are of the 3×3 Hermitian matrix. It has three real diagonal elements and three complex off-diagonal elements. Therefore, the number of real-valued independent parameters is nine $(3 + 6 = 9)$.

- **Ensemble Average Kennaugh Matrix**

$$\langle [K] \rangle =$$

$$\begin{bmatrix} \dfrac{\langle |S_{HH}|^2 + 2|S_{HV}|^2 + |S_{VV}|^2 \rangle}{2} & \dfrac{\langle |S_{HH}|^2 - |S_{VV}|^2 \rangle}{2} & \langle \mathrm{Re}\{(S_{HH} + S_{VV})S_{HV}^*\} \rangle & \langle \mathrm{Im}\{(S_{HH} - S_{VV})S_{HV}^*\} \rangle \\ \dfrac{\langle |S_{HH}|^2 - |S_{VV}|^2 \rangle}{2} & \dfrac{\langle |S_{HH}|^2 - 2|S_{HV}|^2 + |S_{VV}|^2 \rangle}{2} & \langle \mathrm{Re}\{(S_{HH} - S_{VV})S_{HV}^*\} \rangle & \langle \mathrm{Im}\{(S_{HH} + S_{VV})S_{HV}^*\} \rangle \\ \langle \mathrm{Re}\{(S_{HH} + S_{VV})S_{HV}^*\} \rangle & \langle \mathrm{Re}\{(S_{HH} - S_{VV})S_{HV}^*\} \rangle & \langle |S_{HV}|^2 + \mathrm{Re}\{S_{HH}S_{VV}^*\} \rangle & \langle \mathrm{Im}\{S_{HH}S_{VV}^*\} \rangle \\ \langle \mathrm{Im}\{(S_{HH} - S_{VV})S_{HV}^*\} \rangle & \langle \mathrm{Im}\{(S_{HH} + S_{VV})S_{HV}^*\} \rangle & \langle \mathrm{Im}\{S_{HH}S_{VV}^*\} \rangle & \langle |S_{HV}|^2 - \mathrm{Re}\{S_{HH}S_{VV}^*\} \rangle \end{bmatrix}$$

$$(4.4.5)$$

Kennaugh matrix is a 4×4 real-valued and symmetric matrix. The number of independent elements looks like $4 + 3 + 2 + 1 = 10$; however, there exists a relation that

$$k_{00} = k_{11} + k_{22} + k_{33} \quad (4.4.6)$$

Therefore, the number of independent parameters is nine.

It can be understood that the number of independent parameters is nine, even if the matrix form is different. The covariance matrix, coherency matrix, Kennaugh matrix, etc., retain the second-order statistics of polarimetric information and can be used for statistical analysis as averaging. They are connected to each other. For example, if the rotation operation is applied to the coherency matrix to minimize the T_{33} component, we have mathematically $\langle \mathrm{Re}\{(S_{HH} - S_{VV})S_{HV}^*\} \rangle = 0$. The independent parameters in the coherency matrix become eight. This condition makes the Kennaugh matrix as

$$2\langle [K] \rangle = \begin{bmatrix} TP & \langle |S_{HH}|^2 - |S_{VV}|^2 \rangle & P_{od} & P_h \\ \langle |S_{HH}|^2 - |S_{VV}|^2 \rangle & \langle |S_{HH}|^2 - 2|S_{HV}|^2 + |S_{VV}|^2 \rangle & 0 & P_{cd} \\ P_{od} & 0 & 2\langle |S_{HV}|^2 + \mathrm{Re}\{S_{HH}S_{VV}^*\} \rangle & 2\langle \mathrm{Im}\{S_{HH}S_{VV}^*\} \rangle \\ P_h & P_{cd} & 2\langle \mathrm{Im}\{S_{HH}S_{VV}^*\} \rangle & 2\langle |S_{HV}|^2 - \mathrm{Re}\{S_{HH}S_{VV}^*\} \rangle \end{bmatrix}$$

where TP, P_h, P_{od}, and P_{cd} represent scattering powers in the six-component scattering power decomposition (see Chapter 8). Since this is a symmetric matrix, the independent number becomes eight again.

4.5 THEORETICAL COVARIANCE MATRIX BY INTEGRATION

In this section, the ensemble-averaged covariance matrix is derived for the purpose of creating physical scattering models. Ensemble averaging is carried out by integration over rotation angles as shown in Figure 4.5. At first, a scattering matrix in the HV polarization basis is rotated around the radar line of sight, then the corresponding covariance matrix is calculated. The elements of the covariance matrix are weighted by the probability density function and integrated over angles. The integral result is used as a theoretical covariance matrix for the scattering model.

4.5.1 COVARIANCE MATRIX IN THE LINEAR HV BASIS

To simplify expressions, we put the scattering matrix in the HV basis as

$$[S(HV)] = \begin{bmatrix} S_{HH} & S_{HV} \\ S_{VH} & S_{VV} \end{bmatrix} = \begin{bmatrix} a & c \\ c & b \end{bmatrix} \tag{4.5.1}$$

where $S_{HH} = a$, $S_{VV} = b$, and $S_{HV} = c$. Rotation θ around the radar line of sight yields, referring to (4.2.9),

$$[S(\theta)] = \begin{bmatrix} S_{hh} & S_{hv} \\ S_{vh} & S_{hh} \end{bmatrix} = \begin{bmatrix} \cos\theta & \sin\theta \\ -\sin\theta & \cos\theta \end{bmatrix} \begin{bmatrix} a & c \\ c & b \end{bmatrix} \begin{bmatrix} \cos\theta & -\sin\theta \\ \sin\theta & \cos\theta \end{bmatrix} \tag{4.5.2}$$

$$\begin{cases} S_{hh} = a\cos^2\theta + b\sin^2\theta + c\sin 2\theta \\ S_{vv} = a\sin^2\theta + b\cos^2\theta - c\sin 2\theta \\ 2S_{hv} = 2c\cos 2\theta - (a-b)\sin 2\theta \end{cases}$$

We can create the covariance matrix $[C(\theta)]$ based on (4.5.2).

$$[C(\theta)] = \begin{bmatrix} |S_{hh}|^2 & \sqrt{2}\,S_{hh}S_{hv}^* & S_{hh}S_{vv}^* \\ \sqrt{2}\,S_{hv}S_{hh}^* & 2|S_{hv}|^2 & \sqrt{2}\,S_{hv}S_{vv}^* \\ S_{vv}S_{hh}^* & \sqrt{2}\,S_{vv}S_{hv}^* & |S_{vv}|^2 \end{bmatrix} = [C(hv)] \tag{4.5.3}$$

The ensemble-averaged covariance matrix is derived from integration over angles weighted by the probability density function $p(\theta)$.

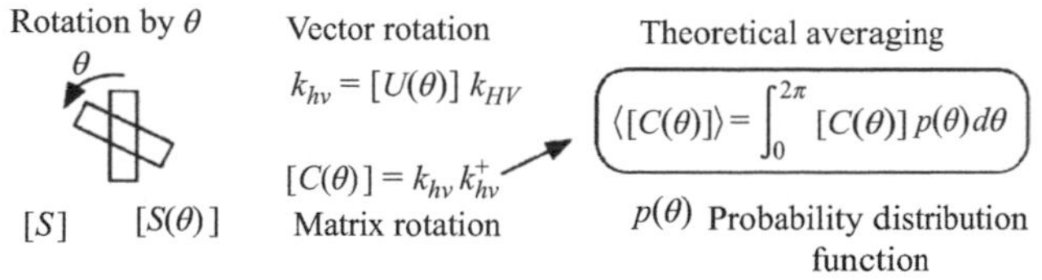

Rotation by θ Vector rotation   Theoretical averaging

$$k_{hv} = [U(\theta)]\, k_{HV}$$

$$\langle [C(\theta)] \rangle = \int_0^{2\pi} [C(\theta)]\, p(\theta)\, d\theta$$

$$[C(\theta)] = k_{hv}\, k_{hv}^+$$

$[S]$ $[S(\theta)]$ Matrix rotation $p(\theta)$ Probability distribution function

FIGURE 4.5 Theoretical averaging by integration.

$$\left\langle\left[C(\theta)\right]\right\rangle^{HV} = \int_0^{2\pi}\left[C(\theta)\right]p(\theta)d\theta \tag{4.5.4}$$

where the superscript HV in $\left\langle\left[C(\theta)\right]\right\rangle^{HV}$ indicates the polarization basis. After the integration of equation (4.5.4), the elements of $\left\langle\left[C(\theta)\right]\right\rangle^{HV}$ become as follows [5]:

$$\left\langle|S_{hh}|^2\right\rangle = |a|^2 I_1 + |b|^2 I_2 + |c|^2 I_3 + 2\,\mathrm{Re}\left\{ab^*\right\}I_4 + 2\,\mathrm{Re}\left\{ac^*\right\}I_5 + 2\,\mathrm{Re}\left\{bc^*\right\}I_6 \tag{4.5.4a}$$

$$\left\langle|S_{vv}|^2\right\rangle = |a^2|I_2 + |b|^2 I_1 + |c|^2 I_3 + 2\,\mathrm{Re}\left\{ab^*\right\}I_4 - 2\,\mathrm{Re}\left\{ac^*\right\}I_6 - 2\,\mathrm{Re}\left\{bc^*\right\}I_5 \tag{4.5.4b}$$

$$\left\langle|S_{hv}|^2\right\rangle = \frac{|b-a|^2}{4}I_3 + |c|^2 I_7 + \mathrm{Re}\left\{c^*(b-a)\right\}I_8 \tag{4.5.4c}$$

$$\left\langle S_{hh}S_{vv}^*\right\rangle = \left(|a|^2 + |b|^2\right)I_4 - |c|^2 I_3 + ab^*I_1 + a^*bI_2 + \left(b^*c - ac^*\right)I_5 + \left(a^*c - bc^*\right)I_6 \tag{4.5.4d}$$

$$\left\langle S_{hh}S_{hv}^*\right\rangle = a\frac{(b-a)^*}{2}I_5 + b\frac{(b-a)^*}{2}I_6 + c\frac{(b-a)^*}{2}I_3 + ac^*I_{10} + bc^*I_9 + |c|^2 I_8 \tag{4.5.4e}$$

$$\left\langle S_{hv}S_{vv}^*\right\rangle = a^*\frac{b-a}{2}I_6 + b^*\frac{b-a}{2}I_5 - c^*\frac{b-a}{2}I_3 + a^*cI_9 + b^*cI_{10} - |c|^2 I_8 \tag{4.5.4f}$$

$$\left\langle S_{hh}^*S_{vv}\right\rangle = \left\langle S_{hh}S_{vv}^*\right\rangle^*,\ \left\langle S_{hh}^*S_{hv}\right\rangle = \left\langle S_{hh}S_{hv}^*\right\rangle^*,\ \left\langle S_{hv}^*S_{vv}\right\rangle = \left\langle S_{hv}S_{vv}^*\right\rangle^* \tag{4.5.4g}$$

where

$$I_1 = \int_0^{2\pi}\cos^4\theta\,p(\theta)d\theta \qquad\qquad I_6 = \int_0^{2\pi}\sin^2\theta\sin 2\theta\,p(\theta)d\theta$$

$$I_2 = \int_0^{2\pi}\sin^4\theta\,p(\theta)d\theta \qquad\qquad I_7 = \int_0^{2\pi}\cos^2 2\theta\,p(\theta)d\theta$$

$$I_3 = \int_0^{2\pi}\sin^2 2\theta\,p(\theta)d\theta \qquad\qquad I_8 = \int_0^{2\pi}\sin 2\theta\cos 2\theta\,p(\theta)d\theta \tag{4.5.5}$$

$$I_4 = \int_0^{2\pi}\sin^2\theta\cos^2\theta\,p(\theta)d\theta \qquad\qquad I_9 = \int_0^{2\pi}\sin^2\theta\cos 2\theta\,p(\theta)d\theta$$

$$I_5 = \int_0^{2\pi}\cos^2\theta\sin 2\theta\,p(\theta)d\theta \qquad\qquad I_{10} = \int_0^{2\pi}\cos^2\theta\cos 2\theta\,p(\theta)d\theta$$

where $p(\theta)$ is the probability density function (PDF) or angular distribution function, satisfying

$$\int_0^{2\pi} p(\theta) d\theta = 1 \qquad (4.5.6)$$

The final covariance matrix form is dependent on $p(\theta)$. The PDF $p(\theta)$ is directly related to physical distributions of the object under observation. It is desirable to take an appropriate function considering actual target distributions. For example, tree branches are randomly oriented if seen from the zenith; however, they are rather oriented in the vertical direction if seen from the horizontal direction. Therefore, we choose three kinds (Figure 4.6) of distribution functions as follows [6]:

$$p(\theta) = \frac{1}{2\pi} \quad \text{uniform}, \qquad (4.5.7)$$

$$p(\theta) = \begin{cases} \dfrac{1}{2}\sin\theta & \text{for} \quad 0 < \theta < \pi, \\ 0 & \text{else} \end{cases} \qquad (4.5.8)$$

$$p(\theta) = \begin{cases} \dfrac{1}{2}\cos\theta & \text{for} \quad -\dfrac{\pi}{2} < \theta < \dfrac{\pi}{2} \\ 0 & \text{else} \end{cases} \qquad (4.5.9)$$

- $p(\theta) = \dfrac{1}{2\pi}$

Assuming constant PDF, the integrals (4.5.5) yields

$$I_1 = I_2 = \frac{3}{8}, I_3 = I_7 = \frac{1}{2}, I_4 = \frac{1}{8}, I_5 = I_6 = I_8 = 0, I_9 = -\frac{1}{4}$$

The elements of equation (4.5.4) become

$$\left\langle |S_{hh}|^2 \right\rangle = \left\langle |S_{vv}|^2 \right\rangle = \frac{1}{8}|a+b|^2 + \frac{1}{4}\left(|a|^2 + |b|^2\right) + \frac{1}{2}|c|^2 \quad (\text{real}) \qquad (4.5.10a)$$

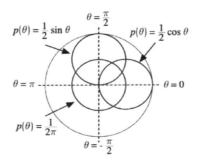

FIGURE 4.6 Probability density function.

$$\left\langle \left| S_{hv} \right|^2 \right\rangle = \frac{1}{8} \left| a - b \right|^2 + \frac{1}{2} \left| c \right|^2 \quad (\text{real}) \tag{4.5.10b}$$

$$\left\langle S_{hh} S_{vv}^* \right\rangle = \left\langle S_{hh}^* S_{vv} \right\rangle = \frac{1}{4} \left| a + b \right|^2 - \frac{1}{8} \left| a - b \right|^2 - \frac{1}{2} \left| c \right|^2 \quad (\text{real}) \tag{4.5.10c}$$

$$\left\langle S_{hh} S_{hv}^* \right\rangle = \left\langle S_{hv} S_{vv}^* \right\rangle = + \frac{j}{2} \mathrm{Im} \left\{ c^* (a - b) \right\} \quad (\text{imaginary}) \tag{4.5.10d}$$

$$\left\langle S_{hh}^* S_{hv} \right\rangle = \left\langle S_{hv}^* S_{vv} \right\rangle = - \frac{j}{2} \mathrm{Im} \left\{ c^* (a - b) \right\} \quad (\text{imaginary}) \tag{4.5.10e}$$

Note that

$$\mathrm{Trace} \left\langle [C] \right\rangle = \left\langle \left| S_{hh} \right|^2 \right\rangle + 2 \left\langle \left| S_{hv} \right|^2 \right\rangle + \left\langle \left| S_{vv} \right|^2 \right\rangle = \left| a \right|^2 + 2 \left| c \right|^2 + \left| b \right|^2 = \mathrm{Span}[S] \tag{4.5.11}$$

and commonly appeared terms are, $\left| a + b \right|^2, \left| a - b \right|^2, \left| c \right|^2, \mathrm{Im} \left\{ c^* (a - b) \right\}$.
For example, if we take the scattering matrix of a flat plate, we substitute $a = b = 1, c = 0$ into equation (4.5.10), yielding

$$\begin{cases} \left\langle \left| S_{hh} \right|^2 \right\rangle = \left\langle \left| S_{vv} \right|^2 \right\rangle = \frac{1}{8} \left| 1 + 1 \right|^2 + \frac{1}{4} (1 + 1) = 1 \\[2mm] \left\langle \left| S_{hv} \right|^2 \right\rangle = 0 \\[2mm] \left\langle S_{hh} S_{vv}^* \right\rangle = \left\langle S_{hh}^* S_{vv} \right\rangle = \frac{1}{4} \left| 2 \right|^2 = 1 \\[2mm] \left\langle S_{hh} S_{hv}^* \right\rangle = \left\langle S_{hv} S_{vv}^* \right\rangle = \left\langle S_{hh}^* S_{hv} \right\rangle = \left\langle S_{hv}^* S_{vv} \right\rangle = 0 \end{cases}$$

Then the ensemble-averaged covariance matrix becomes,

$$\left\langle [C(\theta)] \right\rangle_{\mathrm{plate}}^{HV} = \begin{bmatrix} 1 & 0 & 1 \\ 0 & 0 & 0 \\ 1 & 0 & 1 \end{bmatrix}$$

This $\left\langle [C(\theta)] \right\rangle_{\mathrm{plate}}^{HV}$ becomes a scattering model of a flat plate in the covariance matrix formulation.

- $$p(\theta) = \begin{cases} \dfrac{1}{2} \sin \theta & \text{for} \quad 0 < \theta < \pi \\[2mm] 0 & \text{else} \end{cases}$$

Assuming the above PDF, the integrals (4.5.5) yields

$$I_1 = \frac{3}{15}, \quad I_2 = I_3 = \frac{8}{15}, \quad I_4 = \frac{2}{15}, \quad I_5 = I_6 = I_8 = 0, \quad I_7 = \frac{7}{15}, \quad I_9 = -\frac{6}{15}, \quad I_{10} = \frac{1}{15}$$

The elements of equation (4.5.4) become,

$$\left\langle |S_{hh}|^2 \right\rangle = \frac{3}{15}|a|^2 + \frac{8}{15}|b|^2 + \frac{8}{15}|c|^2 + \frac{4}{15}\mathrm{Re}\left\{ab^*\right\} \quad (\text{real}) \tag{4.5.12a}$$

$$\left\langle |S_{vv}|^2 \right\rangle = \frac{8}{15}|a|^2 + \frac{3}{15}|b|^2 + \frac{8}{15}|c|^2 + \frac{4}{15}\mathrm{Re}\left\{ab^*\right\} \quad (\text{real}) \tag{4.5.12b}$$

$$\left\langle |S_{hv}|^2 \right\rangle = \frac{2}{15}|a-b|^2 + \frac{7}{15}|c|^2 \quad (\text{real}) \tag{4.5.12c}$$

$$\left\langle S_{hh}S_{vv}^* \right\rangle = \frac{2}{15}\left(|a|^2 + |b|^2 - 4|c|^2\right) + \frac{8}{15}a^*b + \frac{3}{15}ab^* \quad (\text{complex}) \tag{4.5.12d}$$

$$\left\langle S_{hh}S_{hv}^* \right\rangle = \frac{4}{15}c(b-a)^* + \frac{1}{15}c^*a - \frac{6}{15}c^*b \quad (\text{complex}) \tag{4.5.12e}$$

$$\left\langle S_{hv}S_{vv}^* \right\rangle = \frac{4}{15}c^*(b-a) + \frac{1}{15}b^*c - \frac{6}{15}a^*c \quad (\text{complex}) \tag{4.5.12f}$$

- $$p(\theta) = \begin{cases} \dfrac{1}{2}\cos\theta & \text{for } -\dfrac{\pi}{2} < \theta < \dfrac{\pi}{2} \\ 0 & \text{else} \end{cases}$$

Assuming the above $p(\theta)$, the integrals (4.5.5) yield

$$I_2 = \frac{3}{15}, \quad I_1 = I_3 = \frac{8}{15}, \quad I_4 = \frac{2}{15}, \quad I_5 = I_6 = I_8 = 0, \quad I_7 = \frac{7}{15}, \quad I_9 = -\frac{1}{15}, \quad I_{10} = \frac{6}{15}$$

The elements of equation (4.5.4) become,

$$\left\langle |S_{hh}|^2 \right\rangle = \frac{8}{15}|a|^2 + \frac{3}{15}|b|^2 + \frac{8}{15}|c|^2 + \frac{4}{15}\mathrm{Re}\left\{ab^*\right\} \quad (\text{real}) \tag{4.5.13a}$$

$$\left\langle |S_{vv}|^2 \right\rangle = \frac{3}{15}|a|^2 + \frac{8}{15}|b|^2 + \frac{8}{15}|c|^2 + \frac{4}{15}\mathrm{Re}\left\{ab^*\right\} \quad (\text{real}) \tag{4.5.13b}$$

$$\left\langle |S_{hv}|^2 \right\rangle = \frac{2}{15}|a-b|^2 + \frac{7}{15}|c|^2 \quad (\text{real}) \tag{4.5.13c}$$

$$\langle S_{hh}S_{vv}^* \rangle = \frac{2}{15}\left(|a|^2+|b|^2-4|c|^2\right)+\frac{3}{15}a^*b+\frac{8}{15}ab^* \quad (\text{complex}) \tag{4.5.13d}$$

$$\langle S_{hh}S_{hv}^* \rangle = \frac{4}{15}c(b-a)^*+\frac{6}{15}c^*a-\frac{1}{15}c^*b \quad (\text{complex}) \tag{4.5.13e}$$

$$\langle S_{hv}S_{vv}^* \rangle = \frac{4}{15}c^*(b-a)+\frac{6}{15}b^*c-\frac{1}{15}a^*c \quad (\text{complex}) \tag{4.5.13f}$$

Equations (4.5.10), (4.5.12), and (4.5.13) generate the general theoretical covariance matrices for any scattering object. Table 4.2 lists covariance matrices of canonical targets based on the preceding equations. These matrices can be used as a theoretical reference in the modeling.

The important feature of the ensemble average matrix is its invariance with respect to target orientations. The scattering matrix is sensitive to the orientation of target. The horizontal dipole has a scattering matrix different from that of the vertical dipole. However, the covariance matrix has the same form for both dipoles. The form of the covariance matrix remains the same regardless of the dipole orientation. This property is important and convenient for detecting and classifying an object in polarimetric observation, especially for airborne PolSAR or spaceborne PolSAR observations. This important property comes from the second-order statistics of polarimetric information contained in the covariance matrix. If the distribution function is different, the elements may change. However, the value itself is close to each other such as 2/8 and 4/15 for C_{22} of dipoles and 2/2 and 16/15 for C_{22} of dihedrals.

4.5.2 Covariance Matrix in the Circular *LR* Basis

It is anticipated that the circular polarization (*LR*) basis is invariant with respect to rotation of the target, that is, roll-invariant. It is worth investigating to see the form of the covariance matrix in the circular polarization basis.

The scattering vector in the circular polarization basis after rotation can be written as,

$$\mathbf{k}_{LR}(\theta)=\left[U_c\right]\left[U_\theta\right]\mathbf{k}_{HV} \tag{4.5.14}$$

$$\begin{bmatrix} S_{LL}(\theta)\\ \sqrt{2}S_{LR}(\theta)\\ S_{RR}(\theta)\end{bmatrix}=\frac{1}{2}\begin{bmatrix} e^{-j2\theta} & j\sqrt{2}e^{-j2\theta} & -e^{-j2\theta}\\ j\sqrt{2} & 0 & j\sqrt{2}\\ -e^{j2\theta} & j\sqrt{2}e^{j2\theta} & e^{j2\theta}\end{bmatrix}\begin{bmatrix} a\\ \sqrt{2}c\\ b\end{bmatrix} \tag{4.5.15}$$

Therefore, the covariance matrix becomes,

$$\left[C(\theta)\right]^{LR}=\mathbf{k}_{LR}(\theta)\mathbf{k}_{LR}^\dagger(\theta)=\left[U_c\right]\left[U_\theta\right]\mathbf{k}_{HV}\mathbf{k}_{HV}^\dagger\left[U_\theta\right]^\dagger\left[U_c\right]^\dagger$$

$$=\begin{bmatrix} |S_{LL}|^2 & \sqrt{2}S_{LL}S_{LR}^*e^{-j2\theta} & S_{LL}S_{RR}^*e^{-j4\theta}\\ \sqrt{2}S_{LR}S_{LL}^*e^{j2\theta} & 2|S_{LR}|^2 & \sqrt{2}S_{LR}S_{RR}^*e^{-j2\theta}\\ S_{RR}S_{LL}^*e^{j4\theta} & \sqrt{2}S_{RR}S_{LR}^*e^{j2\theta} & |S_{RR}|^2\end{bmatrix}, \tag{4.5.16}$$

TABLE 4.2
Theoretical Covariance Matrices of Canonical Targets

Target Type	Scattering Matrix $\begin{bmatrix} S_{HH} & S_{HV} \\ S_{VH} & S_{VV} \end{bmatrix}$	Averaged Covariance Matrix		
		$p(\theta) = \dfrac{1}{2\pi}$	$p(\theta) = \dfrac{1}{2}\sin\theta$	$p(\theta) = \dfrac{1}{2}\cos\theta$
Sphere, Plate	$\begin{bmatrix} 1 & 0 \\ 0 & 1 \end{bmatrix}$	$\begin{bmatrix} 1 & 0 & 1 \\ 0 & 0 & 0 \\ 1 & 0 & 1 \end{bmatrix}$	$\begin{bmatrix} 1 & 0 & 1 \\ 0 & 0 & 0 \\ 1 & 0 & 1 \end{bmatrix}$	$\begin{bmatrix} 1 & 0 & 1 \\ 0 & 0 & 0 \\ 1 & 0 & 1 \end{bmatrix}$
H-dipole	$\begin{bmatrix} 1 & 0 \\ 0 & 0 \end{bmatrix}$	$\dfrac{1}{8}\begin{bmatrix} 3 & 0 & 1 \\ 0 & 2 & 0 \\ 1 & 0 & 3 \end{bmatrix}$	$\dfrac{1}{15}\begin{bmatrix} 3 & 0 & 2 \\ 0 & 4 & 0 \\ 2 & 0 & 8 \end{bmatrix}$	$\dfrac{1}{15}\begin{bmatrix} 8 & 0 & 2 \\ 0 & 4 & 0 \\ 2 & 0 & 3 \end{bmatrix}$
V-dipole	$\begin{bmatrix} 0 & 0 \\ 0 & 1 \end{bmatrix}$	$\dfrac{1}{8}\begin{bmatrix} 3 & 0 & 1 \\ 0 & 2 & 0 \\ 1 & 0 & 3 \end{bmatrix}$	$\dfrac{1}{15}\begin{bmatrix} 8 & 0 & 2 \\ 0 & 4 & 0 \\ 2 & 0 & 3 \end{bmatrix}$	$\dfrac{1}{15}\begin{bmatrix} 3 & 0 & 2 \\ 0 & 4 & 0 \\ 2 & 0 & 8 \end{bmatrix}$
H-dihedral	$\begin{bmatrix} 1 & 0 \\ 0 & -1 \end{bmatrix}$	$\dfrac{1}{2}\begin{bmatrix} 0 & 0 & -1 \\ 0 & 2 & 0 \\ -1 & 0 & 1 \end{bmatrix}$	$\dfrac{1}{15}\begin{bmatrix} 7 & 0 & -7 \\ 0 & 16 & 0 \\ -7 & 0 & 7 \end{bmatrix}$	$\dfrac{1}{15}\begin{bmatrix} 7 & 0 & -7 \\ 0 & 16 & 0 \\ -7 & 0 & 7 \end{bmatrix}$
V-dihedral	$\begin{bmatrix} -1 & 0 \\ 0 & 1 \end{bmatrix}$	$\dfrac{1}{2}\begin{bmatrix} 0 & 0 & -1 \\ 0 & 2 & 0 \\ -1 & 0 & 1 \end{bmatrix}$	$\dfrac{1}{15}\begin{bmatrix} 7 & 0 & -7 \\ 0 & 16 & 0 \\ -7 & 0 & 7 \end{bmatrix}$	$\dfrac{1}{15}\begin{bmatrix} 7 & 0 & -7 \\ 0 & 16 & 0 \\ -7 & 0 & 7 \end{bmatrix}$
Left helix	$\dfrac{1}{2}\begin{bmatrix} 1 & j \\ j & -1 \end{bmatrix}$	$\dfrac{1}{4}\begin{bmatrix} 1 & -j\sqrt{2} & -1 \\ j\sqrt{2} & 2 & -j\sqrt{2} \\ -1 & j\sqrt{2} & 1 \end{bmatrix}$	$\dfrac{1}{4}\begin{bmatrix} 1 & -j\sqrt{2} & -1 \\ j\sqrt{2} & 2 & -j\sqrt{2} \\ -1 & j\sqrt{2} & 1 \end{bmatrix}$	$\dfrac{1}{4}\begin{bmatrix} 1 & -j\sqrt{2} & -1 \\ j\sqrt{2} & 2 & -j\sqrt{2} \\ -1 & j\sqrt{2} & 1 \end{bmatrix}$
Right helix	$\dfrac{1}{2}\begin{bmatrix} 1 & -j \\ -j & -1 \end{bmatrix}$	$\dfrac{1}{4}\begin{bmatrix} 1 & j\sqrt{2} & -1 \\ -j\sqrt{2} & 2 & j\sqrt{2} \\ -1 & -j\sqrt{2} & 1 \end{bmatrix}$	$\dfrac{1}{4}\begin{bmatrix} 1 & j\sqrt{2} & -1 \\ -j\sqrt{2} & 2 & j\sqrt{2} \\ -1 & -j\sqrt{2} & 1 \end{bmatrix}$	$\dfrac{1}{4}\begin{bmatrix} 1 & j\sqrt{2} & -1 \\ -j\sqrt{2} & 2 & j\sqrt{2} \\ -1 & -j\sqrt{2} & 1 \end{bmatrix}$

where

$$\begin{cases} S_{LL} = \dfrac{1}{2}(a-b+j\,2c) \\[2ex] S_{RR} = \dfrac{1}{2}(b-a+j\,2c). \\[2ex] S_{LR} = S_{RL} = \dfrac{j}{2}(a+b) \end{cases} \tag{4.5.17}$$

Assuming constant distribution of $p(\theta)$, we obtain the following result:

$$\langle [C(\theta)]\rangle^{LR} = \frac{1}{2\pi}\int_0^{2\pi}[C(\theta)]^{LR}\,d\theta = \frac{1}{4}\begin{bmatrix} |a-b+j\,2c|^2 & 0 & 0 \\ 0 & 2|a+b|^2 & 0 \\ 0 & 0 & |b-a+j\,2c|^2 \end{bmatrix} \tag{4.5.18}$$

Since the final equation (4.5.18) is of diagonal form, we can see the diagonal elements corresponding to eigenvalues. Therefore, it is convenient for eigenvalue analysis.

On the other hand, the number of independent parameters is three and is less than four in other polarization matrices. Therefore, this form is not so convenient for classifying or identifying a target. For reference, the form of the canonical object becomes as follows:

$$\text{Dipole:}\frac{1}{4}\begin{bmatrix}1&0&0\\0&2&0\\0&0&1\end{bmatrix}\quad \text{Plate:}\begin{bmatrix}0&0&0\\0&2&0\\0&0&0\end{bmatrix}\quad \text{Dihedral:}\begin{bmatrix}1&0&0\\0&0&0\\0&0&1\end{bmatrix}$$

$$\text{L-helix:}\begin{bmatrix}0&0&0\\0&0&0\\0&0&1\end{bmatrix}\quad \text{R-helix:}\begin{bmatrix}1&0&0\\0&0&0\\0&0&0\end{bmatrix}$$

4.6   COHERENCY MATRIX BY INTEGRATION

The coherency matrix has the advantage of mathematical orthogonality and at the same time representing physical scattering mechanisms. It is useful for the interpretation of the PolSAR image. Here we derive the coherency matrix by integration in the same way as the covariance matrix using the scattering vector.

A scattering vector $\mathbf{k}_P$ and its rotation vector $\mathbf{k}_P(\theta)$ can be expressed as

$$\mathbf{k}_P = \frac{1}{\sqrt{2}}\begin{bmatrix}S_{HH}+S_{VV}\\S_{HH}-S_{VV}\\2S_{HV}\end{bmatrix} = \frac{1}{\sqrt{2}}\begin{bmatrix}a+b\\a-b\\2c\end{bmatrix}, \quad \mathbf{k}_P(\theta) = \frac{1}{\sqrt{2}}\begin{bmatrix}S_{hh}+S_{vv}\\S_{hh}-S_{vv}\\2S_{hv}\end{bmatrix} \tag{4.6.1}$$

Based on equation (4.2.9), the rotation relation can be expressed in a simple form as

$$\begin{bmatrix} S_{hh} + S_{vv} \\ S_{hh} - S_{vv} \\ 2S_{hv} \end{bmatrix} = \begin{bmatrix} 1 & 0 & 0 \\ 0 & \cos 2\theta & \sin 2\theta \\ 0 & -\sin 2\theta & \cos 2\theta \end{bmatrix} \begin{bmatrix} S_{HH} + S_{VV} \\ S_{HH} - S_{VV} \\ 2S_{HV} \end{bmatrix}$$

so that

$$\mathbf{k}_P(\theta) = [R_P(\theta)] \mathbf{k}_P \tag{4.6.2}$$

where

$$[R_P(\theta)] = \begin{bmatrix} 1 & 0 & 0 \\ 0 & \cos 2\theta & \sin 2\theta \\ 0 & -\sin 2\theta & \cos 2\theta \end{bmatrix} : \text{Rotation matrix} \tag{4.6.3}$$

Therefore, the coherency matrix $[T(\theta)]$ after rotation is given by

$$[T(\theta)] = \mathbf{k}_P(\theta) \mathbf{k}_P^\dagger(\theta) = [R_P(\theta)] \mathbf{k}_P \mathbf{k}_P^\dagger [R_P(\theta)]^\dagger$$

$$= [R_P(\theta)][T][R_P(\theta)]^\dagger = \begin{bmatrix} T_{11}(\theta) T_{12}(\theta) T_{13}(\theta) \\ T_{21}(\theta) T_{22}(\theta) T_{23}(\theta) \\ T_{31}(\theta) T_{32}(\theta) T_{33}(\theta) \end{bmatrix}. \tag{4.6.4}$$

The elements are:

$$T_{11}(\theta) = T_{11}, \quad T_{12}(\theta) = T_{12} \cos 2\theta + T_{13} \sin 2\theta, \quad T_{13}(\theta) = T_{13} \cos 2\theta - T_{12} \sin 2\theta,$$

$$T_{21}(\theta) = T_{12}^*(\theta), \quad T_{22}(\theta) = T_{22}\cos^2 2\theta + T_{33}\sin^2 2\theta + \mathrm{Re}\{T_{23}\} \sin 4\theta,$$

$$T_{23}(\theta) = \mathrm{Re}\{T_{23}\} \cos 4\theta - \frac{T_{22} - T_{33}}{2} \sin 4\theta + j\mathrm{Im}\{T_{23}\}, \tag{4.6.5}$$

$$T_{31}(\theta) = T_{13}^*(\theta), \quad T_{32}(\theta) = T_{23}^*(\theta), \quad T_{33}(\theta) = T_{33}\cos^2 2\theta + T_{22}\sin^2 2\theta - \mathrm{Re}\{T_{23}\} \sin 4\theta$$

In order to derive ensemble averaging, we carry out the following integration with three kinds of probability density functions.

$$\langle [T(\theta)] \rangle = \int_0^{2\pi} [T(\theta)] p(\theta) d\theta \tag{4.6.6}$$

The integration yields

- for $p(\theta) = \dfrac{1}{2\pi}$

$$\left\langle\left[T(\theta)\right]\right\rangle = \begin{bmatrix} \dfrac{1}{2}\left|a+b\right|^2 & 0 & 0 \\[2mm] 0 & \dfrac{1}{4}\left|a-b\right|^2 + \left|c\right|^2 & j\mathrm{Im}\left\{c^*\left(a-b\right)\right\} \\[2mm] 0 & -j\mathrm{Im}\left\{c^*\left(a-b\right)\right\} & \dfrac{1}{4}\left|a-b\right|^2 + \left|c\right|^2 \end{bmatrix} \qquad (4.6.7)$$

- for $p(\theta) = \dfrac{1}{2}\sin\theta$,

$$\left\langle\left[T(\theta)\right]\right\rangle = \begin{bmatrix} \dfrac{1}{2}\left|a+b\right|^2 & -\dfrac{1}{6}(a+b)(a-b)^* & -\dfrac{1}{3}c^*(a+b) \\[2mm] -\dfrac{1}{6}(a+b)^*(a-b) & \dfrac{7}{30}\left|a-b\right|^2 + \dfrac{16}{15}\left|c\right|^2 & \dfrac{7}{15}c^*(a-b) - \dfrac{8}{15}c(a-b)^* \\[2mm] -\dfrac{1}{3}c(a+b)^* & \dfrac{7}{15}c(a-b)^* - \dfrac{8}{15}c^*(a-b) & \dfrac{8}{30}\left|a-b\right|^2 + \dfrac{14}{15}\left|c\right|^2 \end{bmatrix}$$

$$(4.6.8)$$

- for $p(\theta) = \dfrac{1}{2}\cos\theta$,

$$\left\langle\left[T(\theta)\right]\right\rangle = \begin{bmatrix} \dfrac{1}{2}\left|a+b\right|^2 & \dfrac{1}{6}(a+b)(a-b)^* & \dfrac{1}{3}c^*(a+b) \\[2mm] -\dfrac{1}{6}(a+b)^*(a-b) & \dfrac{7}{30}\left|a-b\right|^2 + \dfrac{16}{15}\left|c\right|^2 & \dfrac{7}{15}c^*(a-b) - \dfrac{8}{15}c(a-b)^* \\[2mm] \dfrac{1}{3}c(a+b)^* & \dfrac{7}{15}c(a-b)^* - \dfrac{8}{15}c^*(a-b) & \dfrac{8}{30}\left|a-b\right|^2 + \dfrac{14}{15}\left|c\right|^2 \end{bmatrix}$$

$$(4.6.9)$$

Note that $\mathrm{Trace}\left\langle\left[T(\theta)\right]\right\rangle = \left|a\right|^2 + 2\left|c\right|^2 + \left|b\right|^2 = \mathrm{Span}\left[S\right]$ applies to all coherency matrices (4.6.7–4.6.9).

It is understood that four terms, $\left|a+b\right|^2$, $\left|a-b\right|^2$, $\left|c\right|^2$, and $\mathrm{Im}\left\{c^*(a-b)\right\}$, appear as independent parameters in the coherency matrix in the same way as in the covariance matrix. These four terms are important polarimetric indexes.

Canonical targets are listed in Table 4.3. Plate, sphere, and helix have the same form regardless of probability functions. Dipoles and dihedrals have slightly different forms; however, the element values are close to each other as regards to PDF. Table 4.3 shows the basic scattering models in the scattering power decomposition (Chapters 7 and 8).

The terminology is given to coherency matrix whose constitutes are as follows:

Reflection symmetry   Rotation symmetry   Azimuthal symmetry

$$\left\langle[T]\right\rangle = \begin{bmatrix} x & x & 0 \\ x & x & 0 \\ 0 & 0 & x \end{bmatrix} \qquad \begin{bmatrix} x & 0 & 0 \\ 0 & x & x \\ 0 & x & x \end{bmatrix} \qquad \begin{bmatrix} x & 0 & 0 \\ 0 & x & 0 \\ 0 & 0 & x \end{bmatrix}$$

TABLE 4.3

Coherency Matrix of Canonical Objects by Integration

Target	Vector $\mathbf{k}_P(\theta)$	Coherency Matrix	Normalized Coherency Matrix by Integration $p(\theta)=\frac{1}{2\pi}$	$p(\theta)=\frac{1}{2}\sin\theta$	$p(\theta)=\frac{1}{2}\cos\theta$
Sphere, Plate	$\begin{bmatrix} 1 \\ 0 \\ 0 \end{bmatrix}$	$\begin{bmatrix} 1 & 0 & 0 \\ 0 & 0 & 0 \\ 0 & 0 & 0 \end{bmatrix}$	$\begin{bmatrix} 1 & 0 & 0 \\ 0 & 0 & 0 \\ 0 & 0 & 0 \end{bmatrix}$	$\begin{bmatrix} 1 & 0 & 0 \\ 0 & 0 & 0 \\ 0 & 0 & 0 \end{bmatrix}$	$\begin{bmatrix} 1 & 0 & 0 \\ 0 & 0 & 0 \\ 0 & 0 & 0 \end{bmatrix}$
H-dipole	$\begin{bmatrix} 1 \\ \cos 2\theta \\ -\sin 2\theta \end{bmatrix}$	$\begin{bmatrix} 1 & \cos 2\theta & -\sin 2\theta \\ \cos 2\theta & \cos^2 2\theta & \frac{\sin 4\theta}{2} \\ -\sin 2\theta & \frac{\sin 4\theta}{2} & \sin^2 2\theta \end{bmatrix}$	$\frac{1}{4}\begin{bmatrix} 2 & 0 & 0 \\ 0 & 1 & 0 \\ 0 & 0 & 1 \end{bmatrix}$	$\frac{1}{30}\begin{bmatrix} 15 & 5 & 0 \\ -5 & 7 & 0 \\ 0 & 0 & 8 \end{bmatrix}$	$\frac{1}{30}\begin{bmatrix} 15 & 5 & 0 \\ 5 & 7 & 0 \\ 0 & 0 & 8 \end{bmatrix}$
V-dipole	$\begin{bmatrix} 1 \\ -\cos 2\theta \\ \sin 2\theta \end{bmatrix}$	$\begin{bmatrix} 1 & -\cos 2\theta & \sin 2\theta \\ -\cos 2\theta & \cos^2 2\theta & \frac{\sin 4\theta}{2} \\ \sin 2\theta & \frac{\sin 4\theta}{2} & \sin^2 2\theta \end{bmatrix}$	$\frac{1}{4}\begin{bmatrix} 2 & 0 & 0 \\ 0 & 1 & 0 \\ 0 & 0 & 1 \end{bmatrix}$	$\frac{1}{30}\begin{bmatrix} 15 & 5 & 0 \\ 5 & 7 & 0 \\ 0 & 0 & 8 \end{bmatrix}$	$\frac{1}{30}\begin{bmatrix} 15 & -5 & 0 \\ -5 & 7 & 0 \\ 0 & 0 & 8 \end{bmatrix}$
H-dihedral	$\begin{bmatrix} 1 \\ \cos 2\theta \\ -\sin 2\theta \end{bmatrix}$	$\begin{bmatrix} 0 & 0 & 0 \\ 0 & \cos^2 2\theta & \frac{\sin 4\theta}{2} \\ 0 & \frac{\sin 4\theta}{2} & \sin^2 2\theta \end{bmatrix}$	$\frac{1}{2}\begin{bmatrix} 0 & 0 & 0 \\ 0 & 1 & 0 \\ 0 & 0 & 1 \end{bmatrix}$	$\frac{1}{30}\begin{bmatrix} 0 & 0 & 0 \\ 0 & 14 & 0 \\ 0 & 0 & 16 \end{bmatrix}$	$\frac{1}{30}\begin{bmatrix} 0 & 0 & 0 \\ 0 & 14 & 0 \\ 0 & 0 & 16 \end{bmatrix}$
V-dihedral	$\begin{bmatrix} 1 \\ -\cos 2\theta \\ \sin 2\theta \end{bmatrix}$	$\begin{bmatrix} 0 & 0 & 0 \\ 0 & \cos^2 2\theta & \frac{\sin 4\theta}{2} \\ 0 & \frac{\sin 4\theta}{2} & \sin^2 2\theta \end{bmatrix}$	$\frac{1}{2}\begin{bmatrix} 0 & 0 & 0 \\ 0 & 1 & 0 \\ 0 & 0 & 1 \end{bmatrix}$	$\frac{1}{30}\begin{bmatrix} 0 & 0 & 0 \\ 0 & 14 & 0 \\ 0 & 0 & 16 \end{bmatrix}$	$\frac{1}{30}\begin{bmatrix} 0 & 0 & 0 \\ 0 & 14 & 0 \\ 0 & 0 & 16 \end{bmatrix}$
Left helix	$e^{j2\theta}\begin{bmatrix} 0 \\ 1 \\ j \end{bmatrix}$	$\begin{bmatrix} 0 & 0 & 0 \\ 0 & 1 & -j \\ 0 & j & 1 \end{bmatrix}$	$\frac{1}{2}\begin{bmatrix} 0 & 0 & 0 \\ 0 & 1 & -j \\ 0 & j & 1 \end{bmatrix}$	$\frac{1}{2}\begin{bmatrix} 0 & 0 & 0 \\ 0 & 1 & -j \\ 0 & j & 1 \end{bmatrix}$	$\frac{1}{2}\begin{bmatrix} 0 & 0 & 0 \\ 0 & 1 & -j \\ 0 & j & 1 \end{bmatrix}$
Right helix	$e^{-j2\theta}\begin{bmatrix} 0 \\ 1 \\ -j \end{bmatrix}$	$\begin{bmatrix} 0 & 0 & 0 \\ 0 & 1 & j \\ 0 & -j & 1 \end{bmatrix}$	$\frac{1}{2}\begin{bmatrix} 0 & 0 & 0 \\ 0 & 1 & j \\ 0 & -j & 1 \end{bmatrix}$	$\frac{1}{2}\begin{bmatrix} 0 & 0 & 0 \\ 0 & 1 & j \\ 0 & -j & 1 \end{bmatrix}$	$\frac{1}{2}\begin{bmatrix} 0 & 0 & 0 \\ 0 & 1 & j \\ 0 & -j & 1 \end{bmatrix}$

4.7 THEORETICAL KENNAUGH MATRIX

Using a rotated scattering matrix (4.5.2), the corresponding Kennaugh matrix $\left[K(\theta)\right]$ becomes,

$$\left[K(\theta)\right] = \begin{bmatrix} \dfrac{|S_{hh}|^2 + 2|S_{hv}|^2 + |S_{vv}|^2}{2} & \dfrac{|S_{hh}|^2 - |S_{vv}|^2}{2} & \mathrm{Re}\left\{(S_{hh}+S_{vv})S_{hv}^*\right\} & \mathrm{Im}\left\{(S_{hh}-S_{vv})S_{hv}^*\right\} \\ \dfrac{|S_{hh}|^2 - |S_{vv}|^2}{2} & \dfrac{|S_{hh}|^2 - 2|S_{hv}|^2 + |S_{vv}|^2}{2} & \mathrm{Re}\left\{(S_{hh}-S_{vv})S_{hv}^*\right\} & \mathrm{Im}\left\{(S_{hh}+S_{vv})S_{hv}^*\right\} \\ \mathrm{Re}\left\{(S_{hh}+S_{vv})S_{hv}^*\right\} & \mathrm{Re}\left\{(S_{hh}-S_{vv})S_{hv}^*\right\} & |S_{hv}|^2 + \mathrm{Re}\left\{S_{hh}S_{vv}^*\right\} & \mathrm{Im}\left\{S_{hh}S_{vv}^*\right\} \\ \mathrm{Im}\left\{(S_{hh}-S_{vv})S_{hv}^*\right\} & \mathrm{Im}\left\{(S_{hh}+S_{vv})S_{hv}^*\right\} & \mathrm{Im}\left\{S_{hh}S_{vv}^*\right\} & |S_{hv}|^2 - \mathrm{Re}\left\{S_{hh}S_{vv}^*\right\} \end{bmatrix}$$

$$(4.7.1)$$

Assuming a uniform PDF, the Kennaugh matrix by integration is given by

$$\left\langle\left[K(\theta)\right]\right\rangle = \frac{1}{2\pi}\int_0^{2\pi}\left[K(\theta)\right]d\theta \tag{4.7.2}$$

By using equation (4.5.9), it is written as

$$\left\langle\left[K(\theta)\right]\right\rangle = \begin{bmatrix} \dfrac{1}{2}\left(|a|^2 + 2|c|^2 + |b|^2\right) & 0 & 0 & \mathrm{Im}\left\{c^*(a-b)\right\} \\ 0 & \dfrac{1}{4}|a+b|^2 & 0 & 0 \\ 0 & 0 & \dfrac{1}{4}|a+b|^2 & 0 \\ \mathrm{Im}\left\{c^*(a-b)\right\} & 0 & 0 & |c|^2 - \mathrm{Re}\left\{ab^*\right\} \end{bmatrix} \tag{4.7.3}$$

Since $\mathrm{Re}\left\{ab^*\right\} = \frac{1}{4}\left(|a+b|^2 - |a-b|^2\right)$, we can consider $|a+b|^2$, $|a-b|^2$, $|c|^2$, and $\mathrm{Im}\left\{c^*(a-b)\right\}$ as four independent parameters again.

4.8 POLARIZATION MATRICES OF CANONICAL TARGETS

Canonical targets expressed by the covariance matrix, coherency matrix, and Kennaugh matrix are listed and compared in Table 4.4. They are normalized so that the trace becomes 1.

It is interesting to compare the forms and check which matrix is suitable for the classification of objects. For example, coherency formulation has the simplest form for a plate or sphere. For a helix target, the coherency matrix formulation gives pure imaginary for T_{23}, and its sign indicates the sense of rotation. For dihedrals, the C_{13} component of a covariance matrix yields negative values, which are easily found. The dipole expression is a sum of a plate and dihedral, etc. These matrix forms are important references for target classification and identification.

TABLE 4.4

Canonical Target Expressed in Various Polarization Matrices

Target	Covariance $\langle[C(HV)]\rangle$	Covariance $\langle[C(LR)]\rangle$	Coherency $\langle[T]\rangle$	Kennaugh $\langle[K]\rangle$
Plate, Sphere	$\dfrac{1}{2}\begin{bmatrix}1&0&1\\0&0&0\\1&0&1\end{bmatrix}$	$\dfrac{1}{2}\begin{bmatrix}0&0&0\\0&2&0\\0&0&0\end{bmatrix}$	$\begin{bmatrix}1&0&0\\0&0&0\\0&0&0\end{bmatrix}$	$\begin{bmatrix}1&0&0&0\\0&1&0&0\\0&0&1&0\\0&0&0&-1\end{bmatrix}$
Dihedral	$\dfrac{1}{4}\begin{bmatrix}1&0&-1\\0&2&0\\-1&0&1\end{bmatrix}$	$\dfrac{1}{2}\begin{bmatrix}1&0&0\\0&0&0\\0&0&1\end{bmatrix}$	$\dfrac{1}{2}\begin{bmatrix}0&0&0\\0&1&0\\0&0&1\end{bmatrix}$	$\begin{bmatrix}1&0&0&0\\0&0&0&0\\0&0&0&0\\0&0&0&1\end{bmatrix}$
Dipole	$\dfrac{1}{8}\begin{bmatrix}3&0&1\\0&2&0\\1&0&3\end{bmatrix}$	$\dfrac{1}{4}\begin{bmatrix}1&0&0\\0&2&0\\0&0&1\end{bmatrix}$	$\dfrac{1}{4}\begin{bmatrix}2&0&0\\0&1&0\\0&0&1\end{bmatrix}$	$\dfrac{1}{4}\begin{bmatrix}2&0&0&0\\0&1&0&0\\0&0&1&0\\0&0&0&0\end{bmatrix}$
L-Helix	$\dfrac{1}{4}\begin{bmatrix}1&-j\sqrt{2}&-1\\j\sqrt{2}&2&-j\sqrt{2}\\-1&j\sqrt{2}&1\end{bmatrix}$	$\begin{bmatrix}0&0&0\\0&0&0\\0&0&1\end{bmatrix}$	$\dfrac{1}{2}\begin{bmatrix}0&0&0\\0&1&-j\\0&j&1\end{bmatrix}$	$\dfrac{1}{2}\begin{bmatrix}1&0&0&-1\\0&0&0&0\\0&0&0&0\\-1&0&0&-1\end{bmatrix}$
R-Helix	$\dfrac{1}{4}\begin{bmatrix}1&j\sqrt{2}&1\\-j\sqrt{2}&2&j\sqrt{2}\\-1&-j\sqrt{2}&1\end{bmatrix}$	$\begin{bmatrix}1&0&0\\0&0&0\\0&0&0\end{bmatrix}$	$\dfrac{1}{2}\begin{bmatrix}0&0&0\\0&1&j\\0&-j&1\end{bmatrix}$	$\dfrac{1}{2}\begin{bmatrix}0&0&0&1\\0&0&0&0\\0&0&0&0\\1&0&0&1\end{bmatrix}$

4.9 MUTUAL TRANSFORMATION OF POLARIZATION MATRICES AND SUMMARY

4.9.1 RELATION BETWEEN COVARIANCE MATRIX AND COHERENCY MATRIX

These two matrices are 3×3 complex valued and semi-definite matrices. Since they are frequently used in the data analysis, the relation is explained again. As shown in Section 4.3, the transformation is carried out as,

$$\left[C(HV)\right] = \mathbf{k}_{HV}\mathbf{k}_{HV}^{\dagger}, \quad \left[T\right] = \mathbf{k}_{P}\mathbf{k}_{P}^{\dagger} \tag{4.9.1}$$

$$\mathbf{k}_{P} = \left[U_{P}\right]\mathbf{k}_{HV}, \quad \left[U_{P}\right] = \frac{1}{\sqrt{2}}\begin{bmatrix}1&0&1\\1&0&-1\\0&\sqrt{2}&0\end{bmatrix} \tag{4.9.2}$$

$$\left[T\right] = \mathbf{k}_{P}\mathbf{k}_{P}^{\dagger} = \left[U_{P}\right]\mathbf{k}_{HV}\mathbf{k}_{HV}^{\dagger}\left[U_{P}\right]^{\dagger} = \left[U_{P}\right]\left[C(HV)\right]\left[U_{P}\right]^{\dagger} \tag{4.9.3}$$

Since $[U_P]$ is a unitary matrix, the covariance matrix and coherency matrix are equivalent mathematically. Therefore, the information contained inside is the same. This also indicates the eigenvalues of both matrices are the same.

$$[T] = [U_P][C][U_P]^\dagger, [C] = [U_P]^\dagger[T][U_P] \tag{4.9.4}$$

In the same way, covariance in the HV basis can be transformed to that in the circular LR basis.

$$\mathbf{k}_{LR} = [U_c]\mathbf{k}_{HV}, \quad [C(LR)] = \mathbf{k}_{LR}\mathbf{k}_{LR}^\dagger \tag{4.9.5}$$

$$[U_c] = \begin{bmatrix} 1 & j\sqrt{2} & -1 \\ j\sqrt{2} & 0 & j\sqrt{2} \\ -1 & j\sqrt{2} & 1 \end{bmatrix} \tag{4.9.6}$$

$$[C(LR)] = \mathbf{k}_{LR}\mathbf{k}_{LR}^\dagger = [U_c]\mathbf{k}_{HV}\mathbf{k}_{HV}^\dagger[U_c]^\dagger = [U_c][C(HV)][U_c]^\dagger \tag{4.9.7}$$

So far, various polarization matrices are introduced and compared. As a summary, the mutual relations and transformations can be visualized as shown in Figure 4.7. These matrices are connected by unitary transformation. Once the scattering matrix is obtained, all polarization matrices can be derived by unitary transformation as shown in Figure 4.4. The number of independent parameters is nine, even if the matrix form is different. The four key parameters $|a+b|^2, |a-b|^2, |c|^2$, and $\text{Im}\{c^*(a-b)\}$ appear in the theoretically averaged matrices.

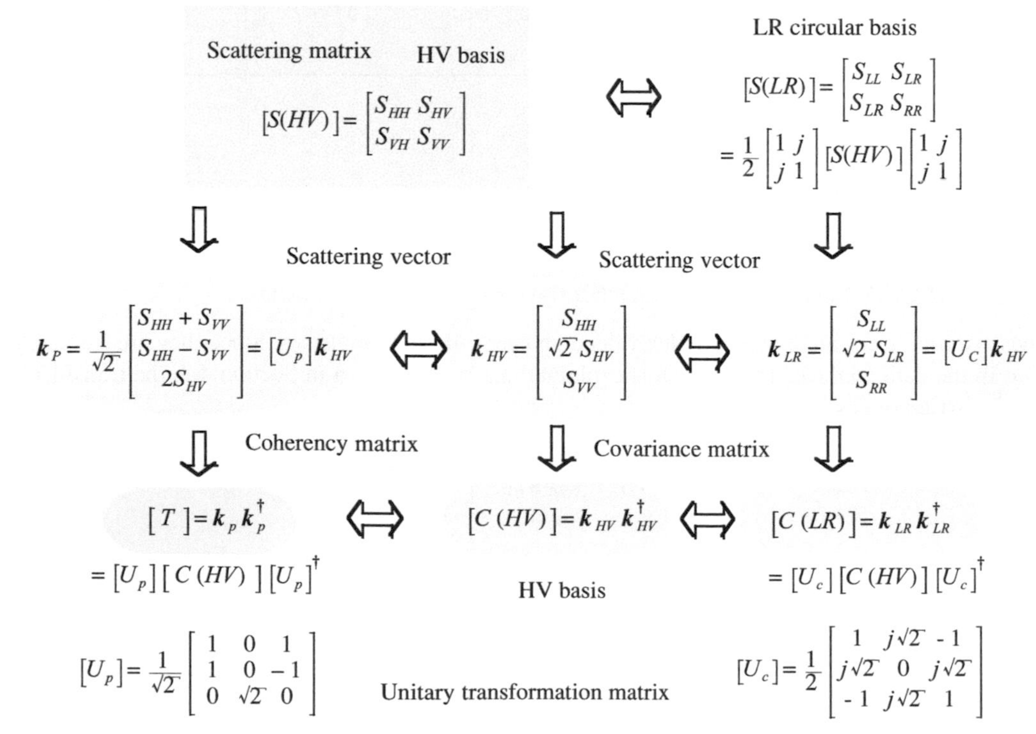

FIGURE 4.7 Mutual transformation of polarization matrices. (From Yamaguchi, Y., *Radar Polarimetry from Basics to Applications: Radar Remote Sensing Using Polarimetric Information (in Japanese)*, IEICE, 2007.)

APPENDIX

A4.1 ROTATION OF COHERENCY MATRIX WITH MINIMIZATION OF T_{33} AND MAXIMIZATION OF T_{22}

Rotation of the coherency matrix around the radar line of sight is carried out by the following equation (Figure A4.1):

$$[T(\theta)] = [R_P(\theta)][T][R_P(\theta)]^\dagger = \begin{bmatrix} T_{11}(\theta) T_{12}(\theta) T_{13}(\theta) \\ T_{21}(\theta) T_{22}(\theta) T_{23}(\theta) \\ T_{31}(\theta) T_{32}(\theta) T_{33}(\theta) \end{bmatrix} \tag{A4.1}$$

$$[R_P(\theta)] = \begin{bmatrix} 1 & 0 & 0 \\ 0 & \cos 2\theta & \sin 2\theta \\ 0 & -\sin 2\theta & \cos 2\theta \end{bmatrix} : \text{Rotation matrix} \tag{A4.2}$$

More explicitly, it can be written as

$$\begin{bmatrix} T_{11}(\theta) T_{12}(\theta) T_{13}(\theta) \\ T_{21}(\theta) T_{22}(\theta) T_{23}(\theta) \\ T_{31}(\theta) T_{32}(\theta) T_{33}(\theta) \end{bmatrix} = \begin{bmatrix} 1 & 0 & 0 \\ 0 & \cos 2\theta & \sin 2\theta \\ 0 & -\sin 2\theta & \cos 2\theta \end{bmatrix} \begin{bmatrix} T_{11} & T_{12} & T_{13} \\ T_{21} & T_{22} & T_{23} \\ T_{31} & T_{32} & T_{33} \end{bmatrix} \begin{bmatrix} 1 & 0 & 0 \\ 0 & \cos 2\theta & -\sin 2\theta \\ 0 & \sin 2\theta & \cos 2\theta \end{bmatrix}$$

$$\tag{A4.3}$$

The element becomes,

$$T_{11}(\theta) = T_{11}, \, T_{12}(\theta) = T_{12} \cos 2\theta + T_{13} \sin 2\theta, \, T_{13}(\theta) = T_{13} \cos 2\theta - T_{12} \sin 2\theta,$$

$$T_{22}(\theta) = T_{22}\cos^2 2\theta + T_{33}\sin^2 2\theta + \text{Re}\{T_{23}\} \sin 4\theta,$$

$$T_{23}(\theta) = \text{Re}\{T_{23}\} \cos 4\theta - \frac{T_{22} - T_{33}}{2} \sin 4\theta + j\text{Im}\{T_{23}\},$$

$$T_{33}(\theta) = T_{33}\cos^2 2\theta + T_{22}\sin^2 2\theta - \text{Re}\{T_{23}\} \sin 4\theta$$

To find the minimum value of T_{33}, we search θ by its derivative = 0

$$\frac{dT_{33}(\theta)}{d\theta} = 2(T_{22} - T_{33}) \sin 4\theta - 4\text{Re}\{T_{23}\}\cos 4\theta = 0 \tag{A4.4}$$

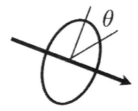

FIGURE A4.1 Rotation around the radar line of sight.

The same equation can be obtained for maximizing T_{22}.

$$\frac{dT_{22}(\theta)}{d\theta} = -2(T_{22} - T_{33})\ \sin 4\theta + 4\text{Re}\{T_{23}\}\cos 4\theta = 0$$

Therefore, the rotation angle can be obtained as

$$\tan 4\theta = \frac{2\text{Re}\{T_{23}\}}{T_{22} - T_{33}}, 2\theta = \frac{1}{2}\tan^{-1}\left(\frac{2\text{Re}\{T_{23}\}}{T_{22} - T_{33}}\right) \tag{A4.5}$$

By this rotation, T_{33} is minimized, and T_{22} is maximized. T_{23} becomes pure imaginary. This situation is a perfect fit for modeling of helix scattering.

$$T_{23}(\theta) = j\text{Im}\{T_{23}\}, \text{Re}\{T_{23}(\theta)\} = 0 \tag{A4.6}$$

After the rotation, the element of coherency matrix becomes,

$$T_{11}(\theta) = T_{11},\ \ T_{12}(\theta) = T_{12}\ \cos 2\theta + T_{13}\ \sin 2\theta,\ \ T_{13}(\theta) = T_{13}\ \cos 2\theta - T_{12}\ \sin 2\theta$$

$$T_{21}(\theta) = T_{12}^*(\theta),\ \ T_{22}(\theta) = T_{22}\cos^2 2\theta + T_{33}\sin^2 2\theta + \text{Re}\{T_{23}\}\ \sin 4\theta,\ \ T_{23}(\theta) = j\text{Im}\{T_{23}\},$$

$$T_{31}(\theta) = T_{13}^*(\theta),\ \ T_{32}(\theta) = T_{23}^*(\theta),\ \ T_{33}(\theta) = T_{33}\cos^2 2\theta + T_{22}\sin^2 2\theta - \text{Re}\{T_{23}\}\ \sin 4\theta \tag{A4.7}$$

T_{22} increases by the amount $\text{Re}\{T_{23}\}\ \sin 4\theta$, whereas T_{33} decreases by the same amount, which contributes to the reduction of the volume-scattering power and the increase of the double-bounce scattering power.

If the rotation angle is $\theta = 45°$, the positions of T_{33} and T_{22} are mutually interchanged. T_{12} and T_{13} also change their positions as in the following equation:

$$\left[T\left(\frac{\pi}{4}\right)\right] = \left[R_P\left(\frac{\pi}{4}\right)\right]\begin{bmatrix} T_{11} & T_{12} & T_{13} \\ T_{21} & T_{22} & T_{23} \\ T_{31} & T_{32} & T_{33} \end{bmatrix}\left[R_P\left(\frac{\pi}{4}\right)\right]^\dagger = \begin{bmatrix} T_{11} & T_{13} & -T_{12} \\ T_{31} & T_{33} & -T_{32} \\ -T_{21} & -T_{23} & T_{22} \end{bmatrix} \tag{A4.8}$$

A4.2 UNITARY TRANSFORMATION OF THE COHERENCY MATRIX WITH MINIMIZATION OF T_{33} AND MAXIMIZATION OF T_{22}

This transformation is intended to reduce T_{33} by mathematical operations [7]. The unitary transformation below is not physically realizable. However, this complex transform also minimizes the T_{33} element.

$$\left[T(\varphi)\right] = \left[U(\varphi)\right]\left[T\right]\left[U(\varphi)\right]^\dagger = \begin{bmatrix} T_{11}(\varphi) & T_{12}(\varphi) & T_{13}(\varphi) \\ T_{21}(\varphi) & T_{22}(\varphi) & T_{23}(\varphi) \\ T_{31}(\varphi) & T_{32}(\varphi) & T_{33}(\varphi) \end{bmatrix} \tag{A4.9}$$

$$\text{If we choose:} \left[U(\varphi) \right] = \begin{bmatrix} 1 & 0 & 0 \\ 0 & \cos 2\varphi & j\sin 2\varphi \\ 0 & j\sin 2\varphi & \cos 2\varphi \end{bmatrix} \qquad \text{(A4.10)}$$

then the elements become

$$T_{11}(\varphi) = T_{11}, \; T_{12}(\varphi) = T_{12}\cos 2\varphi - jT_{13}\sin 2\varphi, \; T_{13}(\varphi) = T_{13}\cos 2\varphi - jT_{12}\sin 2\varphi$$

$$T_{22}(\varphi) = T_{22}\cos^2 2\varphi + T_{33}\sin^2 2\varphi + \text{Im}\{T_{23}\} \; \sin 4\varphi$$

$$T_{23}(\varphi) = \text{Re}\{T_{23}\} + \frac{j}{2}\left[2\text{Im}\{T_{23}\}\cos 4\varphi - (T_{22} - T_{33}) \; \sin 4\varphi \right]$$

$$T_{33}(\varphi) = T_{33}\cos^2 2\varphi + T_{22}\sin^2 2\varphi - \; \text{Im}\{T_{23}\}\sin 4\varphi$$

The minimum T_{33} and the maximum T_{22} can be searched by the following equations:

$$\frac{dT_{33}(\varphi)}{d\varphi} = 2(T_{22} - T_{33}) \; \sin 4\varphi - 4\text{Im}\{T_{23}\} \; \cos 4\varphi = 0 \qquad \text{(A4.11)}$$

$$\frac{dT_{22}(\varphi)}{d\varphi} = -2(T_{22} - T_{33}) \; \sin 4\varphi + 4\text{Im}\{T_{23}\} \; \cos 4\varphi = 0$$

Therefore, the angle can be obtained as

$$\tan 4\varphi = \frac{2\text{Im}\{T_{23}\}}{T_{22} - T_{33}}, \; 2\varphi = \frac{1}{2}\tan^{-1}\left(\frac{2\text{Im}\{T_{23}\}}{T_{22} - T_{33}} \right) \qquad \text{(A4.12)}$$

Under this angle, T_{23} becomes a real value, and the imaginary part of T_{23} vanishes.

$$T_{33}(\varphi) = \text{Re}\{T_{23}\}, \; \text{Im}\{T_{23}(\varphi)\} = 0$$

The element of the coherency matrix after the unitary transformation becomes

$$T_{11}(\varphi) = T_{11}, \; T_{12}(\varphi) = T_{12}\cos 2\varphi - jT_{13}\sin 2\varphi, \; T_{13}(\varphi) = T_{13}\cos 2\varphi - jT_{12}\sin 2\varphi$$

$$T_{21}(\varphi) = T_{12}^*(\varphi), \; T_{22}(\varphi) = T_{22}\cos^2 2\varphi + T_{33}\sin^2 2\varphi + \text{Im}\{T_{23}\} \; \sin 4\varphi, \; T_{33}(\varphi) = \text{Re}\{T_{23}\}$$

$$T_{31}(\varphi) = T_{13}^*(\varphi), \; T_{32}(\varphi) = T_{23}^*(\varphi), \; T_{33}(\varphi) = T_{33}\cos^2 2\varphi + T_{22}\sin^2 2\varphi - \; \text{Im}\{T_{23}\}\sin 4\varphi \quad \text{(A4.13)}$$

If this angle is chosen as 45°, the positions of the element change as follows:

$$\left[U_\varphi\left(\frac{\pi}{4}\right) \right]\begin{bmatrix} T_{11} & T_{12} & T_{13} \\ T_{21} & T_{22} & T_{23} \\ T_{31} & T_{32} & T_{33} \end{bmatrix}\left[U_\varphi\left(\frac{\pi}{4}\right) \right]^\dagger = \begin{bmatrix} T_{11} & -jT_{13} & -jT_{12} \\ jT_{31} & T_{33} & T_{32} \\ jT_{21} & T_{23} & T_{22} \end{bmatrix} \qquad \text{(A4.14)}$$

A4.3   DUAL POL DATA MATRIX

Dual pol data has four independent polarimetric parameters as shown in the following equation. From the 3×3 covariance matrix expression in the HV polarization basis, we eliminate the VV component. Then the covariance matrix by $HH + HV$ becomes as follows:

$$
\begin{bmatrix}
\left\langle |S_{HH}|^2 \right\rangle & \sqrt{2}\left\langle S_{HH}S_{HV}^* \right\rangle & \left\langle S_{HH}S_{VV}^* \right\rangle \\
\sqrt{2}\left\langle S_{HV}S_{HH}^* \right\rangle & 2\left\langle |S_{HV}|^2 \right\rangle & \sqrt{2}\left\langle S_{HV}S_{VV}^* \right\rangle \\
\left\langle S_{VV}S_{HH}^* \right\rangle & \sqrt{2}\left\langle S_{VV}S_{HV}^* \right\rangle & \left\langle |S_{VV}|^2 \right\rangle
\end{bmatrix}
\Rightarrow
\begin{bmatrix}
\left\langle |S_{HH}|^2 \right\rangle & \sqrt{2}\left\langle S_{HH}S_{HV}^* \right\rangle & 0 \\
\sqrt{2}\left\langle S_{HV}S_{HH}^* \right\rangle & 2\left\langle |S_{HV}|^2 \right\rangle & 0 \\
0 & 0 & 0
\end{bmatrix}
\Rightarrow
\begin{bmatrix}
\left\langle |S_{HH}|^2 \right\rangle & \sqrt{2}\left\langle S_{HH}S_{HV}^* \right\rangle \\
\sqrt{2}\left\langle S_{HV}S_{HH}^* \right\rangle & 2\left\langle |S_{HV}|^2 \right\rangle
\end{bmatrix}
$$

There are two real diagonal terms and one complex-valued off-diagonal term in the 2×2 covariance matrix. We have four real polarimetric parameters in total, which is much less than nine in the quad pol case.

For compact pol data in which the left-handed circular wave transmission and H and V channel reception are assumed, the scattering equation and the covariance matrix become,

$$
\begin{bmatrix} E_H^r \\ E_V^r \end{bmatrix} = \begin{bmatrix} S_{HH} S_{HV} \\ S_{VH} S_{VV} \end{bmatrix} \frac{1}{\sqrt{2}} \begin{bmatrix} 1 \\ j \end{bmatrix} = \frac{1}{\sqrt{2}} \begin{bmatrix} S_{HH} + jS_{HV} \\ S_{VH} + jS_{VV} \end{bmatrix}
$$

$$
\left\langle \begin{bmatrix} E_H^r \\ E_V^r \end{bmatrix} \begin{bmatrix} E_H^{r*} E_V^{r*} \end{bmatrix} \right\rangle = \begin{bmatrix} \left\langle |E_H^r|^2 \right\rangle \left\langle E_H^r E_V^{r*} \right\rangle \\ \left\langle E_V^r E_H^{r*} \right\rangle \left\langle |E_V^r|^2 \right\rangle \end{bmatrix}
$$

$$
= \frac{1}{2} \begin{bmatrix} \left\langle |S_{HH} + jS_{HV}|^2 \right\rangle & \left\langle (S_{HH} + jS_{HV})(S_{VH} + jS_{VV})^* \right\rangle \\ \left\langle (S_{VH} + jS_{VV})(S_{HH} + jS_{HV})^* \right\rangle & \left\langle |S_{VH} + jS_{VV}|^2 \right\rangle \end{bmatrix}
$$

From this expression, we can see the matrix is the same form as the preceding 2×2 matrix, and that four real parameters are available for the compact pol. Since it is impossible to retrieve S_{HH}, S_{HV}, S_{VV} from this expression, the compact pol cannot be substituted by the quad pol [8].

REFERENCES

1. W.-M. Boerner et al., eds., *Direct and Inverse Methods in Radar Polarimetry*, Parts 1 and 2, NATO ASI Series, Mathematical and Physical Sciences, vol. 350, Kluwer Academic Publishers, the Netherlands, 1988.
2. F. M. Henderson and A. J. Lewis, *Principles & Applications of Imaging Radar*, Manual of Remote Sensing, 3rd ed., vol. 2, ch. 5, pp. 271–357, John Wiley & Sons, New York, 1998.
3. S. R. Cloude and E. Pottier, "A review of target decomposition theorems in radar polarimetry," *IEEE Trans. Geosci. Remote Sens.*, vol. 34, no. 2, pp. 498–518, March 1996.
4. J. S. Lee and E. Pottier, *Polarimetric Radar Imaging from Basics to Applications*, CRC Press, 2009.
5. Y. Yamaguchi, M. Ishido, T. Moriyama, and H. Yamada, "Four-component scattering model for polarimetric SAR image decomposition," *IEEE Trans. Geosci. Remote Sens.*, vol. 43, no. 8, pp. 1699–1706, 2005.
6. Y. Yamaguchi, *Radar Polarimetry from Basics to Applications: Radar Remote Sensing using Polarimetric Information (in Japanese)*, IEICE, Tokyo, December 2007. ISBN: 978-4-88554-227-7.

7. G. Singh, Y. Yamaguchi, and S.-E. Park, "General four-component scattering power decomposition with unitary transformation of coherency matrix," *IEEE Trans. Geosci. Remote Sens.*, vol. 51, no. 5, pp. 3014–3022, 2013.

8. J. S. Lee, M. R. Grunes, and E. Pottier, "Quantitative comparison of classification capability: Fully polarimetric versus dual and single-polarization SAR," *IEEE Trans. Geosci. Remote Sens.*, vol. 39, no. 11, pp. 2343–2351, 2001.

5 H/A/ᾱ Polarimetric Decomposition

Eigenvalue/eigenvector analysis has mathematical universality. Mathematically, the covariance matrix and coherency matrix are 3 × 3 positive definite Hermitian matrices. There is a unitary transformation relation between the covariance matrix and coherency matrix. Therefore, they have the same three non-negative eigenvalues. Since three eigenvectors belonging to three eigenvalues are orthogonal to each other and represent scattering mechanisms, they are convenient for polarimetric data analyses. Using these merits, Cloude and Pottier proposed a method [1,2] of extracting average parameters from averaged coherency matrices, which retain the second-order statistics of polarimetric information. This method is well-known as H/A/ᾱ decomposition in the radar polarimetry community. In this chapter, we briefly review the eigenvalue/eigenvector method for PolSAR data analysis. The details should be referred to [2–4].

5.1 EIGENVALUES, ENTROPY, AND MEAN ALPHA ANGLE

The coherency matrix $\langle [T] \rangle$ after ensemble averaging of polarimetric data can be diagonalized by a 3 × 3 orthogonal unitary matrix $[U_3]$.

$$\langle [T] \rangle = \frac{1}{n} \sum_{p}^{n} \mathbf{k}_p \mathbf{k}_p^{\dagger} \tag{5.1.1}$$

$$= [U_3] \begin{bmatrix} \lambda_1 & 0 & 0 \\ 0 & \lambda_2 & 0 \\ 0 & 0 & \lambda_3 \end{bmatrix} [U_3]^{\dagger} = \sum_{i=1}^{3} \lambda_i \mathbf{e}_i \mathbf{e}_i^{\dagger} \tag{5.1.2}$$

where, $\lambda_1, \lambda_2, \lambda_3$ are eigenvalues satisfying $\lambda_1 \geq \lambda_2 \geq \lambda_3$, and $\mathbf{e}_1, \mathbf{e}_2, \mathbf{e}_3$ are the corresponding eigenvectors. $[U_3]$ can be expressed as

$$[U_3] = [\mathbf{e}_1 \ \mathbf{e}_2 \ \mathbf{e}_3] = \begin{bmatrix} \cos\alpha_1 & \cos\alpha_2 & \cos\alpha_3 \\ \sin\alpha_1 \cos\beta_1 \, e^{j\delta_1} & \sin\alpha_2 \cos\beta_2 \, e^{j\delta_2} & \sin\alpha_3 \cos\beta_3 \, e^{j\delta_3} \\ \sin\alpha_1 \cos\beta_1 \, e^{j\gamma_1} & \sin\alpha_2 \cos\beta_2 \, e^{j\gamma_2} & \sin\alpha_3 \cos\beta_3 \, e^{j\gamma_3} \end{bmatrix} \tag{5.1.3}$$

Equation (5.1.1) shows statistical averaging, while equation (5.1.2) shows a mathematical decomposition by eigenvalues/eigenvectors. This statistical model is represented by the Bernoulli process, and the scattering process can be expressed by $\mathbf{e}_1, \mathbf{e}_2, \mathbf{e}_3$ with the probability P_i such that

$$P_i = \frac{\lambda_i}{\lambda_1 + \lambda_2 + \lambda_3} \quad (i = 1, 2, 3) \tag{5.1.4}$$

The scattering process is represented by a sum of three independent scattering mechanisms. Entropy H and mean alpha angle $\bar{\alpha}$ are defined using this probability [2–4].

$$H = -\sum_{i=1}^{3} P_i \log_3 P_i \quad (0 \le H \le 1) \tag{5.1.5}$$

$$\bar{\alpha} = \sum_{i=1}^{3} P_i \, \alpha_i \left(0° \le \bar{\alpha} \le 90°\right) \tag{5.1.6}$$

Now, we look into the meaning of these parameters from the user's point of view.

5.2   SOME PARAMETERS BY EIGENVALUES

Let's consider eigenvalues and what we know in the previous chapter. The eigenvalue derived from a finite number of measured data, and the eigenvalue derived from integration with probability distribution are different. If we look for theoretical eigenvalues, the expression of the covariance matrix (4.5.18) in the circular polarization basis provides the formula as follows:

$$\begin{cases} \lambda_1 = \dfrac{1}{2}|a+b|^2 \\[2mm] \lambda_2 = \dfrac{1}{4}|a-b+j2c|^2 \\[2mm] \lambda_3 = \dfrac{1}{4}|a-b-j2c|^2 \end{cases} \tag{5.2.1}$$

where we assume

$$\lambda_1 \ge \lambda_2 \ge \lambda_3 \tag{5.2.2}$$

Once eigenvalues are given, **total power** and **anisotropy** parameters can be defined as

$$\textbf{Total power} \qquad TP = \lambda_1 + \lambda_2 + \lambda_3 \tag{5.2.3}$$

$$\textbf{Anisotropy} \qquad A = \frac{\lambda_2 - \lambda_3}{\lambda_2 + \lambda_3} \tag{5.2.4}$$

Now let's take an example of these parameters. Eigenvalue expansion has the form of

$$\langle [T] \rangle = [U_3] \begin{bmatrix} \lambda_1 & 0 & 0 \\ 0 & \lambda_2 & 0 \\ 0 & 0 & \lambda_3 \end{bmatrix} [U_3]^\dagger.$$

If $\langle [T] \rangle$ is a diagonal matrix from the outset, then the unitary matrix should be $[U_3] = \begin{bmatrix} 1 & 0 & 0 \\ 0 & 1 & 0 \\ 0 & 0 & 1 \end{bmatrix}$

Then we have,

$$
\begin{bmatrix} 1 & 0 & 0 \\ 0 & 1 & 0 \\ 0 & 0 & 1 \end{bmatrix} = \begin{bmatrix} \cos\alpha_1 & \cos\alpha_2 & \cos\alpha_3 \\ \sin\alpha_1 \cos\beta_1\, e^{j\delta_1} & \sin\alpha_2 \cos\beta_2\, e^{j\delta_2} & \sin\alpha_3 \cos\beta_3\, e^{j\delta_3} \\ \sin\alpha_1 \cos\beta_1\, e^{j\gamma_1} & \sin\alpha_2 \cos\beta_2\, e^{j\gamma_2} & \sin\alpha_3 \cos\beta_3\, e^{j\gamma_3} \end{bmatrix}
$$

Therefore, the following equation should apply, $\cos\alpha_1 = 1$, $\cos\alpha_2 = 0$, $\cos\alpha_3 = 0$, leading to

$$
\alpha_1 = 0, \quad \alpha_2 = \frac{\pi}{2}, \quad \alpha_3 = \frac{\pi}{2}. \tag{5.2.5}
$$

A cloud of randomly oriented dipoles may be a good example. The coherency matrix and eigenvalues of a cloud of dipoles are given as

$$
\langle [T] \rangle = \frac{1}{4} \begin{bmatrix} 2 & 0 & 0 \\ 0 & 1 & 0 \\ 0 & 0 & 1 \end{bmatrix}, \quad \begin{cases} \lambda_1 = 1/2 \\ \lambda_2 = 1/4 \\ \lambda_3 = 1/4 \end{cases}
$$

Therefore, polarimetric parameters can be determined by

$$
P_1 = \frac{1/2}{1/2 + 1/4 + 1/4} = \frac{1}{2}, \quad P_2 = \frac{1/4}{1/2 + 1/4 + 1/4} = \frac{1}{4}, \quad P_3 = \frac{1}{4}
$$

$$
H = -\frac{1}{2}\log_3 \frac{1}{2} - \frac{1}{4}\log_3 \frac{1}{4} - \frac{1}{4}\log_3 \frac{1}{4} = 0.95
$$

$$
\bar{\alpha} = \sum_{i=1}^{3} P_i \alpha_i = \frac{1}{2}0 + \frac{1}{4}\frac{\pi}{2} + \frac{1}{4}\frac{\pi}{2} = \frac{\pi}{4}
$$

In a similar way, using eigenvalues of canonical targets in the diagonal matrix forms and α_i of equation (5.2.5), the entropy H and mean alpha angle $\bar{\alpha}$ are listed in Table 5.1.

These targets can be plotted on the 2D H-$\bar{\alpha}$ map as shown in Figure 5.1, referring to Table 5.1. We can understand the following items from the definition of parameters and from Figure 5.1.

- Entropy H ranges from 0 to 1. $H = 0$ corresponds to single eigenvalue case $(\lambda_1 = 1, \lambda_2 = \lambda_3 = 0)$. This physically means there is single scattering mechanism only and is caused by the surface scattering. $H = 1$ corresponds to the situation of $(\lambda_1 = \lambda_2 = \lambda_3 = 1/3)$. Three scattering mechanisms occur simultaneously for $H = 1$. This shows completely random scattering like randomly distributed objects. As H increases from 0 to 1, it shows increasing randomness of the scattering mechanism. Therefore, H can be regarded as the index of randomness.
- The mean alpha angle $\bar{\alpha}$ shows polarimetric scattering dependency ranging from 0° to 90° as shown in Figures 5.1 and 5.2. $\bar{\alpha} = 0°$ corresponds to the surface scattering caused by sea surface or bare soil, $\bar{\alpha} = 45°$ to the dipole scattering, and $\bar{\alpha} = 90°$ corresponds to dihedrals or helix scattering.

TABLE 5.1

Eigenvalues, *H*, and $\bar{\alpha}$ of Canonical Target

	Scattering Matrix	$\begin{bmatrix} \lambda_1 & 0 & 0 \\ 0 & \lambda_2 & 0 \\ 0 & 0 & \lambda_3 \end{bmatrix}$	Eigenvalues	Angle $\bar{\alpha}$	Entropy *H*
Plate, sphere	$\begin{bmatrix} 1 & 0 \\ 0 & 1 \end{bmatrix}$	$\begin{bmatrix} 1 & 0 & 0 \\ 0 & 0 & 0 \\ 0 & 0 & 0 \end{bmatrix}$	$\lambda_1 = 1$ $\lambda_2 = 0$ $\lambda_3 = 0$	0	0
Dipole	$\begin{bmatrix} 1 & 0 \\ 0 & 0 \end{bmatrix}$	$\dfrac{1}{4}\begin{bmatrix} 2 & 0 & 0 \\ 0 & 1 & 0 \\ 0 & 0 & 1 \end{bmatrix}$	$\lambda_1 = 1/2$ $\lambda_2 = 1/4$ $\lambda_3 = 1/4$	$\dfrac{\pi}{4}$	0.95
Dihedral	$\begin{bmatrix} 1 & 0 \\ 0 & -1 \end{bmatrix}$	$\dfrac{1}{2}\begin{bmatrix} 0 & 0 & 0 \\ 0 & 1 & 0 \\ 0 & 0 & 1 \end{bmatrix}$	$\lambda_1 = 0$ $\lambda_2 = 1/2$ $\lambda_3 = 1/2$	$\dfrac{\pi}{2}$	0.63
Helix	$\dfrac{1}{2}\begin{bmatrix} 1 & j \\ j & -1 \end{bmatrix}$	$\begin{bmatrix} 0 & 0 & 0 \\ 0 & 0 & 0 \\ 0 & 0 & 1 \end{bmatrix}$	$\lambda_1 = 0$ $\lambda_2 = 0$ $\lambda_3 = 1$	$\dfrac{\pi}{2}$	0
Random		$\begin{bmatrix} 1/3 & 0 & 0 \\ 0 & 1/3 & 0 \\ 0 & 0 & 1/3 \end{bmatrix}$	$\lambda_1 = 1/3$ $\lambda_2 = 1/3$ $\lambda_3 = 1/3$	$\dfrac{\pi}{3}$	1

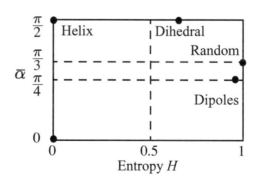

FIGURE 5.1   Canonical target locations on the H$\bar{\alpha}$ map.

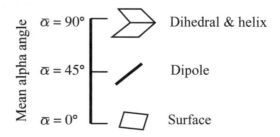

FIGURE 5.2   Mean alpha angle and scatterer. (From Pottier, E. and Lee, J.S., Application of the <H/A/α> polarimetric decomposition theorem for unsupervised classification of fully polarimetric SAR data on the Whishart distribution, in *Proceedings of EUSAR*, Germany, 2000.)

The 2D plot of Figure 5.1 is convenient for interpreting target-scattering mechanisms. According to the value of H, we can understand the complexity of scattering, that is, multiple and random scattering occur in the large H region, whereas simple scattering occurs in the small H region. Furthermore, the information of $\bar{\alpha}$ indicates the type of scattering objects. According to this plot, the forest locates near ($H = 1$, $\bar{\alpha} = 45°$) and the sea surface locates around ($H = 0$, $\bar{\alpha} = 0°$).

It should be noted that eigenvalues used here are derived from equation (5.2.1) and not from real data. For real applications, the eigenvalue plot should be derived from real data.

5.3 SIMPLE CALCULATION METHOD OF ENTROPY

In this section, a simple calculation method of entropy is presented, without log execution or eigenvalue analysis. This method has been proposed by Yang [5] and is useful for real-time application by a personal computer. The brief explanation is given by a covariance formulation.

Covariance matrix is given as
$$[C] = \begin{bmatrix} C_{11} & C_{12} & C_{13} \\ C_{21} & C_{22} & C_{23} \\ C_{31} & C_{32} & C_{33} \end{bmatrix} \tag{5.3.1}$$

The eigenvalue equation becomes,
$$[C]\mathbf{x} = \lambda\mathbf{x} \tag{5.3.2}$$

This leads to
$$\lambda^3 + a_2\lambda^2 + a_1\lambda + a_0 = 0 \tag{5.3.3}$$

where

$$a_0 = c_{11}c_{23}c_{32} + c_{22}c_{13}c_{31} + c_{33}c_{12}c_{21} - c_{11}c_{22}c_{33} - c_{12}c_{23}c_{31} - c_{13}c_{32}c_{21}$$

$$a_1 = c_{11}c_{22} + c_{22}c_{33} + c_{33}c_{11} - c_{12}c_{21} - c_{23}c_{32} - c_{13}c_{31} \tag{5.3.4}$$

$$a_2 = -c_{11} - c_{22} - c_{33}$$

The eigenvalues can be obtained by solving equation (5.3.2). However, the entropy is in the following form without eigenvalue expression:

$$H = -\sum_{i=1}^{3} P_i \log_3 P_i \tag{5.3.5}$$

Approximate entropy (AH) to equation (5.3.5) can be expressed in a series expansion as,

$$AH = 2.3506 - 5.7613 \sum_{i=1}^{3} k_i^2 + 6.0611 \sum_{i=1}^{3} k_i^3 - 2.6504 \sum_{i=1}^{3} k_i^4 \tag{5.3.6}$$

Based on Vieta's theorem, AH can be further simplified as

$$\text{If } \frac{a_1}{a_2^2} \leq 0.0481, \frac{a_0}{a_2^3} \leq 0.0006 \Rightarrow AH = 5.2819\frac{a_1}{a_2^2} + 54.8584\frac{a_0}{a_2^3} - 35.6980\frac{a_1^2}{a_2^4}. \tag{5.3.7}$$

$$\text{If } \frac{a_1}{a_2^2} \geq 0.0481, \frac{a_0}{a_2^3} \geq 0.0006 \Rightarrow AH = 3.9408\frac{a_1}{a_2^2} + 7.5818\frac{a_0}{a_2^3} - 5.3008\frac{a_1^2}{a_2^4}. \tag{5.3.8}$$

The error of approximate entropy of equations (5.3.7) and (5.3.8) is less than 5%.

5.4   APPLICATION TO CLASSIFICATION

The most attractive and effective application of $H/\bar{\alpha}$ method is classification of PolSAR data. A 2D $H/\bar{\alpha}$ plane as shown in Figure 5.3 is used to classify targets based on the mean-scattering mechanism [3]. The $H/\bar{\alpha}$ plane is subdivided into nine zones of target class with different scattering characteristics. The location of the boundaries is set based on the general properties of the scattering mechanisms.

Z9: Low entropy surface scatterer
   In this zone, low entropy scattering processes with $\bar{\alpha} < 42.5°$ occur. Examples are water-surface, very smooth land surface, etc.
Z8: Low entropy dipole scattering
   Strongly correlated *HH* and *VV* returns occur in this zone.
Z7: Low entropy multiple scattering events
   Isolated dihedral scatters are examples in this low entropy region.
Z6: Medium entropy surface scatterer
   The zone includes surface scatter with small roughness or vegetation.
Z5: Medium entropy vegetation scattering
   Vegetation on a rough surface falls in this zone.
Z4: Medium entropy multiple scattering
   Sparse forest has the double-bounce scattering and moderate entropy.
Z3: None
Z2: High entropy vegetation scattering
   Trees and forests are typical scatterers in this zone.
Z1: High entropy multiple scattering
   It is often observed in forests.

Since $H/\bar{\alpha}$ method is based on the eigenvalue of a 3×3 coherency matrix, all information on fully polarimetric data is included. This classification scheme has been supported by many researchers. One problem for the $H/\bar{\alpha}$ approach is non-uniqueness of the eigenvalue combination, which derives the same H. For example, both combinations give the same value of H.

$$\left(\lambda_1 = 1, \lambda_2 = 1, \lambda_3 = 0.3\right) \Rightarrow P_1 = \frac{1}{1+1+0.3} = \frac{1}{2.3}, P_2 = \frac{1}{2.3}, P_3 = \frac{0.3}{2.3}$$

$$H = -\frac{1}{2.3}\log_3\frac{1}{2.3} - \frac{1}{2.3}\log_3\frac{1}{2.3} - \frac{0.3}{2.3}\log_3\frac{0.3}{2.3} = 0.9$$

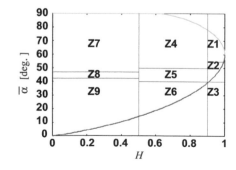

FIGURE 5.3   Scattering zones in $H/\bar{\alpha}$ plane. (From Pottier, E. and Lee, J.S., Application of the <H/A/α> polarimetric decomposition theorem for unsupervised classification of fully polarimetric SAR data on the Whishart distribution, in *Proceedings of EUSAR*, Germany, 2000.)

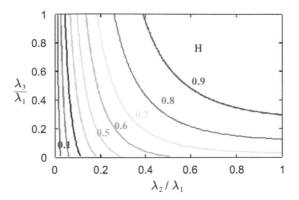

FIGURE 5.4 Combination of eigenvalues, which gives the same H value. (From Pottier, E. and Lee, J.S., Application of the <H/A/α> polarimetric decomposition theorem for unsupervised classification of fully polarimetric SAR data on the Whishart distribution, in *Proceedings of EUSAR*, Germany, 2000.)

$$\left(\lambda_1 = 1, \lambda_2 = 0.4, \lambda_3 = 0.4\right) \Rightarrow P_1 = \frac{1}{1+0.4+0.4} = \frac{1}{1.8}, \quad P_2 = \frac{0.4}{18}, \quad P_3 = \frac{0.4}{18}$$

$$H = -\frac{1}{1.8}\log_3\frac{1}{1.8} - \frac{0.4}{1.8}\log_3\frac{0.4}{1.8} - \frac{0.4}{1.8}\log_3\frac{0.4}{1.8} = 0.9$$

In this case, misclassification occurs due to the same H, even if eigenvalues are different. The combination of eigenvalues that give the same H are plotted in Figure 5.4.

In order to avoid such a misclassification problem, Pottier [3] introduced an anisotropy parameter to discriminate the situation.

$$\text{Anisotropy} \qquad A = \frac{\lambda_2 - \lambda_3}{\lambda_2 + \lambda_3} \qquad\qquad (5.2.4)$$

This parameter gives $A = 0.54$ for $\left(\lambda_1 = 1, \lambda_2 = 1, \lambda_3 = 0.3\right)$ and $A = 0$ for $\left(\lambda_1 = 1, \lambda_2 = 0.4, \lambda_3 = 0.4\right)$.
It becomes possible to distinguish these two scatterings by the anisotropy.
Anisotropy is a roll-invariant parameter and has the following properties:

$A = 0 \Rightarrow$ if $\lambda_2 = \lambda_3 = 0$, there is single scattering mechanism

             if $\lambda_2 = \lambda_3 \neq 0$, there are two scattering mechanisms with same contribution

$A \neq 0 \Rightarrow$ if $\lambda_2 \neq 0, \lambda_3 = 0$, there are two scattering mechanisms

             if $\lambda_2 \neq 0, \lambda_3 \neq 0$, there are three scattering mechanisms

Therefore, it would be useful to add an anisotropy axis to the $H/\bar{\alpha}$ plane to make 3D space for classification of objects. A 3D classification will be the extension of the $H/\bar{\alpha}$ scheme and may better serve fully polarimetric data classification.

If the anisotropy axis is replaced by total power (TP), then power information can be included in the classification as shown in Figure 5.5. In the real data analysis, the region with low entropy has noise so that the anisotropy parameter also behaves in a noisy fashion. In such a case, power information may better serve classification [6].

Further extensions can be made, taking into account various polarimetric parameters [4],

$$(1-H)(1-A), H(1-A), HA, (1-H)A, \quad A_{12} = \frac{\lambda_1 - \lambda_2}{\lambda_1 + \lambda_2}, \quad A_{13} = \frac{\lambda_1 - \lambda_3}{\lambda_1 + \lambda_3}$$

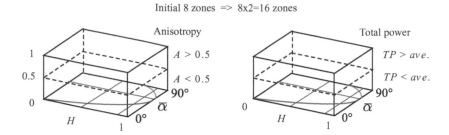

FIGURE 5.5   3D classification space by H/$\overline{\alpha}$/A and H/$\overline{\alpha}$/TP. (From Pottier, E. and Lee, J.S., Application of the <H/A/α> polarimetric decomposition theorem for unsupervisedclassification of fully polarimetric SAR data on the Whishart distribution, in *Proceedings of EUSAR*, Germany, 2000.)

5.5   CLASSIFICATION RESULTS

Some classification results of the San Francisco area using AIRSAR data are shown in the following figures.

From the data analysis, we can find that

- $H/\overline{\alpha}$ classifies terrain precisely.
- $H/\overline{\alpha}/A$ method: Fine and accurate classification is performed when polarimetric scattering property is distinct. But accuracy decreases when scattering property is similar or for mixed targets because A is developed for high entropy.
- $H/\overline{\alpha}/TP$ classifies vegetation area precisely owing to power information. TP from vegetation and urban area is different (Figure 5.6).

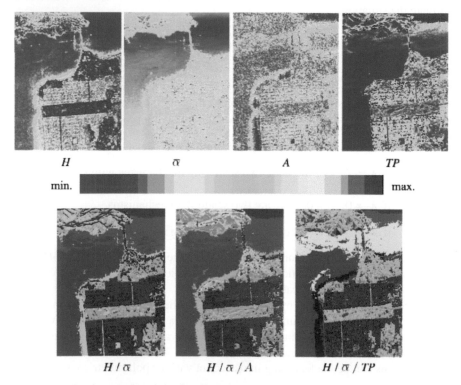

FIGURE 5.6   Classification results of the San Francisco area by AIRSAR. (From Pottier, E. and Lee, J.S., Application of the <H/A/α> polarimetric decomposition theorem for unsupervised classification of fully polarimetric SAR data on the Whishart distribution, in Proceedings of EUSAR, Germany, 2000.)

5.6 REMARKS

As is well-known, eigenvalue/eigenvector analysis has mathematical universality. The eigenvalues of the matrix represent scattering powers. The largest eigenvalue has the largest power, and the smallest eigenvalue corresponds to the smallest power. Hence, there is a one-to-one correspondence between the magnitude of the eigenvalue and the power itself. If we choose the largest eigenvalue in the PolSAR image, it always shows the largest power throughout the image regardless of the scattering mechanism. Therefore, combinations of eigenvalues such as H, mean alpha angle A, and TP are preferred rather than using a single eigenvalue [7].

REFERENCES

1. S. R. Cloude and E. Pottier, "A review of target decomposition theorems in radar polarimetry," *IEEE Trans. Geosci. Remote Sens.*, vol. 34, no. 2, pp. 498–518, 1996.
2. S. R. Cloude and E. Pottier, "An entropy based classification scheme for land applications of polarimetric SAR," *IEEE Trans. Geosci. Remote Sens.*, vol. 35, no. 1, pp. 68–78, 1997.
3. E. Pottier and J. S. Lee, "Application of the <<H/A/α>> polarimetric decomposition theorem for unsupervised classification of fully polarimetric SAR data on the Whishart distribution," *Proceedings of EUSAR*, Germany, 2000.
4. J. S. Lee and E. Pottier, *Polarimetric Radar Imaging from Basics to Applications*, CRC Press, 2009.
5. J. Yang, Y. Chen, Y. Peng, Y. Yamaguchi, and H. Yamada, "New formula of the polarization entropy," *IEICE Trans. Commun.*, vol. E89-B, no. 3, pp. 1033–1035, 2006.
6. K. Kimura, Y. Yamaguchi, and H. Yamada, "Unsupervised land classification using H/alpha/TP space applied to POLSAR image analysis," *IEICE Trans. Commun.*, vol. E87-B, no. 6, pp. 1639–1647, 2004.
7. Y. Yamaguchi, *Radar Polarimetry from Basics to Applications: Radar Remote Sensing using Polarimetric Information (in Japanese)*, IEICE, Tokyo, December 2007.

6 Compound Scattering Matrix

6.1 INTRODUCTION

This chapter is devoted to compound scattering matrix composed of dipoles. Any type of scattering matrix can be created by a combination of dipoles, which can be the basis of a physical scattering model. Dipole is a thin metallic wire which is long enough when compared to wavelength. The coherent scattering is assumed.

Radar range resolution ΔR is determined by the bandwidth B of the transmitting signal, namely $\Delta R = \frac{c}{2B}$, where c is the speed of light. The azimuth resolution is determined by radar antenna size by the definition of synthetic aperture radar (SAR). These radar resolutions are determined by radar hardware design.

On the other hand, the object size is independent of radar resolutions. Some objects are much larger than range or azimuth resolutions, and others may be smaller than resolutions. If we pay our attention to the size relation, there may be various cases for scattering phenomena according to object size, wavelength, and radar resolutions, as shown in Figure 6.1. In this chapter, we consider the situation where resolution > object size > wavelength. This situation happens in typical radar measurements.

Consider the case where multiple objects exist in range or azimuth directions as shown in Figure 6.2. Each scatterer is assumed to have a scattering matrix [S]. How is the total scattering matrix for these cases? Is it a direct coherent sum or not? If the coherent sum applies, the scattering matrix can be obtained directly and convenient for further analysis.

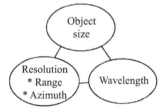

FIGURE 6.1   Size relations in radar sensing.

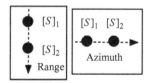

FIGURE 6.2   Scattering matrix of multiple objects.

To be more specific, we consider the two targets #1 and #2 are separated by distance d along the range direction as shown in Figure 6.3.

We assume that the spacing d is less than the radar range resolution ΔR and also less than the wavelength. This assumption applies to almost all radar measurement situation. Target #1 has radar cross section (RCS) σ_1, and target #2 has σ_2. Then the compound or total scattering matrix by two targets becomes the coherent sum,

$$\sqrt{\sigma_{total}}\,[S]_{total} = \sqrt{\sigma_1}\,[S]_1 + \sqrt{\sigma_2}\,[S]_2 \, \exp\left(-j\frac{4\pi d}{\lambda}\right) \qquad (6.1.1)$$

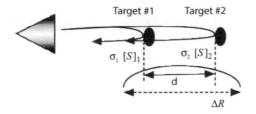

FIGURE 6.3 Total scattering matrix when the spacing d is less than the range resolution ΔR.

If the same RCS targets are aligned, that is, $\sigma_1 = \sigma_2 = \sigma_{total}$, then the scattering matrix will be,

$$[S]_{total} = [S]_1 + [S]_2 \exp\left(-j\frac{4\pi d}{\lambda}\right) = [S]_1 + [S]_2 P(d), \qquad (6.1.2)$$

$$\text{where} \quad P(d) = \exp\left(-j\frac{4\pi d}{\lambda}\right) \qquad (6.1.3)$$

is the phase function by the spacing d. The phase function has the following characteristics:

$$d = 0 \qquad P(0) = \exp(-j\,0) = 1 \qquad (6.1.4)$$

$$d = \frac{\lambda}{8} \qquad P\left(\frac{1}{8}\lambda\right) = \exp\left(-j\frac{1}{2}\pi\right) = -j \qquad (6.1.5)$$

$$d = \frac{2}{8}\lambda \qquad P\left(\frac{2}{8}\lambda\right) = \exp(-j\pi) = -1 \qquad (6.1.6)$$

$$d = \frac{3}{8}\lambda \qquad P\left(\frac{3}{8}\lambda\right) = \exp\left(-j\frac{3}{2}\pi\right) = j \qquad (6.1.7)$$

$$d = \frac{4}{8}\lambda \qquad P\left(\frac{4}{8}\lambda\right) = \exp(-j2\pi) = 1 \qquad (6.1.8)$$

$$P(d) = P\left(d + n\frac{\lambda}{2}\right) \qquad n = 1, 2, 3, \ldots \qquad (6.1.9)$$

The purpose of this chapter is to confirm the applicability of equation (6.1.2) by theoretical and experimental investigations (Figure 6.4). For theoretical verification, the finite difference time

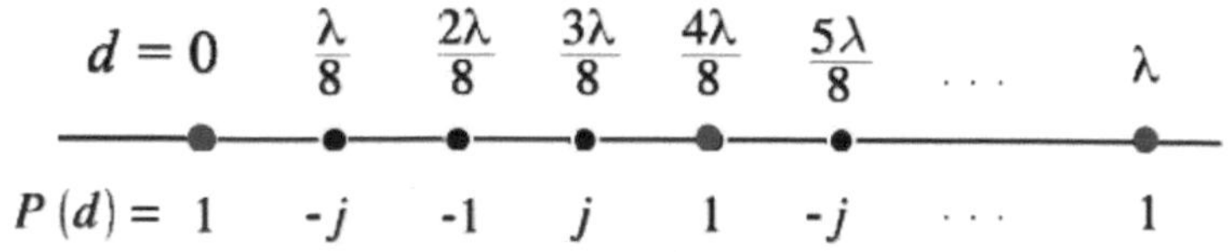

$$d = 0 \qquad \frac{\lambda}{8} \qquad \frac{2\lambda}{8} \qquad \frac{3\lambda}{8} \qquad \frac{4\lambda}{8} \qquad \frac{5\lambda}{8} \qquad \cdots \qquad \lambda$$

$$P(d) = \quad 1 \qquad -j \qquad -1 \qquad j \qquad 1 \qquad -j \qquad \cdots \qquad 1$$

FIGURE 6.4 Phase function $P(d)$.

domain (FDTD) method is adopted to deal with complex structures [1,2]. If the FDTD simulation works, the result can be beneficial to

- Validation of experimental data
- Analysis for unknown data

For experimental validation, we measured the compound-scattering matrix directly in an anechoic chamber. Since experimental data confirm the physical scattering mechanism, the result proofs the evidence. Although experimental verification is valid, the experimental situation is rather limited, that is, all situations cannot be confirmed. Therefore, a combined approach is employed in this chapter.

6.2 COMPOUND SCATTERING MATRIX

Based on Equations (6.1.2) and (6.1.3), any scattering matrix can be created using an elementary scatterer such as a dipole or dihedral corner reflector. It should be noted that equation (6.1.2) applies only for the same RCS target. If the mixture of a dipole and dihedral corner reflector is used, we have to use equation (6.1.1), accounting for the RCS value. The matrix created by multiple targets as shown in Figure 6.3 is called a compound matrix. For simplicity, we choose the dipole for creating various scattering matrices. If dihedrals are employed, it becomes impossible to locate them together within the wavelength range. In addition, the shadowing effect by the first dihedral would cause the distortion of the second dihedral. Therefore, a dipole is the best elementary target to deal with.

Four dipoles in Figure 6.5 are the elements of the compound matrix. For simplicity, we denote the 45°-oriented dipole as $[S]_1$ and the −45°-oriented dipole as $[S]_2$,

$$[S]_1 = [S]_{dipole}^{45°} = \frac{1}{2}\begin{bmatrix} 1 & 1 \\ 1 & 1 \end{bmatrix}, \quad [S]_2 = [S]_{dipole}^{-45°} = \frac{1}{2}\begin{bmatrix} 1 & -1 \\ -1 & 1 \end{bmatrix}$$

6.2.1 EXAMPLES OF COMPOUND SCATTERING MATRICES

If we increase the spacing d between horizontal and vertical dipoles, the coherent sum becomes

$$d = \frac{\lambda}{8} \Rightarrow \quad [S]_{dipole}^{H} + [S]_{dipole}^{V} P\left(\frac{\lambda}{8}\right) = \begin{bmatrix} 1 & 0 \\ 0 & 0 \end{bmatrix} - j\begin{bmatrix} 0 & 0 \\ 0 & 1 \end{bmatrix} = \begin{bmatrix} 1 & 0 \\ 0 & -j \end{bmatrix} \quad : \text{model 1}$$

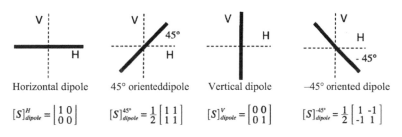

FIGURE 6.5   Basic dipole configuration and scattering matrix.

$$d = \frac{2}{8}\lambda \Rightarrow \quad [S]_{dipole}^{H} + [S]_{dipole}^{V} P\left(\frac{2\lambda}{8}\right) = \begin{bmatrix} 1 & 0 \\ 0 & 0 \end{bmatrix} - \begin{bmatrix} 0 & 0 \\ 0 & 1 \end{bmatrix} = \begin{bmatrix} 1 & 0 \\ 0 & -1 \end{bmatrix} \quad :\text{dihedral}$$

$$d = \frac{3}{8}\lambda \Rightarrow \quad [S]_{dipole}^{H} + [S]_{dipole}^{V} P\left(\frac{3\lambda}{8}\right) = \begin{bmatrix} 1 & 0 \\ 0 & 0 \end{bmatrix} + j\begin{bmatrix} 0 & 0 \\ 0 & 1 \end{bmatrix} = \begin{bmatrix} 1 & 0 \\ 0 & j \end{bmatrix} \quad :\text{model 2}$$

$$d = \frac{4}{8}\lambda \Rightarrow \quad [S]_{dipole}^{H} + [S]_{dipole}^{V} P\left(\frac{4\lambda}{8}\right) = \begin{bmatrix} 1 & 0 \\ 0 & 0 \end{bmatrix} + \begin{bmatrix} 0 & 0 \\ 0 & 1 \end{bmatrix} = \begin{bmatrix} 1 & 0 \\ 0 & 1 \end{bmatrix} \quad :\text{flat plate}$$

These corresponding models are shown in Figure 6.6 with illustrations. This result indicates an interesting phenomenon, that is, the VV component changes according to the spacing d. The value of S_{VV} can be real, complex, and pure imaginary. This change is caused by the phase delay between H and V dipoles. From this fact, we can deduce that the complex numbers in the scattering matrix and in other matrices are related to phase delay. Although the reflection or scattering itself on the object may be real-valued, the distance of scattering center causes a phase delay in the measured scattering matrix.

Next, we employ the 45°-oriented dipole $[S]_1$ and the −45°-oriented dipole $[S]_2$. The non-orthogonal combination yields a complicated double-bounce scattering because of $\text{Re}\{S_{HH}S_{VV}^*\} < 0$. They can be found in oriented urban areas. We call them models 3 and 4.

$$\text{model 3:} \quad [S]_1 + [S]_{dipole}^{V} P\left(\frac{\lambda}{4}\right) = \frac{1}{2}\begin{bmatrix} 1 & 1 \\ 1 & 1 \end{bmatrix} - \begin{bmatrix} 0 & 0 \\ 0 & 1 \end{bmatrix} = \frac{1}{2}\begin{bmatrix} 1 & 1 \\ 1 & -1 \end{bmatrix}$$

$$\text{model 4:} \quad [S]_2 + [S]_{dipole}^{V} P\left(\frac{\lambda}{4}\right) = \frac{1}{2}\begin{bmatrix} 1 & -1 \\ -1 & 1 \end{bmatrix} - \begin{bmatrix} 0 & 0 \\ 0 & 1 \end{bmatrix} = \frac{1}{2}\begin{bmatrix} 1 & -1 \\ -1 & -1 \end{bmatrix}$$

Orthogonal combination yields a flat plate or HV reflector by an appropriate spacing.

$$\text{flat plate:} \quad [S]_1 + [S]_2 P\left(\frac{\lambda}{2}\right) = \frac{1}{2}\begin{bmatrix} 1 & 1 \\ 1 & 1 \end{bmatrix} + \frac{1}{2}\begin{bmatrix} 1 & -1 \\ -1 & 1 \end{bmatrix} = \begin{bmatrix} 1 & 0 \\ 0 & 1 \end{bmatrix}$$

$$HV \text{ reflector:} \quad [S]_1 + [S]_2 P\left(\frac{\lambda}{4}\right) = \frac{1}{2}\begin{bmatrix} 1 & 1 \\ 1 & 1 \end{bmatrix} - \frac{1}{2}\begin{bmatrix} 1 & -1 \\ -1 & 1 \end{bmatrix} = \begin{bmatrix} 0 & 1 \\ 1 & 0 \end{bmatrix}$$

This HV reflector has the same scattering matrix as the 45°-oriented dihedral.

The preceding models are shown in Figure 6.7.

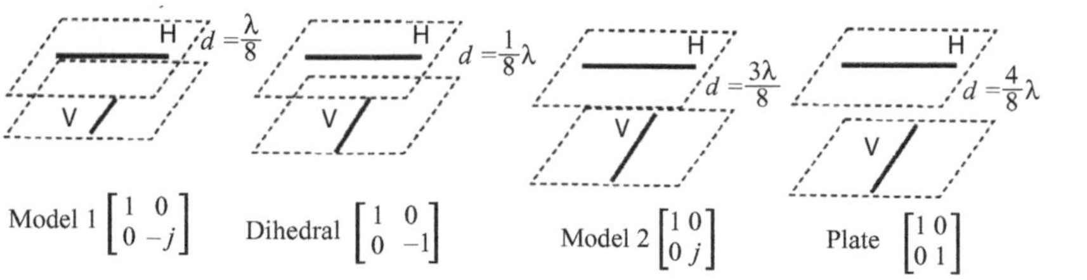

FIGURE 6.6   Compound scattering matrix by H- and V-dipoles.

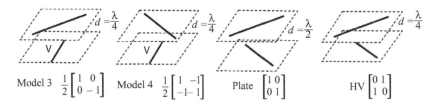

$$\text{Model 3} \quad \frac{1}{2}\begin{bmatrix} 1 & 0 \\ 0 & -1 \end{bmatrix} \qquad \text{Model 4} \quad \frac{1}{2}\begin{bmatrix} 1 & -1 \\ -1 & 1 \end{bmatrix} \qquad \text{Plate} \quad \begin{bmatrix} 1 & 0 \\ 0 & 1 \end{bmatrix} \qquad \text{HV} \quad \begin{bmatrix} 0 & 1 \\ 1 & 0 \end{bmatrix}$$

FIGURE 6.7 Double-bounce and other targets.

Multiple dipoles can create various scattering matrices as

$$\text{model 5:} \quad [S]_{dipole}^H + [S]_{dipole}^V P\left(\frac{\lambda}{4}\right) + [S]_{dipole}^H P\left(\frac{\lambda}{2}\right) = \begin{bmatrix} 1 & 0 \\ 0 & 0 \end{bmatrix} - \begin{bmatrix} 0 & 0 \\ 0 & 1 \end{bmatrix} + \begin{bmatrix} 1 & 0 \\ 0 & 0 \end{bmatrix} = \begin{bmatrix} 2 & 0 \\ 0 & -1 \end{bmatrix}$$

$$\text{model 6:} \quad [S]_{dipole}^H + [S]_{dipole}^V P\left(\frac{\lambda}{4}\right) + [S]_{dipole}^H P\left(\frac{3\lambda}{4}\right) = \begin{bmatrix} 1 & 0 \\ 0 & 0 \end{bmatrix} - \begin{bmatrix} 0 & 0 \\ 0 & 1 \end{bmatrix} - \begin{bmatrix} 0 & 0 \\ 0 & 1 \end{bmatrix} = \begin{bmatrix} 1 & 0 \\ 0 & -2 \end{bmatrix}$$

These specific double-bounce models are shown in Figure 6.8.

If we place four dipoles as

$$[S]_{ex1}^{total} = [S]_1 + [S]_2 P(0) + [S]_1 P\left(\frac{\lambda}{8}\right) + [S]_2 P\left(\frac{3\lambda}{8}\right)$$

then the compound scattering matrix becomes

$$= \frac{1}{2}\begin{bmatrix} 1 & 1 \\ 1 & 1 \end{bmatrix} + \frac{1}{2}\begin{bmatrix} 1 & -1 \\ -1 & 1 \end{bmatrix} - \frac{j}{2}\begin{bmatrix} 1 & 1 \\ 1 & 1 \end{bmatrix} + \frac{j}{2}\begin{bmatrix} 1 & -1 \\ -1 & 1 \end{bmatrix} = \begin{bmatrix} 1 & -j \\ -j & 1 \end{bmatrix}$$

Mathematically, the preceding equation holds, but it is difficult to install two dipoles in the same plane $(d = 0)$ or in the very near plane $(d = \lambda / 8)$ in the measurement. In such a case, we can use the periodic property of the phase function $P(d) = P\left(d + n\frac{\lambda}{2}\right)$. The preceding combination can be rewritten as

$$= \frac{1}{2}\begin{bmatrix} 1 & 1 \\ 1 & 1 \end{bmatrix} - \frac{j}{2}\begin{bmatrix} 1 & 1 \\ 1 & 1 \end{bmatrix} + \frac{j}{2}\begin{bmatrix} 1 & -1 \\ -1 & 1 \end{bmatrix} + \frac{1}{2}\begin{bmatrix} 1 & -1 \\ -1 & 1 \end{bmatrix} = \begin{bmatrix} 1 & -j \\ -j & 1 \end{bmatrix}$$

This combination corresponds to a different dipole arrangement

$$[S]_{ex1}^{total} = [S]_1 + [S]_1 P\left(\frac{7\lambda}{8}\right) + [S]_2 P\left(\frac{13\lambda}{8}\right) + [S]_2 P\left(\frac{20\lambda}{8}\right)$$

$$\text{Model 5} \quad \begin{bmatrix} 2 & 0 \\ 0 & -1 \end{bmatrix} \qquad\qquad \text{Model 6} \quad \begin{bmatrix} 1 & 0 \\ 0 & -2 \end{bmatrix}$$

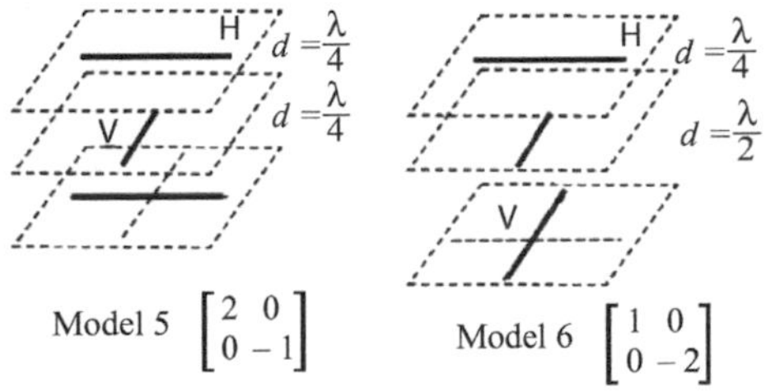

FIGURE 6.8 Double-bounce scattering with different *HH* and *VV* magnitudes.

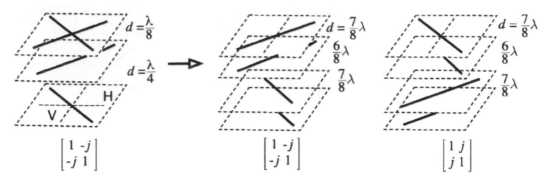

FIGURE 6.9 Four-dipole configurations.

This arrangement is easy to install dipoles at different planes as shown in Figure 6.9. Similarly, we can create a similar scattering matrix with $S_{HV} = j$.

$$[S]_{ex2}^{total} = [S]_2 + [S]_2 P\left(\frac{7\lambda}{8}\right) + [S]_1 P\left(\frac{13\lambda}{8}\right) + [S]_1 P\left(\frac{20\lambda}{8}\right)$$

$$= \frac{1}{2}\begin{bmatrix} 1 & -1 \\ -1 & 1 \end{bmatrix} - \frac{j}{2}\begin{bmatrix} 1 & -1 \\ -1 & 1 \end{bmatrix} + \frac{j}{2}\begin{bmatrix} 1 & 1 \\ 1 & 1 \end{bmatrix} + \frac{1}{2}\begin{bmatrix} 1 & 1 \\ 1 & 1 \end{bmatrix} = \begin{bmatrix} 1 & j \\ j & 1 \end{bmatrix}$$

Helix scattering can be obtained by four dipoles as shown in Figure 6.10. The spacing can be adjusted according to $P(d) = P\left(d + n\frac{\lambda}{2}\right)$. By the appropriate spacing, complete helix target can be created.

$$\text{Left-helix:}\quad [S]_{l-helix}^{total} = [S]_{dipole}^{H} + [S]_2 P\left(\frac{\lambda}{8}\right) + [S]_{dipole}^{V} P\left(\frac{2\lambda}{8}\right) + [S]_1 P\left(\frac{3\lambda}{8}\right)$$

$$= \begin{bmatrix} 1 & 0 \\ 0 & 0 \end{bmatrix} - \frac{j}{2}\begin{bmatrix} 1 & -1 \\ -1 & 1 \end{bmatrix} - \begin{bmatrix} 0 & 0 \\ 0 & 1 \end{bmatrix} + \frac{j}{2}\begin{bmatrix} 1 & 1 \\ 1 & 1 \end{bmatrix} = \begin{bmatrix} 1 & j \\ j & -1 \end{bmatrix}$$

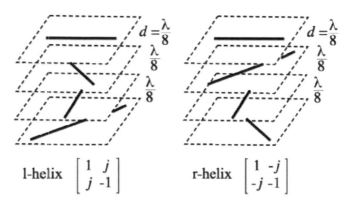

FIGURE 6.10 Helix scattering by four dipoles.

Right-helix: $\left[S \right]_{r-helix}^{total} = \left[S \right]_{dipole}^{H} + \left[S \right]_{1} P\left(\dfrac{\lambda}{8} \right) + \left[S \right]_{dipole}^{V} P\left(\dfrac{2\lambda}{8} \right) + \left[S \right]_{2} P\left(\dfrac{3\lambda}{8} \right)$

$$= \begin{bmatrix} 1 & 0 \\ 0 & 0 \end{bmatrix} - \frac{j}{2}\begin{bmatrix} 1 & 1 \\ 1 & 1 \end{bmatrix} - \begin{bmatrix} 0 & 0 \\ 0 & 1 \end{bmatrix} + \frac{j}{2}\begin{bmatrix} 1 & -1 \\ -1 & 1 \end{bmatrix} = \begin{bmatrix} 1 & -j \\ -j & -1 \end{bmatrix}$$

6.3 FDTD ANALYSIS

FDTD simulation on the coherent sum in equation (6.1.2) has been applied to investigate the scattering phenomena from compound dipoles. A Gaussian pulse was illuminated to targets, and the reflected wave was collected to create scattering matrix elements. FDTD parameters are listed in Table 6.1.

6.3.1 COMPOUND SCATTERING MATRIX IN THE RANGE DIRECTION

Two orthogonal wires and its rotation by 45° were chosen for test objects to calculate scattering matrix. After the scattering calculation, a polarization signature for each spacing distance was derived as shown in Figure 6.11. The symbol on the top of Figure 6.11 indicates two orthogonal wires viewed from the top. It is seen that the polarization signature changes from the flat plate to the dihedral, and goes back to the flat plate. It changes in a periodical fashion.

From Figure 6.11, we can see the spacing between dipoles plays a very important role in the resultant scattering matrix. Two orthogonal dipoles with appropriate spacing can be any kinds of scatterers. This theoretical result is very important in interpreting the scattering matrix acquired by the real PolSAR system.

The next target is a helix model for the configuration in Figure 6.10. The FDTD analyses on left- and right-helix models were performed, and the polarization signatures were calculated as shown in the upper side of Figure 6.12. The calculated polarization signatures are a little bit distorted due to electromagnetic coupling between close dipoles and the numerical error; however, the overall signatures are those of the helix. They return the circular polarization power.

This simulation result was confirmed by the PolSAR measurement in an anechoic chamber. The measured polarization signatures are very similar to those of FDTD simulations (Figure 6.12). Hence, we can confirm the applicability of the coherent sum equation (6.1.2).

6.3.2 COMPOUND SCATTERING MATRIX IN THE AZIMUTH DIRECTION

If two targets are in the same range and aligned in the azimuth direction, then the coherent sum may be written as

$$\sqrt{\sigma_{total}}\left[S \right]_{total} = \sqrt{\sigma_1}\left[S \right]_1 + \sqrt{\sigma_2}\left[S \right]_2 \qquad (6.3.1)$$

TABLE 6.1
FDTD Parameters

Time step	5.7 ps
Grid size	3 mm
Number of cells	$100 \times 100 \times 100$
Frequency	10 GHz
Plane wave	Gaussian pulse
Target	Wires, plate, dihedral

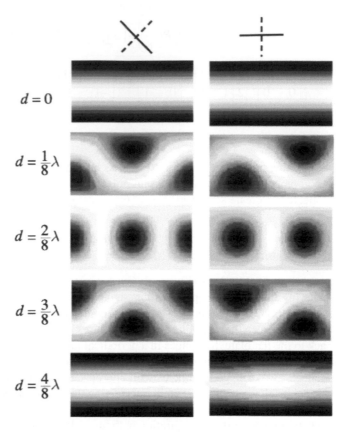

FIGURE 6.11   Polarization signature of crossed dipole with various spacings.

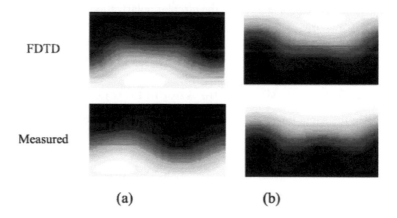

FIGURE 6.12   Polarization signature by simulation and measurement: (a) left helix and (b) right helix.

This property was simulated by the FDTD method as shown in Figure 6.13. In these figures, the target arrangement is depicted first, and the corresponding FDTD polarization signature as well as the coherent sum polarization signature are compared. If we decompose the compound scattering matrix by the Ks, Kd, Kh method by E. Krogager [3], the constitution ratio becomes as listed in Table 6.2. The validity of the coherent sum equation (6.3.1) is confirmed in this figure.

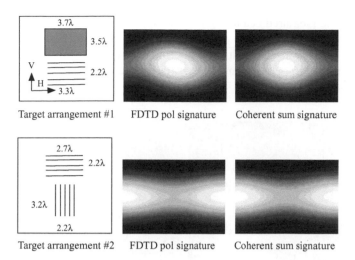

FIGURE 6.13   Composite scattering matrix by two targets aligned in the same range.

TABLE 6.2
Component Ratio Derived by the Decomposition

Component Ratio		Plate (%)	Dihedral (%)
Arrangement #1	FDTD	83	17
	Coherent sum	82	18
Arrangement #2	FDTD	93	7
	Coherent sum	92	8

6.4   COMPOUND SCATTERING MATRIX MEASUREMENT

For validation of compound scattering matrix formation, a scattering matrix acquisition was carried out in a well-controlled anechoic chamber at Niigata University. The measurement specification is listed in Table 6.3. The range resolution was chosen to be 15 cm so that the length of the compound dipoles in the range direction becomes smaller than 15 cm. A dipole 3 mm in diameter and 96 cm long can be considered a long wire object for the center frequency of 10 GHz (the wavelength 3 cm). The experimental scene is shown in Figure 6.14.

After the PolSAR acquisition, compound scattering matrices were derived. They are compared with theoretical values by the polarization signature as shown in Table 6.4. It is understood and surprising to see that measured signatures are very close to theoretical polarization ones.

TABLE 6.3
Radar Measurement Specifications

Center frequency	10 GHz
Bandwidth	1 GHz
Range resolution	15 cm
Dipole size	3 mm × 96 cm

Some errors in the measured scattering matrix element come from misalignment and multiple scattering of dipoles. When multiple and non-orthogonal dipoles organize a specific combined matrix, the errors come from the multiple scattering and mutual coupling effect. Although there are some measurement errors, the compound scattering matrix can be achieved by a combination of oriented dipoles.

FIGURE 6.14 Compound scattering matrix measurement in an anechoic chamber.

TABLE 6.4
Comparison of Theoretical and Measured Polarization Signature

Compound Model	Theoretical	Measured

Plate

$[S] = \begin{bmatrix} 1 & 0 \\ 0 & 1 \end{bmatrix}$

$\begin{bmatrix} 1.0 & 0.009\,j \\ 0.009\,j & 0.988 - 0.046\,j \end{bmatrix}$

$\begin{bmatrix} 1 & 0 \\ 0 & 1 \end{bmatrix} = \begin{bmatrix} 0 & 0 \\ 0 & 1 \end{bmatrix} + \begin{bmatrix} 1 & 0 \\ 0 & 0 \end{bmatrix}$

Dihedral

$[S] = \begin{bmatrix} 1 & 0 \\ 1 & -1 \end{bmatrix}$

$\begin{bmatrix} 1.0 & -0.027 - 0.011\,j \\ -0.027 - 0.011\,j & -0.963 + 0.235\,j \end{bmatrix}$

$\begin{bmatrix} 1 & 0 \\ 0 & -1 \end{bmatrix} = \begin{bmatrix} 1 & 0 \\ 0 & 0 \end{bmatrix} - \begin{bmatrix} 0 & 0 \\ 0 & 1 \end{bmatrix}$

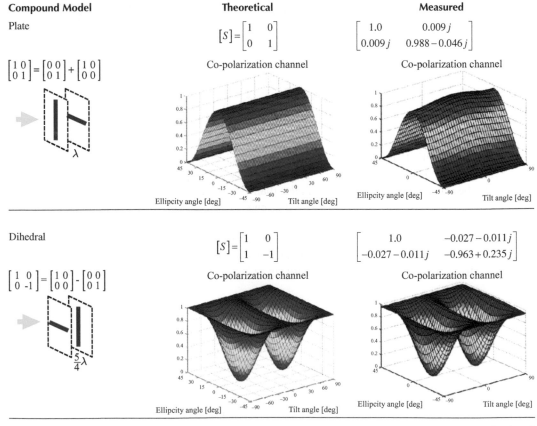

(Continued)

TABLE 6.4 (*Continued*)
Comparison of Theoretical and Measured Polarization Signature

Compound Model	Theoretical	Measured
45°-oriented dipole	$2[S] = \begin{bmatrix} 1 & 1 \\ 1 & 1 \end{bmatrix}$	$\begin{bmatrix} 1.0 & 0.842 - 0.074\,j \\ 0.842 - 0.074\,j & 0.969 + 0.108\,j \end{bmatrix}$

$[S] = \frac{1}{2}\begin{bmatrix} 1 & 1 \\ 1 & 1 \end{bmatrix}$

Co-polarization channel

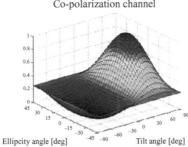

Co-polarization channel

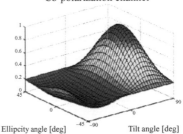

−45°-oriented dipole

$2[S] = \begin{bmatrix} 1 & -1 \\ -1 & 1 \end{bmatrix}$

$\begin{bmatrix} 1.0 & -1.087 - 0.061\,j \\ -1.087 - 0.061\,j & 0.985 + 0.242\,j \end{bmatrix}$

$[S] = \frac{1}{2}\begin{bmatrix} 1 & -1 \\ -1 & 1 \end{bmatrix}$

Co-polarization channel

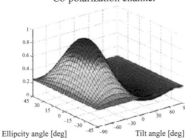

Co-polarization channel

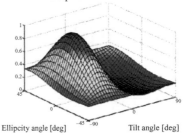

Compound 1

$[S] = \begin{bmatrix} 1 & -j \\ -j & 1 \end{bmatrix}$

$\begin{bmatrix} 1.0 & 0.038 - 1.031\,j \\ 0.038 - 1.031\,j & 0.897 + 0.159\,j \end{bmatrix}$

$\begin{bmatrix} 1 & -j \\ -j & 1 \end{bmatrix} = \frac{1}{2}\begin{bmatrix} 1 & -1 \\ -1 & 1 \end{bmatrix} + \frac{j}{2}\begin{bmatrix} 1 & -1 \\ -1 & 1 \end{bmatrix}$
$- \frac{j}{2}\begin{bmatrix} 1 & 1 \\ 1 & 1 \end{bmatrix} + \frac{1}{2}\begin{bmatrix} 1 & 1 \\ 1 & 1 \end{bmatrix}$

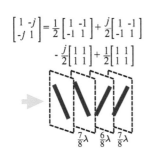

$\frac{7}{8}\lambda \quad \frac{6}{8}\lambda \quad \frac{7}{8}\lambda$

Co-polarization channel

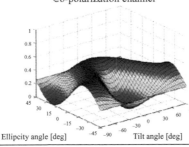

Co-polarization channel

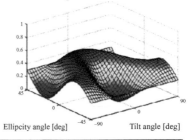

Compound 2

$[S] = \begin{bmatrix} 1 & j \\ j & 1 \end{bmatrix}$

$\begin{bmatrix} 1.0 & -0.045 + 1.135\,j \\ -0.045 + 1.135\,j & 1.006 - 0.186\,j \end{bmatrix}$

(*Continued*)

TABLE 6.4 (*Continued*)

Comparison of Theoretical and Measured Polarization Signature

Compound Model	Theoretical	Measured

Compound Model

$$\begin{bmatrix} 1 & j \\ j & 1 \end{bmatrix} = \frac{1}{2}\begin{bmatrix} 1 & 1 \\ 1 & 1 \end{bmatrix} + \frac{j}{2}\begin{bmatrix} 1 & 1 \\ 1 & 1 \end{bmatrix}$$

$$-\frac{j}{2}\begin{bmatrix} 1 & -1 \\ -1 & 1 \end{bmatrix} + \frac{1}{2}\begin{bmatrix} 1 & -1 \\ -1 & 1 \end{bmatrix}$$

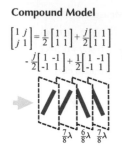

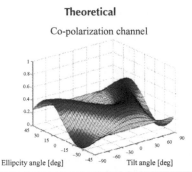

Co-polarization channel

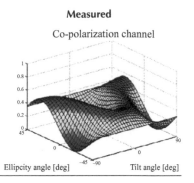

Co-polarization channel

Left helix

$$[S] = \begin{bmatrix} 1 & j \\ j & -1 \end{bmatrix}$$

$$\begin{bmatrix} 1.0 & 0.293 + 1.07j \\ 0.293 + 1.07j & -0.866 + 0.584j \end{bmatrix}$$

$$\begin{bmatrix} 1 & j \\ j & -1 \end{bmatrix} = \begin{bmatrix} 1 & 0 \\ 0 & 0 \end{bmatrix} - \frac{j}{2}\begin{bmatrix} 1 & -1 \\ -1 & 1 \end{bmatrix}$$

$$-\frac{1}{2}\begin{bmatrix} 0 & 0 \\ 0 & 1 \end{bmatrix} + \frac{j}{2}\begin{bmatrix} 1 & 1 \\ 1 & 1 \end{bmatrix}$$

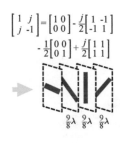

Co-polarization channel

Co-polarization channel

Model 2

$$[S] = \begin{bmatrix} 1 & 0 \\ 0 & j \end{bmatrix}$$

$$\begin{bmatrix} 1.0 & 0.014 + 0.012j \\ 0.014 + 0.012j & -0.056 + 1.085j \end{bmatrix}$$

$$\begin{bmatrix} 1 & 0 \\ 0 & j \end{bmatrix} = \begin{bmatrix} 1 & 0 \\ 0 & 0 \end{bmatrix} + j\begin{bmatrix} 0 & 0 \\ 0 & 1 \end{bmatrix}$$

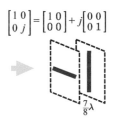

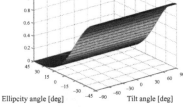

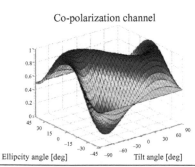

Co-polarization channel

Model 3

$$2[S] = \begin{bmatrix} 1 & 1 \\ 1 & -1 \end{bmatrix}$$

$$\begin{bmatrix} 1.0 & 1.637 + 0.097j \\ 1.637 + 0.097j & -1.182 - 0.037j \end{bmatrix}$$

(*Continued*)

TABLE 6.4 (*Continued*)

Comparison of Theoretical and Measured Polarization Signature

Compound Model	Theoretical	Measured

$$\frac{1}{2}\begin{bmatrix} 1 & 1 \\ 1 & -1 \end{bmatrix} = \frac{1}{2}\begin{bmatrix} 1 & 1 \\ 1 & 1 \end{bmatrix} - \begin{bmatrix} 0 & 0 \\ 0 & 1 \end{bmatrix}$$

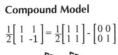

Co-polarization channel

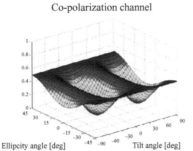

Co-polarization channel

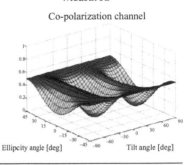

Model 4

$$2[S] = \begin{bmatrix} 1 & -1 \\ -1 & -1 \end{bmatrix}$$

$$\begin{bmatrix} 1.0 & -1.393 - 0.359\,j \\ -1.393 + 0.359\,j & -1.068 + 0.621\,j \end{bmatrix}$$

$$\frac{1}{2}\begin{bmatrix} 1 & -1 \\ -1 & -1 \end{bmatrix} = \frac{1}{2}\begin{bmatrix} 1 & -1 \\ -1 & 1 \end{bmatrix} - \begin{bmatrix} 0 & 0 \\ 0 & 1 \end{bmatrix}$$

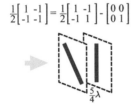

Co-polarization channel

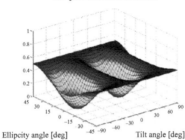

Co-polarization channel

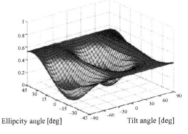

Model 5

$$[S] = \begin{bmatrix} 2 & 0 \\ 0 & -1 \end{bmatrix}$$

$$\begin{bmatrix} 2.0 & 0.044 - 0.068\,j \\ 0.044 - 0.068\,j & -1.075 + 0.445\,j \end{bmatrix}$$

$$\begin{bmatrix} 2 & 0 \\ 0 & -1 \end{bmatrix} = \begin{bmatrix} 1 & 0 \\ 0 & 0 \end{bmatrix} - \begin{bmatrix} 0 & 0 \\ 0 & 1 \end{bmatrix} + \begin{bmatrix} 1 & 0 \\ 0 & 0 \end{bmatrix}$$

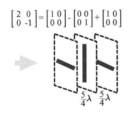

Co-polarization channel

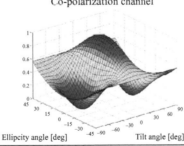

Co-polarization channel

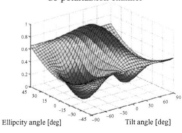

(Continued)

TABLE 6.4 (*Continued*)

Comparison of Theoretical and Measured Polarization Signature

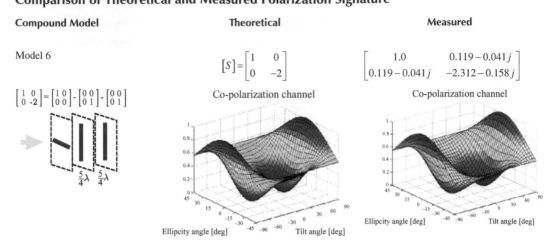

Compound Model	Theoretical	Measured

Model 6

$[S] = \begin{bmatrix} 1 & 0 \\ 0 & -2 \end{bmatrix}$

$\begin{bmatrix} 1.0 & 0.119 - 0.041j \\ 0.119 - 0.041j & -2.312 - 0.158j \end{bmatrix}$

$\begin{bmatrix} 1 & 0 \\ 0 & -2 \end{bmatrix} = \begin{bmatrix} 1 & 0 \\ 0 & 0 \end{bmatrix} - \begin{bmatrix} 0 & 0 \\ 0 & 1 \end{bmatrix} - \begin{bmatrix} 0 & 0 \\ 0 & 1 \end{bmatrix}$

Co-polarization channel

Co-polarization channel

6.5 SUMMARY

Through FDTD analyses and experimental results, we understand that all scattering models can be made by a combination of dipoles. These compound scattering matrices come from a coherent sum of multiple dipole scatterings with certain spacings within radar resolution. The compound matrix configurations are a little bit complex but may be found as combinations of tree branches in forests, jungles, etc. We can interpret these scattering mechanisms simply by polarization signatures. The compound matrix in Table 6.4 can be the basis for the modeling of scattering mechanisms.

REFERENCES

1. K. Kitayama, Y. Yamaguchi, J. Yang, and H. Yamada, "Compound scattering matrix of targets aligned in the range direction," *IEICE Trans. Commun.*, vol. E.84-B, no. 1, pp. 81–88, 2001.
2. K. Kitayama, Y. Takayanagi, Y. Yamaguchi, and H. Yamada, "Polarimetric calibration using a corrugated parallel plate target," *Trans. of IEICE B-II*, vol. J-81, no. 10, pp. 914–921, 1998.
3. E. Krogager and Z. H. Czyz, "Properties of the sphere, diplane, helix (target scattering matrix) decomposition," *Proc. JIPR-3*, pp. 106–114, Nantes, France, 1995.
4. G. Singh, S. Mohanty, Y. Yamaguchi, and Y. Yamazaki, "Physical scattering interpretation of POLSAR coherency matrix by using compound scattering phenomenon," *IEEE Trans. Geosci. Remote Sens.*, vol. 58, no. 4, pp. 2541–2556, 2020.

7 Scattering Mechanisms and Modeling

7.1 INTRODUCTION

This chapter is devoted to the understanding of scattering mechanisms in PolSAR images and modeling of these scattering mechanisms by corresponding matrices. If scattering mechanisms are connected to physically feasible scattering, it would serve immediate interpretation and retrieval of objects in the PolSAR image.

At first, we check the image of San Francisco (Figure 7.1). The corresponding gray-scale images of nine parameters in the covariance matrix [C] and coherency matrix [T] are shown in the consecutive figures to understand the polarimetric information contained in the quad pol data.

We can see in the covariance matrix images of Figure 7.2 that; C_{11} image is the brightest among nine element images. This image represents $|HH|^2$ power in the HH radar channel. Not only in urban areas but also in mountain areas the image is the brightest. The C_{33} image comes from the VV channel power $|VV|^2$ and has the second brightest image. The brightest area is corresponding to the orthogonal building block urban area in the lower-right corner. The orthogonal building block means that buildings are aligned to normal to radar illumination. The C_{22} image is based on the cross-polarized channel power $|HV|^2$ and has different bright areas in an oriented urban block at the lower corner. $\text{Re}\{C_{13}\}$ and $\text{Im}\{C_{13}\}$ images show cross-correlation, look similar ($|\text{Re}\{C_{13}\}|$ > $|\text{Im}\{C_{13}\}|$), and have bright areas around orthogonal building block urban region.

Other components by C_{12} and C_{23} have very small contributions in the whole image. $|\text{Im}\{C_{12}\}|$ and $|\text{Im}\{C_{23}\}|$ images are almost the same. Although there may be nine parameters, some of them are too small compared to others. There are magnitude imbalances among these parameters. The number of contributions by nine parameters is scene dependent.

Similarly, we can see in the coherency matrix images of Figure 7.3 that T_{11} is the brightest, and T_{22} seems to be the second brightest, but their bright locations are different. T_{11} may represent surface scattering, but it also can be seen in the double-bounce urban areas in the lower-right corner. The third brightest is the $|\text{Re}\{T_{12}\}|$ image, which occurs in orthogonal urban areas. On the other hand, T_{33} appears in the oriented urban blocks, which is the same as the C_{22} image. All images have some values in orthogonal urban blocks. Although there may be nine parameters, some of them are too small compared to others. There are magnitude imbalances as well. Contributions of nine parameters are scene dependent.

Now, we try to retrieve polarimetric information(s) by nine parameters and we make use of them for scattering power decomposition. As seen in the preceding figures, there is magnitude imbalance among parameters. Just using each parameter would not serve fruitful results. The next step is to retrieve scattering mechanisms based on the second-order statistics of the signal and experimental evidences.

It has been well-known from the analyses on the huge amount of NASA JPL AIRSAR data sets that there are three major scattering phenomena:

- Surface scattering
- Double-bounce scattering
- Volume (or defuse) scattering

FIGURE 7.1   Google Earth image of San Francisco.

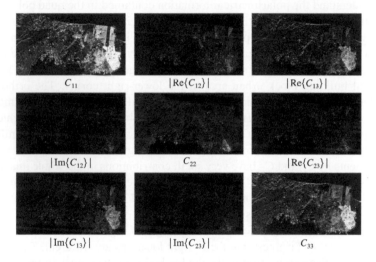

FIGURE 7.2   Covariance matrix images of San Francisco by ALOS2 Quad pol data.

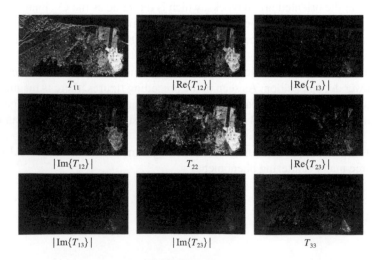

FIGURE 7.3   Coherency matrix images of San Francisco by ALOS2 Quad pol data.

The purpose of this chapter is to explore the modeling of nine parameters including the existing three scattering mechanisms. Here we refer to "modeling" as deriving suitable coherency matrix based on a physically realizable scattering mechanism.

7.2 INTERPRETATION OF NINE PARAMETERS AND MODELING

The coherency matrix is closely related to scattering mechanisms and is convenient and easy to rotate in mathematical formulations. We focus our attention to each element of the ensemble-averaged coherency matrix for modeling of physical scattering.

$$
\langle [T] \rangle = \begin{bmatrix} T_{11} & T_{12} & T_{13} \\ T_{21} & T_{22} & T_{23} \\ T_{31} & T_{32} & T_{33} \end{bmatrix}
$$

$$
= \begin{bmatrix} \dfrac{\left\langle |S_{HH}+S_{VV}|^2 \right\rangle}{2} & \dfrac{\left\langle (S_{HH}+S_{VV})(S_{HH}-S_{VV})^* \right\rangle}{2} & \left\langle (S_{HH}+S_{VV})S_{HV}^* \right\rangle \\[2mm] \dfrac{\left\langle (S_{HH}+S_{VV})^*(S_{HH}-S_{VV}) \right\rangle}{2} & \dfrac{\left\langle |S_{HH}-S_{VV}|^2 \right\rangle}{2} & \left\langle (S_{HH}-S_{VV})S_{HV}^* \right\rangle \\[2mm] \left\langle (S_{HH}+S_{VV})^* S_{HV} \right\rangle & \left\langle (S_{HH}-S_{VV})^* S_{HV} \right\rangle & \left\langle 2|S_{HV}|^2 \right\rangle \end{bmatrix} \quad (7.2.1)
$$

It has been known from experimental evidences that the main contributors are:

T_{11}: surface scattering
T_{22}: double-bounce scattering
T_{33}: volume scattering

Other contributors for scattering are not well-defined physically.

T_{12}: Re$\{T_{12}\}$, Im$\{T_{12}\}$
T_{13}: Re$\{T_{13}\}$, Im$\{T_{13}\}$
T_{23}: Re$\{T_{23}\}$ mainly caused by an oriented surface
Im$\{T_{23}\}$: helix scattering

These contributors can be realized by compound dipoles or oriented surfaces as shown in the next section. We will check each element by referring to Figure 7.3.

7.3 T_{11}: SURFACE SCATTERING

The term can be expanded as

$$
T_{11} = \frac{1}{2}\left\langle |S_{HH}+S_{VV}|^2 \right\rangle = \frac{1}{2}\left\langle |S_{HH}|^2 \right\rangle + \frac{1}{2}\left\langle |S_{VV}|^2 \right\rangle + \left\langle \mathrm{Re}\{S_{HH}S_{VV}^*\} \right\rangle \quad (7.3.1)
$$

If Re$\{S_{HH}S_{VV}^*\} > 0$, then the value of T_{11} becomes maximum. This situation means that S_{HH} and S_{VV} are in-phase, that is, they have the same sign, which corresponds to the surface scattering. T_{11} is the brightest in the whole images of Figure 7.3. This property can be seen not only in bare soil surfaces but also in urban areas.

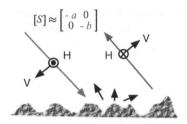

FIGURE 7.4   Surface scattering model.

Figure 7.4 shows a typical surface scattering model. Since the direction of the scattered electric field becomes opposite for both H- and V-polarized incidence waves, the scattering elements have a negative sign ($S_{HH} \approx -a, S_{VV} \approx -b$). This yields that S_{HH} and S_{VV} are in-phase, which can be written as

$$\mathrm{Re}\left\{S_{HH}S_{VV}^*\right\} > 0. \tag{7.3.2}$$

From the experimental data from the sea surface or bare soil, it is known that S_{HV} is negligible ($S_{HV} \approx 0$). For modeling of these surface scatterings, a slight modification in a scattering vector may apply.

$$\mathbf{k}_P = \frac{1}{\sqrt{2}}\begin{bmatrix} S_{HH}+S_{VV} \\ S_{HH}-S_{VV} \\ 2S_{HV} \end{bmatrix} \Rightarrow \quad \mathbf{k}_P = \begin{bmatrix} 1 \\ \beta \\ 0 \end{bmatrix} \tag{7.3.3}$$

where $\beta = \frac{S_{HH}-S_{VV}}{S_{HH}+S_{VV}}$ and $|\beta| < 1$ are assumed. The corresponding coherency matrix becomes

$$\langle [T] \rangle_{surfae} = \begin{bmatrix} 1 & \beta & 0 \\ \beta & |\beta|^2 & 0 \\ 0 & 0 & 0 \end{bmatrix}. \tag{7.3.4}$$

This surface scattering model in equation (7.3.4) indicates that the contribution of the T_{11} element is the largest, while accounting for T_{22}, T_{12}, and T_{21} terms by the complex parameter β satisfying $|\beta| < 1$. β can be determined or estimated afterward. $\beta = 0$ makes equation (7.3.4) to those of a sphere or flat plate.

7.4 T_{22}: DOUBLE-BOUNCE SCATTERING

The T_{22} term can be expanded as

$$T_{22} = \frac{1}{2}\left\langle |S_{HH}-S_{VV}|^2 \right\rangle = \frac{1}{2}\left\langle |S_{HH}|^2 \right\rangle + \frac{1}{2}\left\langle |S_{VV}|^2 \right\rangle - \left\langle \mathrm{Re}\left\{S_{HH}S_{VV}^*\right\} \right\rangle \tag{7.4.1}$$

If $\mathrm{Re}\left\{S_{HH}S_{VV}^*\right\} < 0$, the value of T_{22} becomes maximum. This means that S_{HH} and S_{VV} are out of phase, that is, they have opposite signs, which corresponds to the double-bounce scattering. This property can be seen in right-angle structures, such as a building wall to road surface, manmade structures, and tall trees stems on the ground. It is also known that S_{HV} is negligible ($S_{HV} \approx 0$) in the double-bounce scattering. Figure 7.5 shows the typical double-bounce scattering model.

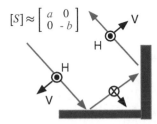

FIGURE 7.5 Double-bounce scattering model.

For the *H*-polarized wave, the direction of the electric field vector changes once per bounce. After the double-bounce, the direction of the reflected wave becomes the same as that of the incidence wave. On the other hand, the direction of the *V*-polarized wave after the double-bounce is opposite to the direction of incidence wave. This results in an opposite sign of *HH* and *VV* element, and characterized by

$$\mathrm{Re}\left\{ S_{HH} S_{VV}^{*} \right\} < 0. \tag{7.4.2}$$

For modeling of these double-bounce scattering, a slight modification may apply in scattering vector,

$$\mathbf{k}_{P} = \frac{1}{\sqrt{2}} \begin{bmatrix} S_{HH} + S_{VV} \\ S_{HH} - S_{VV} \\ 2S_{HV} \end{bmatrix} \Rightarrow \mathbf{k}_{P} = \begin{bmatrix} \alpha \\ 1 \\ 0 \end{bmatrix} \tag{7.4.3}$$

where $\alpha = \frac{S_{HH} + S_{VV}}{S_{HH} - S_{VV}}$ and $|\alpha| < 1$ are assumed. The corresponding coherency matrix becomes a model for the double-bounce scattering,

$$\langle [T] \rangle_{double} = \begin{bmatrix} |\alpha|^{2} & \alpha & 0 \\ \alpha^{*} & 1 & 0 \\ 0 & 0 & 0 \end{bmatrix}. \tag{7.4.4}$$

This scattering model indicates that the contribution of T_{22} is the largest, while accounting for T_{11}, T_{12} and T_{21} terms by complex parameter α with $|\alpha| < 1$. $\alpha = 0$ makes equation (7.4.4) to that of a dihedral.

7.5 T_{33}: VOLUME SCATTERING

$$T_{33} = \left\langle 2|S_{HV}|^{2} \right\rangle \tag{7.5.1}$$

This term represents the cross-polarization power. S_{HV} is generated by forests, trees, vegetations, oriented urban blocks, man-made structures, sloped/oriented surfaces, etc. There is no simple physical model except for the 45°-oriented dihedral to represent T_{33}.

7.5.1 VOLUME SCATTERING BY VEGETATION

Majority of the cross-polarization power comes from trees, forests, and vegetation in PolSAR images. It is known that the reflection symmetry condition applies for natural distributed target.

$$\langle S_{HH} S_{HV}^{*} \rangle \approx 0, \quad \langle S_{VV} S_{HV}^{*} \rangle \approx 0 \tag{7.5.2}$$

FIGURE 7.6   Volume scattering model.

This condition comes from random scattering in natural vegetation as shown in Figure 7.6. There are so many scattering centers inside the vegetation volume. If the contribution of each scattering point is summed up, then the total value becomes zero. This is called the reflection symmetry condition. The corresponding ensemble averaged covariance or coherency matrix comes in the form of

$$
\langle [C] \rangle = \begin{bmatrix} X & 0 & X \\ 0 & X & 0 \\ X & 0 & X \end{bmatrix}, \ \langle [T] \rangle = \begin{bmatrix} X & X & 0 \\ X & X & 0 \\ 0 & 0 & X \end{bmatrix} \tag{7.5.3}
$$

For theoretical modeling of trees or forests, a random cloud dipole model is employed. If forests or trees are seen from the zenith, they look like a completely random cloud of wire or dipoles. If the off-nadir angle becomes oblique as shown in Figure 7.7, the distribution of branches (dipole) becomes not uniform. Considering the physical situations, we employ some distribution functions.

Theoretical averaging is carried out by

$$
\langle [C] \rangle_{vol} = \int_0^\pi \left[C(\theta) \right] p(\theta) \, d\theta, \ \langle [T] \rangle_{vol} = \int_0^\pi \left[T(\theta) \right] p(\theta) \, d\theta \tag{7.5.4}
$$

The integration result becomes as follows:

$$
\text{Uniform PDF} \quad p(\theta) = \frac{1}{2\pi} \quad \langle [C] \rangle_{vol} = \frac{1}{8} \begin{bmatrix} 3 & 0 & 1 \\ 0 & 2 & 0 \\ 1 & 0 & 3 \end{bmatrix}, \ \langle [T] \rangle_{vol} = \frac{1}{4} \begin{bmatrix} 2 & 0 & 0 \\ 0 & 1 & 0 \\ 0 & 0 & 1 \end{bmatrix} \tag{7.5.5}
$$

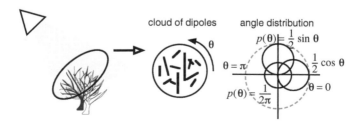

FIGURE 7.7   Dipole distributions seen from radar.

Vertical PDF $\quad p(\theta) = \dfrac{1}{2}\sin\theta \quad \langle[C]\rangle_{vol} = \dfrac{1}{15}\begin{bmatrix} 8 & 0 & 2 \\ 0 & 4 & 0 \\ 2 & 0 & 3 \end{bmatrix}, \quad \langle[T]\rangle_{vol} = \dfrac{1}{30}\begin{bmatrix} 15 & 5 & 0 \\ 5 & 7 & 0 \\ 0 & 0 & 8 \end{bmatrix}$ (7.5.6)

Horizontal PDF $\quad p(\theta) = \dfrac{1}{2}\cos\theta \quad \langle[C]\rangle = \dfrac{1}{15}\begin{bmatrix} 3 & 0 & 2 \\ 0 & 4 & 0 \\ 2 & 0 & 8 \end{bmatrix}, \quad \langle[T]\rangle_{vol} = \dfrac{1}{30}\begin{bmatrix} 15 & -5 & 0 \\ -5 & 7 & 0 \\ 0 & 0 & 8 \end{bmatrix}$ (7.5.7)

The criterion on how to choose the best volume scattering model is the power ratio of $\langle|S_{HH}|^2\rangle$ and $\langle|S_{VV}|^2\rangle$. Since the HH and VV ratio in the covariance matrix becomes $10\log\left(\frac{8}{3}\right) = 4.26$ dB, we set the boundary at ±2 dB as shown in Table 7.1.

There are various volume scattering models proposed. The next mathematical model is proposed by An et al. [1], which gives the maximum entropy,

$$\text{Maximum entropy model} \qquad \langle[T]\rangle_{vol} = \dfrac{1}{3}\begin{bmatrix} 1 & 0 & 0 \\ 0 & 1 & 0 \\ 0 & 0 & 1 \end{bmatrix} \qquad (7.5.8)$$

7.5.2 Volume Scattering by Oriented Surface

If S_{HV} scattering comes from an oriented urban area with respect to radar illumination, it is caused by ground surface to oriented building wall scattering. This scattering may be modeled as oriented dihedral scattering as shown in Figure 7.8 [2]. This coherency matrix can be obtained by the following equation:

$$\langle[T]\rangle_{vol}^{dihedral} = \int_{-\pi/2}^{\pi/2}[T(\theta)]_{dihedral}\dfrac{\cos\theta}{2}d\theta = \dfrac{1}{15}\begin{bmatrix} 0 & 0 & 0 \\ 0 & 7 & 0 \\ 0 & 0 & 8 \end{bmatrix} \qquad (7.5.9)$$

This oriented surface scattering is very strong in the right lower corner (triangular area) of the San Francisco image (Figure 7.2). This phenomenon was actually confirmed in an anechoic chamber.

TABLE 7.1

Choice of Volume Scattering Model for Vegetations

$10\log\dfrac{\langle|S_{VV}|^2\rangle}{\langle|S_{HH}|^2\rangle}$	−4 dB	−2 dB	0 dB	2 dB	4 dB
$\langle[C]\rangle_{vol} =$	$\dfrac{1}{15}\begin{bmatrix} 8 & 0 & 2 \\ 0 & 4 & 0 \\ 2 & 0 & 3 \end{bmatrix}$		$\dfrac{1}{8}\begin{bmatrix} 3 & 0 & 1 \\ 0 & 2 & 0 \\ 1 & 0 & 3 \end{bmatrix}$		$\dfrac{1}{15}\begin{bmatrix} 3 & 0 & 2 \\ 0 & 4 & 0 \\ 2 & 0 & 8 \end{bmatrix}$
$\langle[T]\rangle_{vol} =$	$\dfrac{1}{30}\begin{bmatrix} 15 & 5 & 0 \\ 5 & 7 & 0 \\ 0 & 0 & 8 \end{bmatrix}$		$\dfrac{1}{4}\begin{bmatrix} 2 & 0 & 0 \\ 0 & 1 & 0 \\ 0 & 0 & 1 \end{bmatrix}$		$\dfrac{1}{30}\begin{bmatrix} 15 & -5 & 0 \\ -5 & 7 & 0 \\ 0 & 0 & 8 \end{bmatrix}$

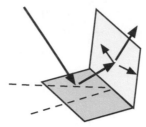

FIGURE 7.8   Oriented surface scattering.

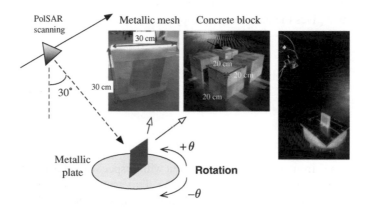

FIGURE 7.9   Polarimetric scattering measurement from oriented surfaces.

The experimental situation is depicted in Figure 7.9, where two typical surface objects are employed. One is a single mesh surface of 30 × 30 cm. The mesh surface is expected to exhibit wide-angle range scattering. The other is two concrete blocks imitating the urban building model.

The center frequency is 15 GHz with 4 GHz bandwidth, resulting in the range resolution of 3.75 cm. The scanning width was chosen as 1.28 m with an incremental interval of 1 cm. A PolSAR measurement was carried out to obtain the scattering matrix of each orientation angle (−40° to 40° with 5° incremental) as shown in Figure 7.9. After polarimetric calibration and synthetic aperture radar (SAR) processing, scattering matrices of each scene have been obtained. Once the quad pol data (scattering matrices) are acquired, they are converted to a coherency matrix form. At first, a four-component scattering power decomposition scheme (Y40 in Chapter 8) was applied to check the scattering mechanisms of each scene from −40° to +40°, as shown in Figure 7.10.

When the surface is orthogonal to radar (0°), the scattering mechanism should be double-bounce. This phenomenon can be clearly seen in Figure 7.10 exhibiting red as the double-bounce. RGB color coding is applied with red for the double-bounce, green for the volume scattering, and blue for the surface scattering. As the orientation angle increases or decreases around 10°, we can see increasing green color. As the surface direction is oriented more than 10° from the azimuth direction, the cross-polarized HV component is generated. This HV component generates the volume scattering (green color). This phenomenon becomes dominant in oriented urban areas. If the surface is oriented more than 30°, the backscattering strength tends to fade. These characteristics are common to both positive and negative orientation angles.

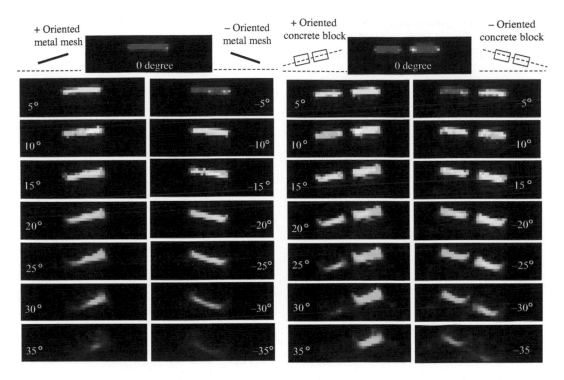

FIGURE 7.10 Four-component scattering power decomposition images of oriented metallic mesh and concrete block. Off-nadir angle is 30°.

The same scattering properties and the angle characteristics can be seen in the concrete block object. Therefore, the oriented surface structure exhibits the same polarization properties regardless of construction materials. We employ equation (7.5.9) as one of the volume scattering models in an oriented urban area.

7.6 Im{T_{23}}: HELIX SCATTERING

Im{T_{23}} term can be written as

$$\text{Im}\left\{T_{23}\right\} = \text{Im}\left\{\left\langle \left(S_{HH} - S_{VV}\right) S_{HV}^{*}\right\rangle\right\} = \text{Im}\left\{\left\langle S_{HH}S_{HV}^{*}\right\rangle + \left\langle S_{HV}S_{VV}^{*}\right\rangle\right\} = \text{Im}\left\{C_{12}\right\} + \text{Im}\left\{C_{23}\right\}. \quad (7.6.1)$$

This means that Im{T_{23}} is the same as the sum of Im{C_{12}} and Im{C_{23}}. This term is rather small compared with other elements as can be seen in Figure 7.3. However, it is a roll-invariant parameter and remains as one of the four important parameters (Chapter 4). This is essentially identical with circular polarization power generated by helix scattering as shown in Figure 7.11.

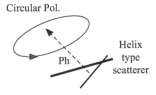

FIGURE 7.11 Helix scattering.

This circular polarization appears in a complex urban scattering scenario and is created by several dipoles aligned in the range direction (Chapter 6 and [3]).

The Im$\{T_{23}\}$ model can be most effectively represented by helix scatterer as

$$\text{For Im}\{T_{23}\} > 0, \quad [T]^r_{helix} = \frac{1}{2}\begin{bmatrix} 0 & 0 & 0 \\ 0 & 1 & j \\ 0 & -j & 1 \end{bmatrix} \quad \text{generated by } [S]^r_{helix} = \frac{1}{2}\begin{bmatrix} 1 & -j \\ -j & -1 \end{bmatrix}$$

$$\text{For Im}\{T_{23}\} < 0, \quad [T]^l_{helix} = \frac{1}{2}\begin{bmatrix} 0 & 0 & 0 \\ 0 & 1 & -j \\ 0 & j & 1 \end{bmatrix} \quad \text{generated by } [S]^l_{helix} = \frac{1}{2}\begin{bmatrix} 1 & j \\ j & -1 \end{bmatrix}$$

$$\text{Resulting in } \langle [T] \rangle_{helix} = \frac{1}{2}\begin{bmatrix} 0 & 0 & 0 \\ 0 & 1 & \pm j \\ 0 & \mp j & 1 \end{bmatrix} \quad (7.6.2)$$

7.7   Re$\{T_{23}\}$: ORIENTED DOUBLE-BOUNCE SCATTERING

The Re$\{T_{23}\}$ element is rather big in an urban area, next to T_{22} and T_{11} components. This fact can be found in the downtown area of San Francisco of Figure 7.3. This bright area of Re$\{T_{23}\}$ is not orthogonal to radar illumination. They are mainly double-bounce structures but are aligned a little bit oblique to the illumination direction. The scattering comes from oriented building walls in urban blocks. This term Re$\{T_{23}\} = \text{Re}\{\langle (S_{HH} - S_{VV}) S^*_{HV} \rangle\}$ indicates the correlation between HV and HH-VV. Therefore, we may call it as oriented double-bounce scattering, or mixed dipole scattering [4].

7.7.1   ALOS2 IMAGE ANALYSIS

Now, we check the value of Re$\{T_{23}\}$ in more detail, which is dependent on the orientation of building blocks. Oriented urban blocks with positive and negative direction with respect to the azimuth direction show the following result in Figure 7.12.

This figure clearly exhibits the dependency on orientation direction of Re$\{T_{23}\}$ with respect to the azimuth or range directions. The sign of Re$\{T_{23}\}$ may become an indicator of the building block direction.

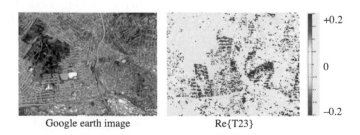

Google earth image   Re{T23}

FIGURE 7.12   Oriented urban blocks with positive and negative direction with respect to azimuth direction in the San Francisco area.

7.7.2 Oriented Surface Measurement in an Anechoic Chamber

In order to confirm this orientation dependency, we measured the scattering property in the anechoic chamber. The experimental situation is the same as Figure 7.9. The $\mathrm{Re}\{T_{23}\}$ images corresponding to Figure 7.10 are created as shown in Figure 7.13. It is seen that the left-hand side images are blue, indicating $\mathrm{Re}\{T_{23}\} < 0$, whereas the right-hand side images are yellow-red with $\mathrm{Re}\{T_{23}\} > 0$. Therefore, the value of $\mathrm{Re}\{T_{23}\}$ is orientation dependent.

From this experimental fact, we may check the surface orientation direction with respect to azimuth direction using the sign of $\mathrm{Re}\{T_{23}\}$. Another object under test is the concrete block, which is a model of oriented urban buildings. The same scattering properties and the angle characteristics can be seen in the concrete block object in Figure 7.13. Therefore, the oriented surface structure exhibits the same polarization properties regardless of construction materials.

The term $\mathrm{Re}\{T_{23}\}$ is eliminated by the rotation operation (minimization of T_{33}, polarization orientation compensation, or deorientation) and has not been accounted for ever before for scattering decomposition. For example, the Y4R or G4U decomposition scheme forces it to zero with redistributing the power into other elements.

Since there is no specific elementary target that directly represents the $\mathrm{Re}\{T_{23}\}$ term, we choose the following double-bounce scattering model (Figure 7.14). The basic characteristics are oblique structure and double-bounce scattering. Since non-orthogonal dipoles with some spacing are used, we may call it the mixed dipole scattering model.

$$[T]_{md} = \frac{1}{2}\begin{bmatrix} 0 & 0 & 0 \\ 0 & 1 & \pm1 \\ 0 & \pm1 & 1 \end{bmatrix} \text{ generated by } \mathbf{k}_p = \frac{1}{\sqrt{2}}\begin{bmatrix} 0 \\ 1 \\ 1 \end{bmatrix} \text{ and } \mathbf{k}_p = \frac{1}{\sqrt{2}}\begin{bmatrix} 0 \\ 1 \\ -1 \end{bmatrix}. \quad (7.7.1)$$

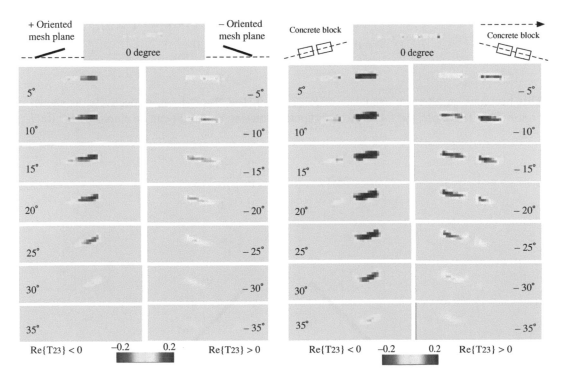

FIGURE 7.13 $\mathrm{Re}\{T_{23}\}$ image of oriented mesh plane and concrete on a metallic plate with off-nadir angle 30°.

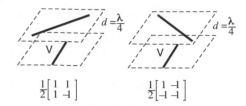

FIGURE 7.14 Example of Re{T_{23}} double-bounce scattering (mixed dipole model).

7.8 Re{T_{13}}: ORIENTED DIPOLE SCATTERING

$$T_{13} = \left\langle \left(S_{HH} + S_{VV} \right) S_{HV}^* \right\rangle = \left\langle S_{HH}S_{HV}^* \right\rangle + \left\langle S_{VV}S_{HV}^* \right\rangle = C_{12} + C_{32} \tag{7.8.1}$$

This term represents the cross-correlation between $\left(S_{HH} + S_{VV} \right)$ and S_{HV}^*. The candidate may be oriented (sloped) surface, which produces the term. However, there is no specific physical model that accounts for this term in the coherency matrix.

Since a ±45°-oriented dipole yields the scattering matrix as

$$\left[S\right]_{dipole}^{45°} = \frac{1}{2}\begin{bmatrix} 1 & 1 \\ 1 & 1 \end{bmatrix}, \quad \left[S\right]_{dipole}^{-45°} = \frac{1}{2}\begin{bmatrix} 1 & -1 \\ -1 & 1 \end{bmatrix},$$

and the corresponding coherency matrix becomes

$$\left[T\right]_{dipole}^{45°} = \frac{1}{2}\begin{bmatrix} 1 & 0 & 1 \\ 0 & 0 & 0 \\ 1 & 0 & 1 \end{bmatrix}, \quad \left[T\right]_{dipole}^{-45°} = \frac{1}{2}\begin{bmatrix} 1 & 0 & -1 \\ 0 & 0 & 0 \\ -1 & 0 & 1 \end{bmatrix}.$$

This becomes an alternate candidate of $\text{Re}\left\{T_{13}\right\} > 0$, and $\text{Re}\left\{T_{13}\right\} < 0$, respectively, as shown in Figure 7.15. Therefore, the ±45°-oriented dipole scattering model [5] can be employed for Re{T_{13}} term.

$$\left[T\right]_{od} = \frac{1}{2}\begin{bmatrix} 1 & 0 & \pm1 \\ 0 & 0 & 0 \\ \pm1 & 0 & 1 \end{bmatrix} \tag{7.8.2}$$

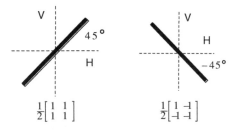

FIGURE 7.15 Example of Re{T_{13}} scattering.

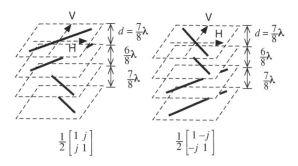

FIGURE 7.16 Example of Im{T_{13}} modeling.

7.9 Im{T_{13}}: COMPOUND DIPOLE SCATTERING

The imaginary part of the scattering matrix or coherency matrix essentially comes from a phase difference caused by multiple scattering centers. In this regard, there is no specific example to represent Im{T_{13}}.

However, if we turn to the compound scattering matrix $\left[S\right]_1^{com} = \dfrac{1}{2}\begin{bmatrix} 1 & -j \\ -j & 1 \end{bmatrix}$, the coherency matrix becomes $\left[T\right]_2^{com} = \dfrac{1}{2}\begin{bmatrix} 1 & 0 & j \\ 0 & 0 & 0 \\ -j & 0 & 1 \end{bmatrix}$, which is convenient for Im$\left\{T_{13}\right\} > 0$. Similarly, the compound matrix $2\left[S\right]_c^{com} = \dfrac{1}{2}\begin{bmatrix} 1 & j \\ j & 1 \end{bmatrix}$ yields $\left[T\right]_2^{com} = \dfrac{1}{2}\begin{bmatrix} 1 & 0 & -j \\ 0 & 0 & 0 \\ j & 0 & 1 \end{bmatrix}$ for Im$\left\{T_{13}\right\} < 0$ [6].

These compound scatterings can be good candidates for Im{T_{13}} modeling (Figure 7.16):

$$\left[T\right]_{cd} = \frac{1}{2}\begin{bmatrix} 0 & 0 & \pm j \\ 0 & 1 & 0 \\ \mp j & 0 & 1 \end{bmatrix} \tag{7.9.1}$$

7.10 Re{T_{12}}: SCATTERING

$$T_{12} = \frac{\left\langle\left(S_{HH} + S_{VV}\right)\left(S_{HH} - S_{VV}\right)^{*}\right\rangle}{2} = \frac{1}{2}\left(\left|S_{HH}\right|^2 - \left|S_{VV}\right|^2\right) + j\,\mathrm{Im}\left\{S_{HH}^{*}S_{VV}\right\} \tag{7.10.1}$$

From this equation, we can recognize that Re{T_{12}} means the power difference of the *HH* and *VV* channel power, and Im{T_{12}} corresponds to the phase difference between *HH* and *VV*. We know the power difference can be found in vegetation modelings. Since this term is not specific, and accounted for in T_{11} surface scattering or T_{22} double-bounce scattering, and also in the vegetation volume scattering, it is difficult to define a good candidate for modeling.

Among various candidates, we can turn to the following case as an example, which can be composed of multiple dipoles with spacing (Figure 7.17).

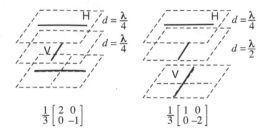

FIGURE 7.17 Example of Re{T_{12}} modeling.

For Re$\{T_{12}\} > 0$, $[S] = \dfrac{1}{3}\begin{bmatrix} 2 & 0 \\ 0 & -1 \end{bmatrix}$, $[T] = \dfrac{1}{10}\begin{bmatrix} 1 & 3 & 0 \\ 3 & 9 & 0 \\ 0 & 0 & 0 \end{bmatrix}$

For Re$\{T_{12}\} < 0$, $[S] = \dfrac{1}{3}\begin{bmatrix} 1 & 0 \\ 0 & -2 \end{bmatrix}$, $[T] = \dfrac{1}{10}\begin{bmatrix} 1 & -3 & 0 \\ -3 & 9 & 0 \\ 0 & 0 & 0 \end{bmatrix}$

7.11 IM{T_{12}}: SCATTERING

Since $\mathrm{Im}\{T_{12}\} = \mathrm{Im}\{S_{HH}^{*}S_{VV}\}$, it is rather difficult to specify targets because there are so many cases. Among various candidates, we can turn to the following case as a very simple example, which can be composed of two dipoles with spacing. The important item is the phase difference between *HH* and *VV*, which can be realized by dipole spacings (Figure 7.18).

For Im$\{T_{12}\} > 0$, $[S] = \begin{bmatrix} 1 & 0 \\ 0 & j \end{bmatrix}$, $[T] = \dfrac{1}{2}\begin{bmatrix} 1 & j & 0 \\ -j & 1 & 0 \\ 0 & 0 & 0 \end{bmatrix}$

For Im$\{T_{12}\} < 0$, $[S] = \dfrac{1}{2}\begin{bmatrix} 1 & 0 \\ 0 & -j \end{bmatrix}$, $[T] = \dfrac{1}{2}\begin{bmatrix} 1 & -j & 0 \\ j & 1 & 0 \\ 0 & 0 & 0 \end{bmatrix}$

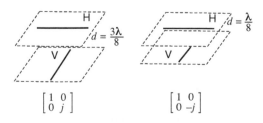

FIGURE 7.18 Example of Im{T_{12}} modeling.

7.12 SCATTERING MODELS FOR NINE PARAMETERS OF COHERENCY MATRIX

Scattering models for nine parameters can be summarized in Table 7.2. The original scattering matrix and the corresponding object are shown in Table 7.2. These scattering models are utilized in the scattering power decompositions in the following chapter.

Note that each object is one example of realizing the scattering matrix. There will be other examples to realize the same scattering matrix. For example, if the spacing between dipoles is a multiple of a half-wavelength, the same scattering matrix, and hence the same coherency matrix, can be obtained. The $\text{Re}\{T_{23}\}$ component can be obtained by an oriented surface and mixed dipoles as shown in Table 7.2.

TABLE 7.2
Scattering Models for Nine Element of Coherency Matrix

Element of [T]	Scattering Model	Original Scattering Matrix	Assumed Object	Remarks
T_{11}	$\begin{bmatrix} 1 & \beta^* & 0 \\ \beta & |\beta|^2 & 0 \\ 0 & 0 & 0 \end{bmatrix}$	$\begin{bmatrix} 1 & 0 \\ 0 & 1 \end{bmatrix}$		Surface scattering
T_{22}	$\begin{bmatrix} |\alpha|^2 & \alpha & 0 \\ \alpha^* & 1 & 0 \\ 0 & 0 & 0 \end{bmatrix}$	$\begin{bmatrix} 1 & 0 \\ 0 & -1 \end{bmatrix}$		Double-bounce scattering
T_{33}	$\frac{1}{4}\begin{bmatrix} 2 & 0 & 0 \\ 0 & 1 & 0 \\ 0 & 0 & 1 \end{bmatrix}, \frac{1}{30}\begin{bmatrix} 15 & \pm 5 & 0 \\ \pm 5 & 7 & 0 \\ 0 & 0 & 8 \end{bmatrix},$ $\frac{1}{15}\begin{bmatrix} 0 & 0 & 0 \\ 0 & 7 & 0 \\ 0 & 0 & 8 \end{bmatrix}$	$\begin{bmatrix} 0 & 1 \\ 1 & 0 \end{bmatrix}$		Volume scattering caused by HV
$\text{Im}\{T_{23}\} > 0$	$\frac{1}{2}\begin{bmatrix} 0 & 0 & 0 \\ 0 & 1 & j \\ 0 & -j & 1 \end{bmatrix}$	$\frac{1}{2}\begin{bmatrix} 1 & -j \\ -j & -1 \end{bmatrix}$	$d = \frac{\lambda}{8}$, $\frac{\lambda}{8}$, $\frac{\lambda}{8}$, $\frac{\lambda}{8}$	Right helix
$\text{Im}\{T_{23}\} < 0$	$\frac{1}{2}\begin{bmatrix} 0 & 0 & 0 \\ 0 & 1 & -j \\ 0 & j & 1 \end{bmatrix}$	$\frac{1}{2}\begin{bmatrix} 1 & j \\ j & -1 \end{bmatrix}$	$d = \frac{\lambda}{8}$, $\frac{\lambda}{8}$, $\frac{\lambda}{8}$, $\frac{\lambda}{8}$	Left helix

(Continued)

TABLE 7.2 (*Continued*)
Scattering Models for Nine Element of Coherency Matrix

Element of [*T*]	Scattering Model	Original Scattering Matrix	Assumed Object	Remarks
$\text{Re}\{T_{12}\} > 0$	$\dfrac{1}{10}\begin{bmatrix} 1 & 3 & 0 \\ 3 & 9 & 0 \\ 0 & 0 & 0 \end{bmatrix}$	$\dfrac{1}{3}\begin{bmatrix} 2 & 0 \\ 0 & -1 \end{bmatrix}$	$d=\dfrac{\lambda}{4}$, $d=\dfrac{\lambda}{4}$	$|HH| > |VV|$
$\text{Re}\{T_{12}\} < 0$	$\dfrac{1}{10}\begin{bmatrix} 1 & -3 & 0 \\ -3 & 9 & 0 \\ 0 & 0 & 0 \end{bmatrix}$	$\dfrac{1}{3}\begin{bmatrix} 1 & 0 \\ 0 & -2 \end{bmatrix}$	$d=\dfrac{\lambda}{4}$, $d=\dfrac{\lambda}{2}$	$|HH| < |VV|$
$\text{Im}\{T_{12}\} > 0$	$\dfrac{1}{2}\begin{bmatrix} 1 & j & 0 \\ -j & 1 & 0 \\ 0 & 0 & 0 \end{bmatrix}$	$\begin{bmatrix} 1 & 0 \\ 0 & j \end{bmatrix}$	$d=\dfrac{3\lambda}{8}$	+ Phase difference between *H*- and *V*-object
$\text{Im}\{T_{12}\} < 0$	$\dfrac{1}{2}\begin{bmatrix} 1 & -j & 1 \\ j & 1 & 0 \\ 0 & 0 & 0 \end{bmatrix}$	$\begin{bmatrix} 1 & 0 \\ 0 & -j \end{bmatrix}$	$d=\dfrac{\lambda}{8}$	− Phase difference between *H*- and *V*-object
$\text{Re}\{T_{13}\} > 0$	$\dfrac{1}{2}\begin{bmatrix} 1 & 0 & 1 \\ 0 & 0 & 0 \\ 1 & 0 & 1 \end{bmatrix}$	$\dfrac{1}{2}\begin{bmatrix} 1 & 1 \\ 1 & 1 \end{bmatrix}$	45°	+45°-oriented dipole
$\text{Re}\{T_{13}\} < 0$	$\dfrac{1}{2}\begin{bmatrix} 1 & 0 & -1 \\ 0 & 0 & 0 \\ -1 & 0 & 1 \end{bmatrix}$	$\dfrac{1}{2}\begin{bmatrix} 1 & -1 \\ -1 & 1 \end{bmatrix}$	- 45°	−45°-oriented dipole
$\text{Im}\{T_{13}\} > 0$	$\dfrac{1}{2}\begin{bmatrix} 1 & 0 & j \\ 0 & 0 & 0 \\ -j & 0 & 1 \end{bmatrix}$	$\dfrac{1}{2}\begin{bmatrix} 1 & j \\ j & 1 \end{bmatrix}$	$d=\dfrac{7}{8}\lambda$, $\dfrac{6}{8}\lambda$, $\dfrac{7}{8}\lambda$	+ Compound dipole

(*Continued*)

TABLE 7.2 (Continued)

Scattering Models for Nine Element of Coherency Matrix

Element of [T]	Scattering Model	Original Scattering Matrix	Assumed Object	Remarks
$\text{Im}\{T_{13}\} < 0$	$\dfrac{1}{2}\begin{bmatrix} 1 & 0 & -j \\ 0 & 0 & 0 \\ j & 0 & 1 \end{bmatrix}$	$\dfrac{1}{2}\begin{bmatrix} 1 & -j \\ -j & 1 \end{bmatrix}$	$d = \frac{7}{8}\lambda$, $\frac{6}{8}\lambda$, $\frac{7}{8}\lambda$	– Compound dipole
$\text{Re}\{T_{23}\} > 0$	$\dfrac{1}{2}\begin{bmatrix} 0 & 0 & 0 \\ 0 & 1 & 1 \\ 0 & 1 & 1 \end{bmatrix}$	$\dfrac{1}{2}\begin{bmatrix} 1 & 1 \\ 1 & -1 \end{bmatrix}$	$d = \frac{\lambda}{4}$	Non-orthogonal dipoles 1 or oriented surface
$\text{Re}\{T_{23}\} > 0$	$\dfrac{1}{2}\begin{bmatrix} 0 & 0 & 0 \\ 0 & 1 & -1 \\ 0 & -1 & 1 \end{bmatrix}$	$\dfrac{1}{2}\begin{bmatrix} 1 & -1 \\ -1 & -1 \end{bmatrix}$	$d = \frac{\lambda}{4}$	Non-orthogonal dipoles 2 or oriented surface

7.13 SUMMARY

This chapter is focused on modeling of scattering mechanisms, that is, the realization of corresponding coherency matrices. Since there are nine independent parameters in the polarization matrix, nine real elements in coherency matrix are chosen to correspond physically to feasible scattering matrices, including positive and negative signs. One-to-one correspondence is established as shown in Table 7.2. The correspondence is not unique because of the periodicity property of the phase function. These modelings would serve an immediate interpretation and retrieval of objects in the PolSAR image.

REFERENCES

1. W. T. An, Y. Cui, and J. Yang, "Three-component model-based decomposition for coherency matrix," *IEEE Trans. Geosci. Remote Sens.*, vol. 48, pp. 2732–2739, 2010.
2. A. Sato, Y. Yamaguchi, G. Singh, and S.-E. Park, "Four-component scattering power decomposition with extended volume scattering model," *IEEE Geosci. Remote Sens. Lett.Ł*, vol. 9, no. 2, pp. 166–170, 2012.
3. Y. Yamaguchi, M. Ishido, T. Moriyama, and H. Yamada, "Four-component scattering model for polarimetric SAR image decomposition," *IEEE Trans. Geosci. Remote Sens.*, vol. 43, no. 8, pp. 1699–1706, 2005.
4. G. Singh, R. Malik, S. Mohanty, V. S. Rathore, K. Yamada, M. Umemura, and Y. Yamaguchi, "Seven-component scattering power decomposition of POLSAR coherency matrix," *IEEE Trans. Geosci. Remote Sens.*, vol. 57, no. 11, pp. 8371–8372, 2019. doi:10.1109/TGRS.2019.2920762.
5. G. Singh and Y. Yamaguchi, "Model-based six-component scattering matrix power decomposition," *IEEE Trans. Geosci. Remote Sens.*, vol. 56, no. 10, pp. 5687–5704, 2018.
6. G. Singh, Y. Yamaguchi, and Y. Yamazaki, "Physical scattering interpretation of POLSAR coherency matrix by using compound scattering phenomenon," *IEEE Trans. Geosci. Remote Sens.*, vol. 58, no. 4, pp. 2541–2556, 2020.

8 Scattering Power Decomposition

This chapter explains the model-based scattering power decomposition method in detail. Scattering power decompositions have been a hot topic in radar polarimetry for more than two decades. Some references [1–24] are listed at the end section of this chapter. There exist nine real independent polarimetric parameters in the 3×3 coherency or covariance matrices. A physical model-based scattering power decomposition tries to account for these polarimetric parameters as much as possible in the decomposition. Based on the scattering models, a measured coherency matrix by ensemble average in an imaging window is expanded by submatrices (scattering models). Finally, the scattering powers are determined by the expansion coefficients.

Once the scattering powers are obtained, it is possible to create a color-coded image. By assigning each power to the RGB color code, we can see the image as shown in Figure 8.1. The resultant image is clear, vivid, and colorful, and seems more than the optical image of Figure 7.1. Since color corresponds to physical scattering power, it is easy to understand and to interpret the scattering mechanism directly.

FIGURE 8.1 Color-coded and six-component scattering power decomposition image of San Francisco. Sequential filter [9] is applied to ALOS2 quad pol data ALOS2229210750-180821.

The original three-component decomposition was developed by Freeman and Durden [10] under the reflection symmetry condition that the cross-correlation between the co- and cross-polarized scattering elements are close to zero for natural distributed objects. Since then, a lot of works on scattering decomposition methods have been carried out by many researchers [12–22] to extend the applicability to man-made areas in the non-reflection symmetry condition, to overcome overestimation of the volume scattering, and to avoid negative power occurrence in the calculation. Brief summaries are given in books [3,6,8]. Our developments on the model-based decomposition, including an addition of the helix scattering (Y4O) [11], rotation of coherency matrix to minimize the T_{33} component (Y4R) [16], discrimination of the HV component by oriented dihedral from vegetation scattering (S4R) [19], unitary transformation (G4U) [20], six-component decomposition with rotation [23], and seven-component without rotation [24] are summarized in Figure 8.2. In this chapter, the advancements of the decomposition scheme are described.

8.1 PREPARATION

Model-based scattering power decomposition needs scattering models that represent physical scattering mechanisms. Since we have already prepared these scattering models in Chapter 6, we use them for the decompositions. The basic idea is to decompose the measured total power (TP) from equation (8.1.2) into a sum of model-based scattering powers as in equation (8.1.3). Since power is a fundamental radar parameter and is rather stable with respect to noise compared with phase information, we can expect stable and reliable results.

$$\textbf{Measured coherency matrix}: \quad \langle [T] \rangle = \begin{bmatrix} T_{11} & T_{12} & T_{13} \\ T_{21} & T_{22} & T_{23} \\ T_{31} & T_{32} & T_{33} \end{bmatrix} \tag{8.1.1}$$

$$\textbf{Total power}: \quad TP = T_{11} + T_{22} + T_{33} \tag{8.1.2}$$

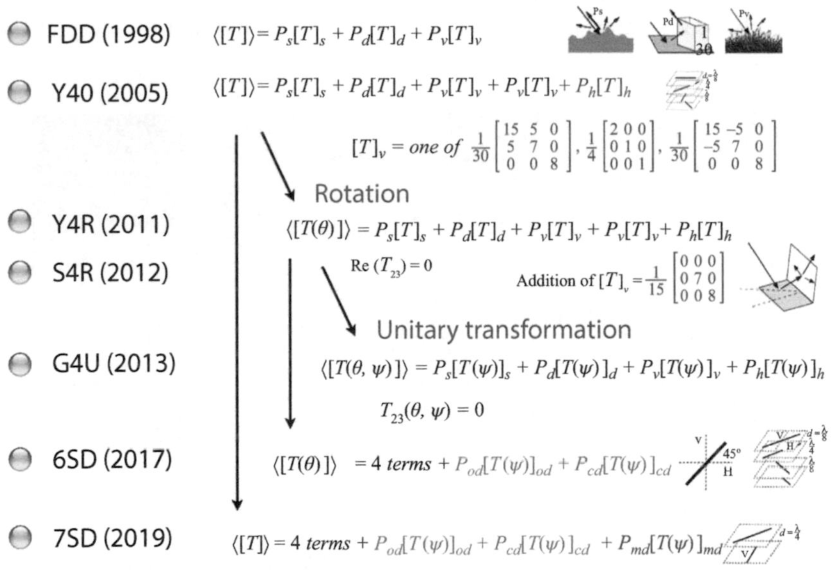

FIGURE 8.2 Development of model-based scattering power decomposition scheme.

$$\textbf{Power decomposition}: \quad TP = P_s + P_d + P_v + P_{md} + P_h + P_{od} + P_{cd} \tag{8.1.3}$$

$$\textbf{Model expansion example}: \langle [T] \rangle = P_s [T]_s + P_d [T]_d + P_v [T]_v + P_{md} [T]_{md}$$
$$+ P_h [T]_h + P_{od} [T]_{od} + P_{cd} [T]_{cd} \tag{8.1.4}$$

where P_s is the surface-scattering power; P_d is the double-bounce scattering power; P_v is the volume-scattering power; P_{md} is the mixed dipole power by $\mathrm{Re}\{T_{23}\}$; P_h is the helix scattering power by $\mathrm{Im}\{T_{23}\}$; P_{od} is the oriented dipole power by $\mathrm{Re}\{T_{13}\}$; and P_{cd} is the compound dipole power by $\mathrm{Im}\{T_{13}\}$.

For model-based decomposition, the coherency matrix is expanded by submatrices in equation (8.1.4). The coherency matrix in equation (8.1.1) is produced by an ensemble average in an imaging window ($M \times N$ pixels) of measurement data. Submatrices represent scattering models and are normalized so that the corresponding coefficient P represents the scattering power itself. According to the data matrix expansion, scattering powers are decided directly.

8.2 FDD 3-COMPONENT DECOMPOSITION

The original decomposition method was developed by Freeman and Durden [10]. Although they have used a covariance matrix approach, the decomposition procedure is the same as the following scheme. The expansion of the first three terms in equation (8.1.4) results in the coherency matrix formulation as

$$
\begin{bmatrix}
T_{11} & T_{12} & T_{13} \\
T_{21} & T_{22} & T_{23} \\
T_{31} & T_{32} & T_{33}
\end{bmatrix}
\Rightarrow
\begin{bmatrix}
T_{11} & T_{12} & 0 \\
T_{21} & T_{22} & 0 \\
0 & 0 & T_{33}
\end{bmatrix}
$$

under the reflection symmetry condition: $\langle S_{HH} S_{HV}^* \rangle \approx 0, \langle S_{VV} S_{HV}^* \rangle \approx 0$.

The first three terms of the expansion become

$$
\begin{bmatrix}
T_{11} & T_{12} & 0 \\
T_{21} & T_{22} & 0 \\
0 & 0 & T_{33}
\end{bmatrix}
= \frac{P_s}{1+|\beta|^2}
\begin{bmatrix}
1 & \beta^* & 0 \\
\beta & |\beta|^2 & 0 \\
0 & 0 & 0
\end{bmatrix}
+ \frac{P_d}{1+|\alpha|^2}
\begin{bmatrix}
|\alpha|^2 & \alpha & 0 \\
\alpha^* & 1 & 0 \\
0 & 0 & 0
\end{bmatrix}
+ \frac{P_v}{4}
\begin{bmatrix}
2 & 0 & 0 \\
0 & 1 & 0 \\
0 & 0 & 1
\end{bmatrix}
\quad (8.2.1)
$$

This yields the following relations:

$$
T_{11} = \frac{P_s}{1+|\beta|^2} + \frac{P_d |\alpha|^2}{1+|\alpha|^2} + \frac{P_v}{2}, \quad
T_{12} = \frac{P_s \beta^*}{1+|\beta|^2} + \frac{P_d \alpha}{1+|\alpha|^2},
$$

$$
T_{22} = \frac{P_s |\beta|^2}{1+|\beta|^2} + \frac{P_d}{1+|\alpha|^2} + \frac{P_v}{4}, \quad
T_{33} = \frac{P_v}{4}.
\quad (8.2.2)
$$

Since $P_v = 4T_{33}$, we have three equations with four unknowns $(P_s, P_d, \alpha, \beta)$.

$$
\begin{cases}
\dfrac{P_s}{1+|\beta|^2} + \dfrac{P_d |\alpha|^2}{1+|\alpha|^2} = S \\[3mm]
\dfrac{P_s |\beta|^2}{1+|\beta|^2} + \dfrac{P_d}{1+|\alpha|^2} = D, \\[3mm]
\dfrac{P_s \beta^*}{1+|\beta|^2} + \dfrac{P_d \alpha}{1+|\alpha|^2} = C
\end{cases}
\qquad
\begin{cases}
S = T_{11} - \dfrac{P_v}{2} = T_{11} - 2T_{33} \\[3mm]
D = T_{22} - \dfrac{P_v}{4} = T_{22} - T_{33} \\[3mm]
C = T_{12}
\end{cases}
\quad (8.2.3)
$$

The three equations can be solved by an approximation $(\alpha = 0 \; or \; \beta = 0)$ using the branch condition C_0.

- Branch condition C_0 (also see Appendix)

The criterion is given as: $C_0 = 2T_{11} - TP$ (8.2.4)

If $C_0 > 0$, the surface scattering is dominant. The details of a minor double-bounce can be neglected. Hence, we put $\alpha = 0$ in the three equations.

$$\alpha = 0 \Rightarrow \beta^* = \frac{C}{S} \Rightarrow P_s = S + \frac{|C|^2}{S}, \quad P_d = D - \frac{|C|^2}{S} \tag{8.2.5}$$

If $C_0 < 0$, the double-bounce scattering is dominant. The details of surface scattering can be neglected. We put $\beta = 0$ in the three equations.

$$\beta = 0 \Rightarrow \alpha = \frac{C}{D} \Rightarrow P_s = S - \frac{|C|^2}{S}, \quad P_d = D + \frac{|C|^2}{S} \tag{8.2.6}$$

We call this criterion the branch condition C_0 hereafter, which discriminates the dominance of P_s or P_d. As seen in equations (8.2.3–8.2.6), three scattering powers, P_s, P_d, and P_v can be calculated directly from the measurement data. Figure 8.3 shows a flow chart of this decomposition algorithm.

Once these powers are obtained, RGB color-coding is applied to each power. The typical and commonly used assignment is: red for the double-bounce power P_d, green for the volume scattering

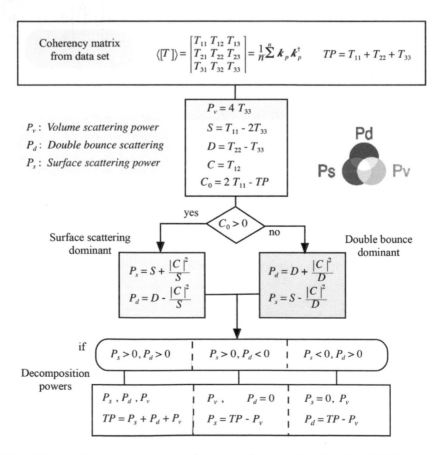

FIGURE 8.3 Original three-component scattering power decomposition algorithm (FDD).

FIGURE 8.4   Freeman and Durden three-component decomposition image of San Francisco.

power P_v, and blue for the surface-scattering power P_s (see Section 8.9). One example of the three-component scattering power decomposition of San Francisco is displayed in Figure 8.4.

Due to the advantages of easy implementation, simple and fast calculation, easy interpretation of the physical scattering mechanism by color, and beautiful color image, this method has attracted a lot of attention for further developments. Five polarization parameters out of nine are used in this decomposition.

When calculating P_s or P_d, we often encounter a negative power occurrence, which is against the physical phenomenon. This is essentially caused by overestimation of $P_v = 4T_{33}$. The right-hand side of equation (8.2.3) sometimes become negative, which causes negative powers in P_s or P_d. Ad hoc constraint that all power should be more than 0 is applied to the decomposition algorithm in Figure 8.3. Another point to notice is its applicability to the non-reflection symmetry area. It cannot be applied to complex urban areas or man-made structure areas by the definition.

8.3   Y40 4-COMPONENT DECOMPOSITION

In order to expand the applicability of the three-component decomposition, the helix scattering is introduced in [11]. Helix is one of the key polarization parameters (i.e., $\mathrm{Im}\{c^*(a-b)\}$ in Chapter 5), and produces circular polarization power. This method (Y40) uses four key parameters in the decomposition. The helix scattering is incorporated in urban area scattering of non-reflection symmetry conditions $\langle S_{HH}S_{HV}^*\rangle \neq 0, \langle S_{VV}S_{HV}^*\rangle \neq 0$. This power also mitigates overestimation of P_v. Y40 also introduced three volume scattering models, taking into account the HH and VV power imbalance. According to the VV/HH power ratio, we choose the most appropriate scattering model for volume scattering. These two items served as improvement in the decomposition results.

- For randomly oriented dipole scattering with uniform distribution (volume scattering with $|\sigma_{HH} - \sigma_{vv}| < 2$ dB), the expansion becomes,

$$
\begin{bmatrix} T_{11} & T_{12} & T_{13} \\ T_{21} & T_{22} & T_{23} \\ T_{31} & T_{32} & T_{33} \end{bmatrix} = \frac{P_s}{1+|\beta|^2}\begin{bmatrix} 1 & \beta^* & 0 \\ \beta & |\beta|^2 & 0 \\ 0 & 0 & 0 \end{bmatrix} + \frac{P_d}{1+|\alpha|^2}\begin{bmatrix} |\alpha|^2 & \alpha & 0 \\ \alpha^* & 1 & 0 \\ 0 & 0 & 0 \end{bmatrix}
$$

$$
+ \frac{P_v}{4}\begin{bmatrix} 2 & 0 & 0 \\ 0 & 1 & 0 \\ 0 & 0 & 1 \end{bmatrix} + \frac{P_h}{2}\begin{bmatrix} 0 & 0 & 0 \\ 0 & 1 & \pm j \\ 0 & \mp j & 1 \end{bmatrix}
$$

(8.3.1)

From this expansion, we can obtain: $P_h = 2\left|\text{Im}\left\{T_{23}\right\}\right|$ (8.3.2)

$$\text{and } P_v = 2\left[2T_{33} - P_h\right]$$ (8.3.3)

The helix scattering power P_h can be retrieved directly from the measured coherency matrix. Since it accounts partially for non-reflection symmetry condition, the applicability of the decomposition is expanded. Although P_h is very small compared to other scattering powers, it also contributes to mitigate overestimation of P_v as shown in equation (8.3.3) and improves the final color image.

Then we have three equations with four unknowns $(P_s, P_d, \alpha, \beta)$ similar to equation (8.2.3).

$$\begin{cases} \dfrac{P_s}{1+|\beta|^2} + \dfrac{P_d |\alpha|^2}{1+|\alpha|^2} = S \\[3mm] \dfrac{P_s |\beta|^2}{1+|\beta|^2} + \dfrac{P_d}{1+|\alpha|^2} = D, \\[3mm] \dfrac{P_s \beta^*}{1+|\beta|^2} + \dfrac{P_d \alpha}{1+|\alpha|^2} = C \end{cases} \qquad \begin{cases} S = T_{11} - \dfrac{P_v}{2} \\[3mm] D = T_{22} - \dfrac{P_v}{4} - \dfrac{P_h}{2} \\[3mm] C = T_{12} \end{cases}$$ (8.3.4)

According to the magnitude imbalance of the HH and VV power, we choose an appropriate volume scattering model for the expansion.

- For $\sigma_{HH} - \sigma_{VV} > 2$ dB, we use $\dfrac{P_v}{30}\begin{bmatrix} 15 & 5 & 0 \\ 5 & 7 & 0 \\ 0 & 0 & 8 \end{bmatrix}$ as the volume scattering model.

$$\begin{bmatrix} T_{11} & T_{12} & T_{13} \\ T_{21} & T_{22} & T_{23} \\ T_{31} & T_{32} & T_{33} \end{bmatrix} = \frac{P_s}{1+|\beta|^2}\begin{bmatrix} 1 & \beta^* & 0 \\ \beta & |\beta|^2 & 0 \\ 0 & 0 & 0 \end{bmatrix} + \frac{P_d}{1+|\alpha|^2}\begin{bmatrix} |\alpha|^2 & \alpha & 0 \\ \alpha^* & 1 & 0 \\ 0 & 0 & 0 \end{bmatrix}$$

$$+ \frac{P_v}{30}\begin{bmatrix} 15 & 5 & 0 \\ 5 & 7 & 0 \\ 0 & 0 & 8 \end{bmatrix} + \frac{P_h}{2}\begin{bmatrix} 0 & 0 & 0 \\ 0 & 1 & \pm j \\ 0 & \mp j & 1 \end{bmatrix}$$ (8.3.5)

- For $\sigma_{VV} - \sigma_{HH} > 2$ dB. we employ $\dfrac{P_v}{30}\begin{bmatrix} 15 & -5 & 0 \\ -5 & 7 & 0 \\ 0 & 0 & 8 \end{bmatrix}$ as the volume scattering model.

$$\begin{bmatrix} T_{11} & T_{12} & T_{13} \\ T_{21} & T_{22} & T_{23} \\ T_{31} & T_{32} & T_{33} \end{bmatrix} = \frac{P_s}{1+|\beta|^2}\begin{bmatrix} 1 & \beta^* & 0 \\ \beta & |\beta|^2 & 0 \\ 0 & 0 & 0 \end{bmatrix} + \frac{P_d}{1+|\alpha|^2}\begin{bmatrix} |\alpha|^2 & \alpha & 0 \\ \alpha^* & 1 & 0 \\ 0 & 0 & 0 \end{bmatrix}$$

$$+ \frac{P_v}{30}\begin{bmatrix} 15 & -5 & 0 \\ -5 & 7 & 0 \\ 0 & 0 & 8 \end{bmatrix} + \frac{P_h}{2}\begin{bmatrix} 0 & 0 & 0 \\ 0 & 1 & \pm j \\ 0 & \mp j & 1 \end{bmatrix}$$ (8.3.6)

TABLE 8.1

Helix Power, Volume-Scattering Power, and Three Equations for Y40 Decomposition Scheme

	$\sigma_{HH} - \sigma_{VV} > 2$ dB	$|\sigma_{HH} - \sigma_{VV}| < 2$ dB	$\sigma_{VV} - \sigma_{HH} > 2$ dB
Helix power	$P_h = 2\|\mathrm{Im}\{T_{23}\}\|$	$P_h = 2\|\mathrm{Im}\{T_{23}\}\|$	$P_h = 2\|\mathrm{Im}\{T_{23}\}\|$
Volume-scattering power	$P_v = \dfrac{15}{8}[2T_{33} - P_h]$	$P_v = 2[2T_{33} - P_h]$	$P_v = \dfrac{15}{8}[2T_{33} - P_h]$

$$\begin{cases} \dfrac{P_s}{1+|\beta|^2} + \dfrac{P_d|\alpha|^2}{1+|\alpha|^2} = S \\[2mm] \dfrac{P_s|\beta|^2}{1+|\beta|^2} + \dfrac{P_d}{1+|\alpha|^2} = D \\[2mm] \dfrac{P_s\beta^*}{1+|\beta|^2} + \dfrac{P_d\,\alpha}{1+|\alpha|^2} = C \end{cases}$$

$$\begin{cases} S = T_{11} - \dfrac{P_v}{2} \\[2mm] D = T_{22} - \dfrac{7}{30}P_v - \dfrac{1}{2}P_h \\[2mm] C = T_{12} - \dfrac{1}{6}P_v \end{cases}$$

$$\begin{cases} S = T_{11} - \dfrac{P_v}{2} \\[2mm] D = T_{22} - \dfrac{P_v}{4} - \dfrac{P_h}{2} \\[2mm] C = T_{12} \end{cases}$$

$$\begin{cases} S = T_{11} - \dfrac{P_v}{2} \\[2mm] D = T_{22} - \dfrac{7}{30}P_v - \dfrac{1}{2}P_h \\[2mm] C = T_{12} + \dfrac{1}{6}P_v \end{cases}$$

The preceding expansions bring similar three equations with unknowns as in Table 8.1. The remaining powers P_s and P_d are given by the aid of the branch condition C_0 (8.2.4–8.2.6). In this case, C_0 becomes,

$$C_0 = 2T_{11} - TP + P_h \tag{8.3.7}$$

The decomposition algorithm of Y40 is displayed in Figure 8.5. In this algorithm, the helix power is extracted first, then according to the *HH* and *VV* power ratio, the volume scattering model is chosen. Finally, the surface or the double-bounce scattering is determined by the branch condition C_0, and the corresponding powers are determined accordingly. Six out of nine parameters are accounted for in this decomposition.

The same data set is used in Figure 8.6 for comparison. Compared to the Freeman Durden Decomposition (FDD) image of Figure 8.4, the yellow color increased a little bit in urban areas near national parks. However, the 40°-oriented urban area (triangular green area in the lower right position) is too green. This is caused by an overestimation of the volume-scattering power.

8.4 Y4R AND S4R

As seen in Figures 8.4 and 8.6, there are bright green-colored areas connected to a bridge in the right lower corner of the image, caused by the volume scattering. This is a highly oriented urban building block area with respect to radar illumination. If the *HV* component is generated, it is automatically assigned to the T_{33} component. If the magnitude ratio of *HV* is rather big in the scattering matrix, the corresponding area tends to show a dominant volume scattering area. Since the area is occupied by non-orthogonal houses to radar illumination and man-made structures, we try to reduce the *HV* component or the corresponding T_{33} component for mitigating the overestimation problem.

8.4.1 MINIMIZATION OF THE *HV* COMPONENT

One compensation idea is to minimize T_{33} mathematically. The minimization of the *HV* component is a fundamental concept in radar polarimetry. This can be carried out easily by rotation operation

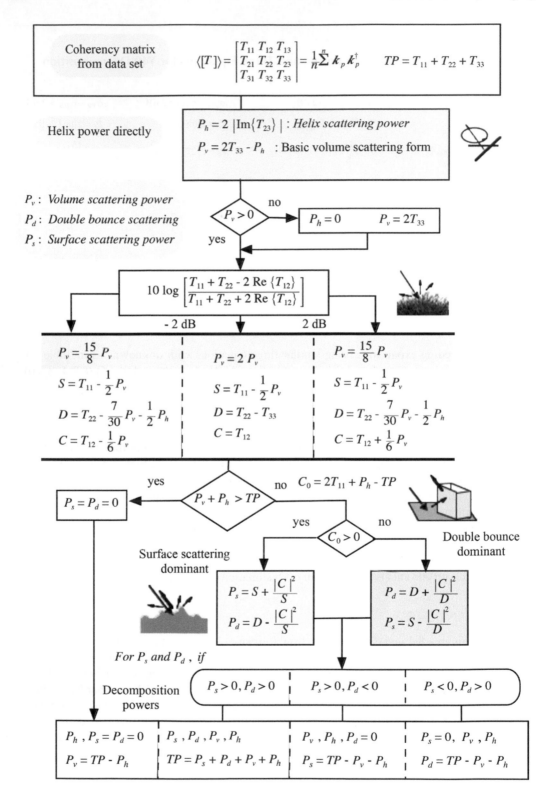

FIGURE 8.5 Four-component scattering power decomposition algorithm (Y40).

FIGURE 8.6   Decomposition image of San Francisco by Y40.

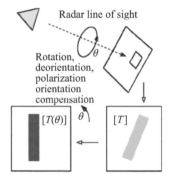

FIGURE 8.7   Rotation around the radar line of sight.

in the coherency matrix. Coherency matrix rotation around the radar line of sight (Figure 8.7) can be expressed as

$$\left[T(\theta)\right]=\left[R(\theta)\right]\left[T\right]\left[R(\theta)\right]^{t} \tag{8.4.1}$$

$$\text{with}\left[R(\theta)\right]=\begin{bmatrix}1 & 0 & 0 \\ 0 & \cos2\theta & \sin2\theta \\ 0 & -\sin2\theta & \cos2\theta\end{bmatrix} \tag{8.4.2}$$

This yields

$$T_{33}(\theta)=T_{33}\cos^{2}2\theta+T_{22}\sin^{2}2\theta-\text{Re}\{T_{23}\}\sin4\theta$$

$$T_{23}(\theta) = \text{Re}\{T_{23}\}\cos 4\theta - \frac{T_{22} - T_{33}}{2}\sin 4\theta + j\,\text{Im}\{T_{23}\}$$

We search the minimum by $\dfrac{dT_{33}(\theta)}{d\theta} = 0$

This leads to
$$\tan 4\theta = \frac{2\text{Re}\{T_{23}\}}{T_{22} - T_{33}} \tag{8.4.3}$$

Therefore the rotation angle can be obtained as: $2\theta = \dfrac{1}{2}\tan^{-1}\left(\dfrac{2\text{Re}\{T_{23}\}}{T_{22} - T_{33}}\right)$ $\tag{8.4.4}$

In addition, $T_{23}(\theta)$ becomes pure imaginary, $T_{23}(\theta) = j\,\text{Im}\{T_{23}\}$ $\tag{8.4.5}$

which is the best fit to the helix scattering modeling. This also means that the rotation makes

$$\text{Re}\{T_{23}(\theta)\} = 0. \tag{8.4.6}$$

Coherency matrix elements after this rotation become:

$$T_{11}(\theta) = T_{11}, \quad T_{12}(\theta) = T_{12}\cos 2\theta + T_{13}\sin 2\theta, \quad T_{13}(\theta) = T_{13}\cos 2\theta - T_{12}\sin 2\theta,$$
$$T_{21}(\theta) = T_{12}^*(\theta), \quad T_{22}(\theta) = T_{22}\cos^2 2\theta + T_{33}\sin^2 2\theta + \text{Re}\{T_{23}\}\sin 4\theta, \quad T_{23}(\theta) = j\text{Im}\{T_{23}\}, \tag{8.4.7}$$
$$T_{31}(\theta) = T_{13}^*(\theta), \quad T_{32}(\theta) = T_{23}^*(\theta), \quad T_{33}(\theta) = T_{33}\cos^2 2\theta + T_{22}\sin^2 2\theta - \text{Re}\{T_{23}\}\sin 4\theta.$$

So, we first rotate the coherency matrix in an imaging window and decompose it according to the matrix expansion method in the next section. Then we move to the neighboring window and repeat the same procedure. The rotation angle will be different window by window as shown in Figure 8.8. The angle is also dependent on the window size (ensemble averaging size). Since this method uses rotation operation, we call it Y4R (rotation of Y4O).

This rotation is also referred to as deorientation [12], or polarization orientation compensation [15]. Equation (8.4.4) is the same as (9.2.8) in the expression of the correlation coefficient in the circular polarization basis. Although the derivation method is different, the rotation angle is the same. Using the rotation or minimization of T_{33}, we expand the measured coherency matrix as follows.

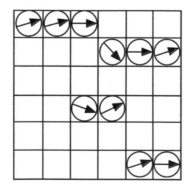

FIGURE 8.8   Rotation angle in an image.

8.4.2 Y4R DECOMPOSITION

For the volume scattering with $|\sigma_{HH} - \sigma_{VV}| < 2$ dB, we expand the measured coherency matrix as follows.

$$
\begin{bmatrix} T_{11}(\theta) & T_{12}(\theta) & T_{13}(\theta) \\ T_{21}(\theta) & T_{22}(\theta) & T_{23}(\theta) \\ T_{31}(\theta) & T_{32}(\theta) & T_{33}(\theta) \end{bmatrix} = \frac{P_s}{1+|\beta|^2} \begin{bmatrix} 1 & \beta^* & 0 \\ \beta & |\beta|^2 & 0 \\ 0 & 0 & 0 \end{bmatrix} + \frac{P_d}{1+|\alpha|^2} \begin{bmatrix} |\alpha|^2 & \alpha & 0 \\ \alpha^* & 1 & 0 \\ 0 & 0 & 0 \end{bmatrix}
$$

$$
+ \frac{P_v}{4} \begin{bmatrix} 2 & 0 & 0 \\ 0 & 1 & 0 \\ 0 & 0 & 1 \end{bmatrix} + \frac{P_h}{2} \begin{bmatrix} 0 & 0 & 0 \\ 0 & 1 & \pm j \\ 0 & \mp j & 1 \end{bmatrix} \tag{8.4.8}
$$

From this expansion, we directly obtain: $P_h = 2\left|\text{Im}\{T_{23}(\theta)\}\right| = 2\left|\text{Im}\{T_{23}\}\right|$ \hfill (8.4.9)

and
$$
P_v = 2\left[2T_{33}(\theta) - P_h\right] \tag{8.4.10}
$$

Then we have three equations with four unknowns $(P_s, P_d, \alpha, \beta)$,

$$
\begin{cases} \dfrac{P_s}{1+|\beta|^2} + \dfrac{P_d|\alpha|^2}{1+|\alpha|^2} = S \\[2ex] \dfrac{P_s|\beta|^2}{1+|\beta|^2} + \dfrac{P_d}{1+|\alpha|^2} = D, \\[2ex] \dfrac{P_s\beta^*}{1+|\beta|^2} + \dfrac{P_d\,\alpha}{1+|\alpha|^2} = C \end{cases} \quad \begin{cases} S = T_{11}(\theta) - \dfrac{P_v}{2} \\[2ex] D = T_{22}(\theta) - T_{33}(\theta) \\[2ex] C = T_{12}(\theta) \end{cases} \tag{8.4.11}
$$

According to the magnitude imbalance of the *HH* and *VV* power, we choose an appropriate volume scattering matrix for submatrix expansion as shown in equations (8.3.5–8.3.6). The results of three equations with unknowns are summarized in Table 8.2. The branch condition in this case remains the same.

$$
C_0 = 2T_{11} - TP + P_h \tag{8.4.12}
$$

Since the rotation makes $\text{Re}\{T_{23}(\theta)\} = 0$, the number of polarization parameters reduces from nine to eight. Therefore, this decomposition scheme accounts for six parameters out of eight. In addition, pure imaginary of $T_{23}(\theta) = j\,\text{Im}\{T_{23}\}$ is the best fit for the helix scattering.

TABLE 8.2
Helix Power, Volume-Scattering Power, and Three Equations for Y4R, S4R Decomposition Scheme

	Vegetation			Dihedral
	$\sigma_{HH} - \sigma_{VV} > 2\,dB$	$\|\sigma_{HH} - \sigma_{VV}\| < 2\,dB$	$\sigma_{VV} - \sigma_{HH} > 2\,dB$	
Helix power	$P_h = 2\|\text{Im}\{T_{23}\}\|$	$P_h = 2\|\text{Im}\{T_{23}\}\|$	$P_h = 2\|\text{Im}\{T_{23}\}\|$	$P_h = 2\|\text{Im}\{T_{23}\}\|$
Volume-scattering power	$P_v = \dfrac{15}{8}\left[2T_{33}(\theta) - P_h\right]$	$P_v = 2\left[2T_{33}(\theta) - P_h\right]$	$P_v = \dfrac{15}{8}\left[2T_{33}(\theta) - P_h\right]$	$P_v = \dfrac{15}{16}\left[2T_{33}(\theta) - P_h\right]$
$\left. \begin{array}{l} \dfrac{P_s}{1+\|\beta\|^2} + \dfrac{P_d\|\alpha\|^2}{1+\|\alpha\|^2} = S \\[2ex] \dfrac{P_s\|\beta\|^2}{1+\|\beta\|^2} + \dfrac{P_d}{1+\|\alpha\|^2} = D \\[2ex] \dfrac{P_s\beta^*}{1+\|\beta\|^2} + \dfrac{P_d\alpha}{1+\|\alpha\|^2} = C \end{array} \right\rbrace$	$S = T_{11}(\theta) - \dfrac{1}{2}P_v$ $D = T_{22}(\theta) - \dfrac{7}{30}P_v - \dfrac{P_h}{2}$ $C = T_{12}(\theta) - \dfrac{1}{6}P_v$	$S = T_{11}(\theta) - \dfrac{1}{2}P_v$ $D = T_{22}(\theta) - T_{33}(\theta)$ $C = T_{12}(\theta)$	$S = T_{11}(\theta) - \dfrac{1}{2}P_v$ $D = T_{22}(\theta) - \dfrac{7}{30}P_v - \dfrac{P_h}{2}$ $C = T_{12}(\theta) + \dfrac{1}{6}P_v$	$S = T_{11}(\theta)$ $D = T_{22}(\theta) - \dfrac{7}{15}P_v - \dfrac{P_h}{2}$ $C = T_{12}(\theta)$

8.4.3   S4R DECOMPOSITION

As can be seen in the oriented urban area of the San Francisco image, the volume scattering is very strong, which exhibits a green color. Since vegetation is the main source of volume scattering, oriented urban scattering is confusing. It is absolutely necessary to distinguish these scatterings caused by the HV component.

The oriented dihedral model is introduced to distinguish these scatterings. The scattering model in the coherency matrix form (Chapter 6 and [19]) is

$$[T]_{dihedral} = \frac{1}{15} \begin{bmatrix} 0 & 0 & 0 \\ 0 & 7 & 0 \\ 0 & 0 & 8 \end{bmatrix} \tag{8.4.13}$$

In order to distinguish this volume scattering model from those of vegetations, the branch condition C_1 is used. This decomposition method is abbreviated as S4R.

- Branch condition C_1 (Appendix)

The criterion is: $C_1 = T_{11}(\theta) - T_{22}(\theta) + \frac{7}{8} T_{33}(\theta) + \frac{1}{16} P_h$ \hfill (8.4.14)

If $C_1 > 0$, the surface scattering is dominant. We use the vegetation scattering model.
If $C_1 < 0$, the dihedral scattering is dominant. We use the preceding scattering model (8.4.13).

The matrix expansion for the dihedral scattering model becomes

$$\begin{bmatrix} T_{11}(\theta) & T_{12}(\theta) & T_{13}(\theta) \\ T_{21}(\theta) & T_{22}(\theta) & T_{23}(\theta) \\ T_{31}(\theta) & T_{32}(\theta) & T_{33}(\theta) \end{bmatrix} = \frac{P_s}{1+|\beta|^2} \begin{bmatrix} 1 & \beta^* & 0 \\ \beta & |\beta|^2 & 0 \\ 0 & 0 & 0 \end{bmatrix} + \frac{P_d}{1+|\alpha|^2} \begin{bmatrix} |\alpha|^2 & \alpha & 0 \\ \alpha^* & 1 & 0 \\ 0 & 0 & 0 \end{bmatrix}$$
$$+ \frac{P_v}{15} \begin{bmatrix} 0 & 0 & 0 \\ 0 & 7 & 0 \\ 0 & 0 & 8 \end{bmatrix} + \frac{P_h}{2} \begin{bmatrix} 0 & 0 & 0 \\ 0 & 1 & \pm j \\ 0 & \mp j & 1 \end{bmatrix} \tag{8.4.15}$$

From the expansion, we can obtain: $P_h = 2\left|\text{Im}\left\{T_{23}(\theta)\right\}\right| = 2\left|\text{Im}\left\{T_{23}\right\}\right|$ \hfill (8.4.16)

and
$$P_v = \frac{16}{15}\left[2T_{33}(\theta) - P_h\right]$$
\hfill (8.4.17)

Then we have three equations with four unknowns $\left(P_s, P_d, \alpha, \beta\right)$,

$$
\begin{cases}
\dfrac{P_s}{1+|\beta|^2} + \dfrac{P_d|\alpha|^2}{1+|\alpha|^2} = S \\[3mm]
\dfrac{P_s|\beta|^2}{1+|\beta|^2} + \dfrac{P_d}{1+|\alpha|^2} = D, \\[3mm]
\dfrac{P_s\beta^*}{1+|\beta|^2} + \dfrac{P_d\,\alpha}{1+|\alpha|^2} = C
\end{cases}
\quad
\begin{cases}
S = T_{11}(\theta) \\[3mm]
D = T_{22}(\theta) - \dfrac{7}{15}P_v - \dfrac{1}{2}P_h \\[3mm]
C = T_{12}(\theta)
\end{cases}
$$
\hfill (8.4.18)

These values are listed in Table 8.2. Hereafter, we can derive P_s and P_v using the same approximation in equations (8.2.4–8.2.6). The flow chart for both Y4R and S4R is shown together in Figure 8.9. The selection of the decomposition method is just putting

$$C_1 = 1 \text{ for Y4R program}$$

$$C_1 = T_{11}(\theta) - T_{22}(\theta) + \frac{7}{8}T_{33}(\theta) + \frac{1}{16}P_h \text{ for S4R program}$$

The Y4R image is shown in Figure 8.10 because both Y4R and S4R methods have yielded the same image at a glance. It is seen that the whole urban areas have become more reddish, which means the double-bounce scattering is increased in man-made structures. The green area remains as a vegetation area. Compared to FDD and the Y40 image, the quality seems improved considerably. In general, the rotation operation with orientation angle less than 25° with respect to radar illumination works well, but for an orientation angle larger than 30°, the method cannot compensate the scattering mechanism.

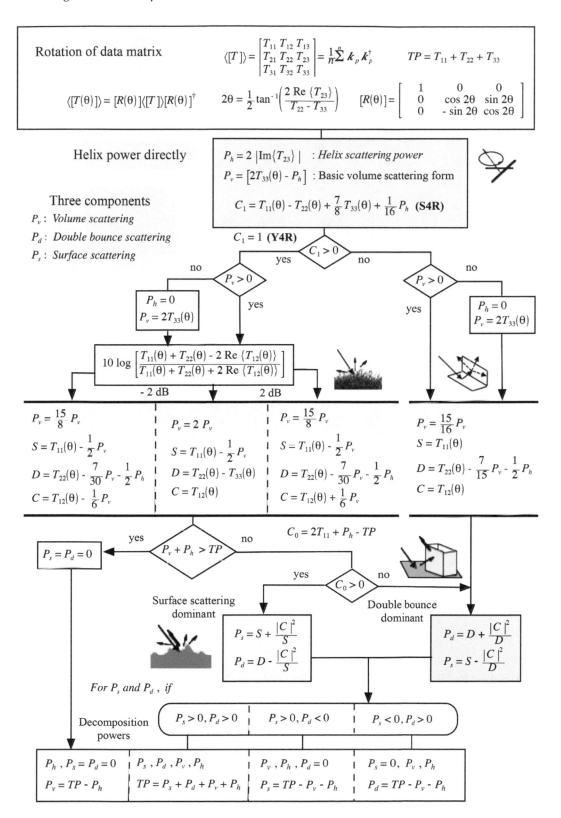

FIGURE 8.9 Flow chart of Y4R and S4R decomposition.

FIGURE 8.10   Decomposition image of San Francisco by Y4R.

8.5   G4U

If we can reduce the number of independent polarization parameters from nine to seven, it is possible to increase the model fitting ratio without increasing the additional scattering model. To be more specific, if we can transform the measured coherency matrix as

$$
\begin{bmatrix} T_{11} & T_{12} & T_{13} \\ T_{21} & T_{22} & T_{23} \\ T_{31} & T_{32} & T_{33} \end{bmatrix} \Rightarrow \begin{bmatrix} T_{11}^{'} & T_{12}^{'} & T_{13}^{'} \\ T_{21}^{'} & T_{22}^{'} & 0 \\ T_{31}^{'} & 0 & T_{33}^{'} \end{bmatrix}
\tag{8.5.1}
$$

then it becomes possible to account for seven parameters out of seven (100%) by the existing four-scattering models in S4R.

It can be achieved by the following double transformations. First, we take rotation transform to minimize T_{33} as in Y4R.

$$
\left[T(\theta) \right] = \left[R(\theta) \right]\left[T \right]\left[R(\theta) \right]^{t} \text{ with } \left[R(\theta) \right] = \begin{bmatrix} 1 & 0 & 0 \\ 0 & \cos 2\theta & \sin 2\theta \\ 0 & -\sin 2\theta & \cos 2\theta \end{bmatrix}
\tag{8.5.2}
$$

This makes: $\operatorname{Re}\left\{ T_{23}(\theta) \right\} = 0$, and $T_{23}(\theta) = j\operatorname{Im}\left\{ T_{23} \right\}$ for $2\theta = \dfrac{1}{2}\tan^{-1}\left(\dfrac{2\operatorname{Re}\left\{ T_{23} \right\}}{T_{22} - T_{33}} \right)$
$\tag{8.5.3}$

Further unitary transform: $\left[T(\psi) \right] = \left[U(\psi) \right]\left[T(\theta) \right]\left[U(\psi) \right]^{t}$
$\tag{8.5.4}$

$$
\text{with} \left[U(\psi) \right] = \begin{bmatrix} 1 & 0 & 0 \\ 0 & \cos 2\psi & j\sin 2\psi \\ 0 & j\sin 2\psi & \cos 2\psi \end{bmatrix}
\tag{8.5.5}
$$

makes $T_{23}(\psi) = \text{Re}\left(j\text{Im}\{T_{23}\}\right) = 0$ for $2\psi = \dfrac{1}{2}\tan^{-1}\left(\dfrac{2\text{Im}\{T_{23}(\theta)\}}{T_{22}(\theta)-T_{33}(\theta)} \right).$ \hfill (8.5.6)

Therefore, the elements of equation (8.5.4) after the double transformations become,

$$T_{11}(\psi) = T_{11}(\theta), T_{23}(\psi) = 0, T_{32}(\psi) = 0,\ T_{21}(\psi) = T_{12}^{*}(\psi), T_{31}(\psi) = T_{13}^{*}(\psi)$$

$$T_{12}(\psi) = T_{12}(\theta)\ \cos 2\psi - jT_{13}(\theta)\ \sin 2\psi,\ \ T_{13}(\psi) = T_{13}(\theta)\ \cos 2\psi - jT_{12}(\theta)\ \sin 2\psi$$

$$T_{22}(\psi) = T_{22}(\theta)\cos^{2}2\psi + T_{33}(\theta)\sin^{2}2\psi + \text{Im}\{T_{23}(\theta)\}\ \sin 4\psi$$

$$T_{33}(\psi) = T_{33}(\theta)\cos^{2}2\psi + T_{22}(\theta)\sin^{2}2\psi - \text{Im}\{T_{23}(\theta)\}\ \sin 4\psi$$

\hfill (8.5.7)

The elements in equation (8.5.7) constitute the same form as in equation (8.5.1). $T(\theta)$ is given by equation (8.4.7). Since unitary transformation does not change any information on the coherency matrix, we can perform scattering power decomposition without loss of generality. Therefore, we use this transformation over the Y4R and S4R formulation model.

$$\left[U(\psi)\right]\left\langle\left[T(\theta)\right]\right\rangle\left[U(\psi)\right]^{t} = \left[U(\psi)\right]\left\{P_{s}[T]_{s} + P_{d}[T]_{d} + P^{v}[T]_{v} + P_{h}[T]_{h}\right\}\left[U(\psi)\right]^{t} \hfill (8.5.8)$$

For the volume scattering with $|\sigma_{HH} - \sigma_{VV}| < 2$ dB, the preceding expansion becomes as follows:

$$
\begin{bmatrix} T_{11}' & T_{12}' & T_{13}' \\ T_{21}' & T_{22}' & 0 \\ T_{31}' & 0 & T_{33}' \end{bmatrix} = \frac{P_{s}}{1+|\beta|^{2}} \begin{bmatrix} 1 & \beta^{*}\cos 2\psi & -j\beta^{*}\sin 2\psi \\ \beta\cos 2\psi & |\beta|^{2}\cos^{2}2\psi & j|\beta|^{2}\dfrac{\sin 4\psi}{2} \\ j\beta\sin 2\psi & j|\beta|^{2}\dfrac{\sin 4\psi}{2} & |\beta|^{2}\sin^{2}2\psi \end{bmatrix}
$$

$$
+ \frac{P_{d}}{1+|\alpha|^{2}} \begin{bmatrix} |\alpha|^{2} & \alpha\cos 2\psi & -j\alpha\sin 2\psi \\ \alpha^{*}\cos 2\psi & \cos^{2}2\psi & -j\dfrac{\sin 4\psi}{2} \\ j\alpha^{*}\sin 2\psi & j\dfrac{\sin 4\psi}{2} & \sin^{2}2\psi \end{bmatrix}
$$

\hfill (8.5.9)

$$
+ \frac{P_{v}}{4}\begin{bmatrix} 2 & 0 & 0 \\ 0 & 1 & 0 \\ 0 & 0 & 1 \end{bmatrix} + \frac{P_{h}}{2}\begin{bmatrix} 0 & 0 & 0 \\ 0 & 1\pm\sin 4\psi & \pm j\cos 4\psi \\ 0 & \mp j\cos 4\psi & 1\mp\sin 4\psi \end{bmatrix}
$$

After arrangement of long calculations, we can derive: $P_h = 2\left|\text{Im}\{T_{23}\}\right|$, $P_v = 2\left[2T_{33}(\theta) - P_h\right]$,

$$\text{Three equations}:\begin{cases}\dfrac{P_s}{1+|\beta|^2}+\dfrac{P_d|\alpha|^2}{1+|\alpha|^2}=S\\[3mm]\dfrac{P_s|\beta|^2}{1+|\beta|^2}+\dfrac{P_d}{1+|\alpha|^2}=D,\\[3mm]\dfrac{P_s\beta^*}{1+|\beta|^2}+\dfrac{P_d\,\alpha}{1+|\alpha|^2}=C\end{cases}\begin{cases}S=T_{11}-\dfrac{1}{2}P_v\\[2mm]D=TP-P_v-P_h-S\\[2mm]C=T_{12}(\theta)+T_{13}(\theta)\end{cases}\tag{8.5.10}$$

Similarly, for $\sigma_{HH}-\sigma_{VV} > 2$ dB, we use $[T]_v = \dfrac{1}{30}\begin{bmatrix}15 & 5 & 0\\ 5 & 7 & 0\\ 0 & 0 & 8\end{bmatrix}$ as the volume scattering model.

For $\sigma_{VV}-\sigma_{HH} > 2$ dB. we employ $[T]_v = \dfrac{1}{30}\begin{bmatrix}15 & -5 & 0\\ -5 & 7 & 0\\ 0 & 0 & 8\end{bmatrix}$ as the volume scattering model.

For the *HV* component from oriented dihedral scattering, we use $[T]_v = \dfrac{1}{15}\begin{bmatrix}0 & 0 & 0\\ 0 & 7 & 0\\ 0 & 0 & 8\end{bmatrix}$.

The results of three equations with unknowns $(P_s, P_d, \alpha, \beta)$ are summarized in Table 8.3. The remaining powers P_s and P_d are given by the aid of the branch condition C_0 (8.2.4–8.2.6).

$$C_0 = 2T_{11} - TP + P_h\tag{8.5.11}$$

Branch condition C_1 becomes,

$$C_1 = T_{11}(\theta) - T_{22}(\theta) + \frac{7}{8}T_{33}(\theta) + \frac{1}{16}P_h\tag{8.5.12}$$

It is interesting to note that the final expression in Table 8.3 becomes quite simple, although the double unitary transformation is applied. The results in Table 8.3 indicate that the decomposition processing can be performed by the values of $T_{21}(\theta), T_{13}(\theta), T_{33}(\theta)$ only. This means that we do not need to calculate any complex expression by the unitary transformation. If we compare Tables 8.1 and 8.2, the difference appears only in S, D, and C expressions using rotation matrix elements. This point is the implicit advantage of unitary transformation.

Since this method reduces the number of parameters from nine to seven and accounts for all parameters in the decomposition, it utilizes 100% of the polarimetric information. The T_{33} element is minimized through rotation and again by unitary transformation. Therefore, the overestimation of volume scattering is reduced (mitigated). Because this method can be applied to reflection symmetry and non-reflection symmetry with reduced volume scattering, the name of the method is called **G**eneral **4**-component scattering decomposition with **U**nitary transformation (**G4U**).

The flow chart of G4U is shown in Figure 8.11. The corresponding San Francisco image is shown in Figure 8.12. The red (the double-bounce scattering) area is expanded compared to FDD,

TABLE 8.3

Helix Power, Volume-Scattering Power, and Three Equations for G4U Decomposition Scheme

	Vegetation			Dihedral
	$\sigma_{HH}-\sigma_{VV}>2$ dB	$\lvert\sigma_{HH}-\sigma_{VV}\rvert<2$ dB	$\sigma_{VV}-\sigma_{HH}>2$ dB	
Helix power	$P_h=2\lvert\mathrm{Im}\{T_{23}\}\rvert$	$P_h=2\lvert\mathrm{Im}\{T_{23}\}\rvert$	$P_h=2\lvert\mathrm{Im}\{T_{23}\}\rvert$	$P_h=2\lvert\mathrm{Im}\{T_{23}\}\rvert$
Volume-scattering power	$P_v=\dfrac{15}{8}\left[2T_{33}(\theta)-P_h\right]$	$P_v=2\left[2T_{33}(\theta)-P_h\right]$	$P_v=\dfrac{15}{8}\left[2T_{33}(\theta)-P_h\right]$	$P_v=\dfrac{15}{16}\left[2T_{33}(\theta)-P_h\right]$
$\left.\begin{array}{l}\dfrac{P_s}{1+\lvert\beta\rvert^2}+\dfrac{P_d\lvert\alpha\rvert^2}{1+\lvert\alpha\rvert^2}=S\\[2ex]\dfrac{P_s\lvert\beta\rvert^2}{1+\lvert\beta\rvert^2}+\dfrac{P_d}{1+\lvert\alpha\rvert^2}=D\\[2ex]\dfrac{P_s\beta^*}{1+\lvert\beta\rvert^2}+\dfrac{P_d\alpha}{1+\lvert\alpha\rvert^2}=C\end{array}\right\}$	$S=T_{11}-\dfrac{1}{2}P_v$ $D=T_{22}-P_v-P_h-S$ $C=T_{12}(\theta)-T_{13}(\theta)-\dfrac{P_v}{6}$	$S=T_{11}-\dfrac{1}{2}P_v$ $D=TP-P_v-P_h-S$ $C=T_{12}(\theta)+T_{12}(\theta)$	$S=T_{11}-\dfrac{1}{2}P_v$ $D=TP-P_v-P_h-S$ $C=T_{12}(\theta)+T_{13}(\theta)+\dfrac{P_v}{6}$	$S=T_{11}$ $D=TP-P_v-P_h-S$ $C=T_{12}(\theta)+T_{13}(\theta)$

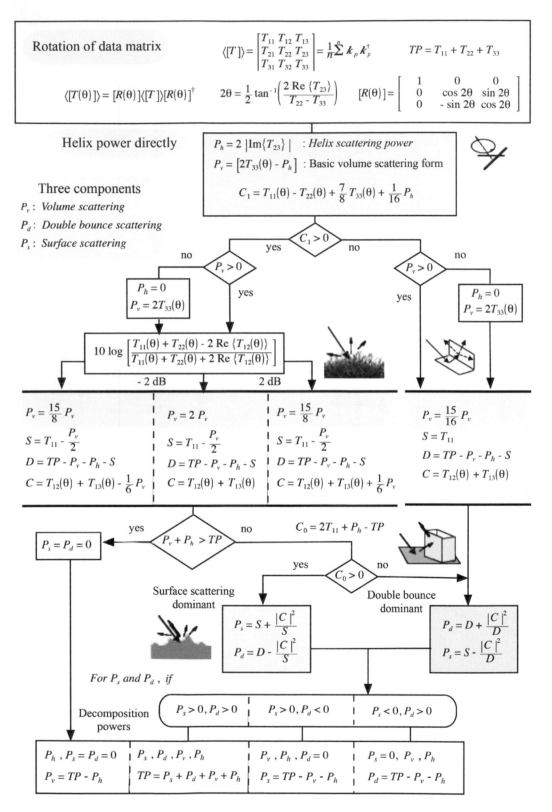

FIGURE 8.11 Flow chart of G4U.

FIGURE 8.12 Decomposition image of San Francisco by G4U.

Y40, and Y4R images. Improvement can be seen in the triangle green urban area in the lower right corner. Although still greenish, it does not look like a vegetation area.

8.6   6SD

No physical model for T_{13} has been employed up to four-component scattering power decompositions. Since T_{13} is rather small or negligible compared to other major terms, no scattering model has been considered yet. We tried to fit a certain model to account for T_{13} term in the six-component decomposition. For the physical modeling of T_{13}, we choose the oriented dipole and the compound dipole models in Chapter 6. These models account for the real and imaginary parts of T_{13}. Then the model expansion can be written as

$$\text{Total power} = P_s + P_d + P_v + P_h + P_{od} + P_{cd} \tag{8.6.1}$$

$$\left\langle \left[T(\theta) \right] \right\rangle = P_s \left\langle [T] \right\rangle_s + P_d \left\langle [T] \right\rangle_d + P_v \left\langle [T] \right\rangle_v + P_h \left[T \right]_h + P_{od} \left[T \right]_{od} + P_{cd} \left[T \right]_{cd} \tag{8.6.2}$$

In addition to the existing four components, we added

- Pod: Oriented dipole power modeled by 45°-oriented dipole,
  and
- Pcd: Compound dipole power by 45°-oriented dipoles with spacing, which is feasible in actual situation.

In the matrix expansion, the left-hand side is the rotated coherency matrix (equation 8.4.7) of measured data. Since $\mathrm{Re}\left\{ T_{23}(\theta) \right\} = 0$, there exist eight parameters in the measured coherency matrix. The right-hand side accounts for eight parameters in total. Therefore, all parameters 8/8 are accounted for in this decomposition (100%).

- For the volume scattering with $|\sigma_{HH} - \sigma_{VV}| < 2$ dB, we expand as follows:

$$
\begin{bmatrix} T_{11}(\theta) & T_{12}(\theta) & T_{13}(\theta) \\ T_{21}(\theta) & T_{22}(\theta) & T_{23}(\theta) \\ T_{31}(\theta) & T_{32}(\theta) & T_{33}(\theta) \end{bmatrix} = \frac{P_s}{1+|\beta|^2} \begin{bmatrix} 1 & \beta^* & 0 \\ \beta & |\beta|^2 & 0 \\ 0 & 0 & 0 \end{bmatrix} + \frac{P_d}{1+|\alpha|^2} \begin{bmatrix} |\alpha|^2 & \alpha & 0 \\ \alpha^* & 1 & 0 \\ 0 & 0 & 0 \end{bmatrix}
$$

$$
+ \frac{P_v}{4} \begin{bmatrix} 2 & 0 & 0 \\ 0 & 1 & 0 \\ 0 & 0 & 1 \end{bmatrix} + \frac{P_h}{2} \begin{bmatrix} 0 & 0 & 0 \\ 0 & 1 & \pm j \\ 0 & \mp j & 1 \end{bmatrix} \tag{8.6.3}
$$

$$
+ \frac{P_{od}}{2} \begin{bmatrix} 1 & 0 & \pm 1 \\ 0 & 0 & 0 \\ \pm 1 & 0 & 1 \end{bmatrix} + \frac{P_{cd}}{2} \begin{bmatrix} 1 & 0 & \pm j \\ 0 & 0 & 0 \\ \mp j & 0 & 1 \end{bmatrix}
$$

From this expansion in equation (8.6.1), we can derive four scattering powers directly.

$$
\text{Helix scattering power, } P_h = 2\left|\text{Im}\left\{T_{23}(\theta)\right\}\right| = 2\left|\text{Im}\left\{T_{23}\right\}\right| \tag{8.6.4}
$$

$$
\text{Oriented dipole power, } P_{od} = 2\left|\text{Re}\left\{T_{13}(\theta)\right\}\right| \tag{8.6.5}
$$

$$
\text{Compound-scattering power, } P_{cd} = 2\left|\text{Im}\left\{T_{13}(\theta)\right\}\right| \tag{8.6.6}
$$

$$
\text{Volume-scattering power, } P_v = 2\left[2T_{33}(\theta) - P_h - P_{od} - P_{cd}\right] \tag{8.6.7}
$$

Then we have three equations with four unknowns $(P_s, P_d, \alpha, \beta)$,

$$
\begin{cases} \dfrac{P_s}{1+|\beta|^2} + \dfrac{P_d|\alpha|^2}{1+|\alpha|^2} = S \\[2ex] \dfrac{P_s|\beta|^2}{1+|\beta|^2} + \dfrac{P_d}{1+|\alpha|^2} = D, \\[2ex] \dfrac{P_s\beta^*}{1+|\beta|^2} + \dfrac{P_d\,\alpha}{1+|\alpha|^2} = C \end{cases} \quad \begin{cases} S = T_{11}(\theta) - \dfrac{P_v}{2} - \dfrac{P_{od}}{2} - \dfrac{P_{cd}}{2} \\[2ex] D = T_{22}(\theta) - \dfrac{P_v}{4} - \dfrac{P_h}{2} \\[2ex] C = T_{12}(\theta) \end{cases} \tag{8.6.8}
$$

According to the magnitude imbalance of the *HH* and *VV* power, we choose an appropriate volume scattering matrix for submatrix expansion as shown in equations (8.3.5–8.3.6). The results of three equations with unknowns are summarized in Table 8.4.

TABLE 8.4

Helix Power, Volume-Scattering Power, Oriented Dipole Power, Compound Dipole Power, and Three Equations for 6SD Decomposition Scheme

	Vegetation			Dihedral
	$\sigma_{HH} - \sigma_{VV} > 2$ dB	$\left|\sigma_{HH} - \sigma_{VV}\right| < 2$ dB	$\sigma_{VV} - \sigma_{HH} > 2$ dB	
Helix power	$P_h = 2\left|\text{Im}\{T_{23}\}\right|$	$P_h = 2\left|\text{Im}\{T_{23}\}\right|$	$P_h = 2\left|\text{Im}\{T_{23}\}\right|$	$P_h = 2\left|\text{Im}\{T_{23}\}\right|$
Oriented dipole power	$P_{od} = 2\left|\text{Re}\{T_{13}(\theta)\}\right|$	$P_{od} = 2\left|\text{Re}\{T_{13}(\theta)\}\right|$	$P_{od} = 2\left|\text{Re}\{T_{13}(\theta)\}\right|$	$P_{od} = 2\left|\text{Re}\{T_{13}(\theta)\}\right|$
Compound dipole power	$P_{cd} = 2\left|\text{Im}\{T_{13}(\theta)\}\right|$	$P_{cd} = 2\left|\text{Im}\{T_{13}(\theta)\}\right|$	$P_{cd} = 2\left|\text{Im}\{T_{13}(\theta)\}\right|$	$P_{cd} = 2\left|\text{Im}\{T_{13}(\theta)\}\right|$
Volume-scattering power	$P_v = \dfrac{15}{8}\left[2T_{33}(\theta) - P_h\right]$	$P_v = 2\left[2T_{33}(\theta) - P_h\right]$	$P_v = \dfrac{15}{8}\left[2T_{33}(\theta) - P_h\right]$	$P_v = \dfrac{15}{16}\left[2T_{33}(\theta) - P_h\right]$
$\dfrac{P_s}{1+\left|\beta\right|^2} + \dfrac{P_d\left|\alpha\right|^2}{1+\left|\alpha\right|^2} = S$	$S = T_{11}(\theta) - \dfrac{P_v}{2} - \dfrac{P_{od}}{2} - \dfrac{P_{cd}}{2}$	$S = T_{11}(\theta) - \dfrac{P_v}{2} - \dfrac{P_{od}}{2} - \dfrac{P_{cd}}{2}$	$S = T_{11}(\theta) - \dfrac{P_v}{2} - \dfrac{P_{od}}{2} - \dfrac{P_{cd}}{2}$	$S = T_{11}(\theta)$
$\dfrac{P_s\left|\beta\right|^2}{1+\left|\beta\right|^2} + \dfrac{P_d}{1+\left|\alpha\right|^2} = D$	$D = T_{22}(\theta) - \dfrac{7}{30}P_v - \dfrac{P_h}{2}$	$D = T_{22}(\theta) - \dfrac{7}{4}P_v - \dfrac{P_h}{2}$	$D = T_{22}(\theta) - \dfrac{7}{30}P_v - \dfrac{P_h}{2}$	$D = T_{22}(\theta) - \dfrac{7}{15}P_v - \dfrac{P_h}{2}$
$\dfrac{P_s\beta^*}{1+\left|\beta\right|^2} + \dfrac{P_d\alpha}{1+\left|\alpha\right|^2} = C$	$C = T_{12}(\theta) - \dfrac{1}{6}P_v$	$C = T_{12}(\theta)$	$C = T_{12}(\theta) + \dfrac{1}{6}P_v$	$C = T_{12}(\theta)$

- For volume scattering caused by oriented dihedral scatter, we expand as

$$
\begin{bmatrix} T_{11}(\theta) & T_{12}(\theta) & T_{13}(\theta) \\ T_{21}(\theta) & T_{22}(\theta) & T_{23}(\theta) \\ T_{31}(\theta) & T_{32}(\theta) & T_{33}(\theta) \end{bmatrix} = \frac{P_s}{1+|\beta|^2} \begin{bmatrix} 1 & \beta^* & 0 \\ \beta & |\beta|^2 & 0 \\ 0 & 0 & 0 \end{bmatrix} + \frac{P_d}{1+|\alpha|^2} \begin{bmatrix} |\alpha|^2 & \alpha & 0 \\ \alpha^* & 1 & 0 \\ 0 & 0 & 0 \end{bmatrix}
$$

$$
+ \frac{P_v}{15} \begin{bmatrix} 0 & 0 & 0 \\ 0 & 7 & 0 \\ 0 & 0 & 8 \end{bmatrix} + \frac{P_h}{2} \begin{bmatrix} 0 & 0 & 0 \\ 0 & 1 & \pm j \\ 0 & \mp j & 1 \end{bmatrix} \tag{8.6.9}
$$

$$
+ \frac{P_{od}}{2} \begin{bmatrix} 1 & 0 & \pm 1 \\ 0 & 0 & 0 \\ \pm 1 & 0 & 1 \end{bmatrix} + \frac{P_{cd}}{2} \begin{bmatrix} 1 & 0 & \pm j \\ 0 & 0 & 0 \\ \mp j & 0 & 1 \end{bmatrix}
$$

This expansion yields, $P_h = 2\left|\text{Im}\{T_{23}\}\right|,$ $P_{od} = 2\left|\text{Re}\{T_{23}(\theta)\}\right|,$

$$
P_{cd} = 2\left|\text{Im}\{T_{13}(\theta)\}\right|, \qquad P_v = \frac{15}{16}\left|2T_{23}(\theta) - P_h - P_{od} - P_{cd}\right|,
$$

$$
\text{Three-equations:} \begin{cases} \dfrac{P_s}{1+|\beta|^2} + \dfrac{P_d|\alpha|^2}{1+|\alpha|^2} = S \\[2mm] \dfrac{P_s|\beta|^2}{1+|\beta|^2} + \dfrac{P_d}{1+|\alpha|^2} = D, \\[2mm] \dfrac{P_s\beta^*}{1+|\beta|^2} + \dfrac{P_d\,\alpha}{1+|\alpha|^2} = C \end{cases} \begin{cases} S = T_{11}(\theta) - \dfrac{P_{od}}{2} - \dfrac{P_{cd}}{2} \\[2mm] D = T_{22}(\theta) - \dfrac{7}{15}P_v - \dfrac{P_h}{2} \\[2mm] C = T_{12}(\theta) \end{cases} \tag{8.6.10}
$$

These expansions bring three equations with unknowns as in Table 8.4. The remaining powers P_s and P_d are given by the aid of branch condition C_0 (8.2.4–8.2.6).

$$
C_0 = 2T_{11} - TP + P_h \tag{8.6.11}
$$

Branch condition C_1 becomes,

$$
C_1 = T_{11}(\theta) - T_{22}(\theta) + \frac{7}{8}T_{33}(\theta) + \frac{1}{16}P_h - \frac{15}{16}(P_{od} + P_{cd}) \tag{8.6.12}
$$

A flow chart of six-component scattering power decomposition is shown in Figure 8.13. The preceding procedure is directly reflected in the decomposition flow.

Rotation of data matrix $\langle[T(\theta)]\rangle = [R(\theta)]\langle[T]\rangle[R(\theta)]^\dagger$ $\langle[T]\rangle = \frac{1}{n}\sum_1^n \mathbf{k}_P \mathbf{k}_P^\dagger = \begin{bmatrix} T_{11} & T_{12} & T_{13} \\ T_{21} & T_{22} & T_{23} \\ T_{31} & T_{32} & T_{33} \end{bmatrix}$

$[R(\theta)] = \begin{bmatrix} 1 & 0 & 0 \\ 0 & \cos 2\theta & \sin 2\theta \\ 0 & -\sin 2\theta & \cos 2\theta \end{bmatrix}$ $2\theta = \frac{1}{2}\tan^{-1}\left(\frac{2\,\mathrm{Re}\,\{T_{23}\}}{T_{22}-T_{33}}\right)$ This rotation makes $\mathrm{Re}\,\{T_{23}\} = 0$

$P_{od} = 2\,|\mathrm{Re}\{T_{13}(\theta)\}|$: Oriented dipole power $\qquad P_h = 2\,|\mathrm{Im}\{T_{23}\}|$: Helix scattering power

$P_{cd} = 2\,|\mathrm{Im}\{T_{13}(\theta)\}|$: Compound dipole power $\qquad TP = T_{11} + T_{22} + T_{33}$: Total Power

$P_v = [2T_{33}(\theta) - P_h - P_{od} - P_{cd}]$: Basic volume scattering form

$C_1 = T_{11}(\theta) - T_{22}(\theta) + \frac{7}{8}T_{33}(\theta) + \frac{1}{16}P_h - \frac{15}{16}P_{od} - \frac{15}{16}P_{cd}$

$C_1 > 0$ — no → $P_v > 0$ — no → $P_v = 0$, $P_{cw} = P_{od} + P_{cd}$

yes ↓ (from $C_1>0$)

$P_v > 0$ — no →

yes ↓

$P_h = P_{od} = P_{cd} = 0$
$P_v = 2T_{33}(\theta)$

$P_{cw} = 2T_{33}(\theta) - P_h$ $\qquad P_h > P_{cw}$ — no → $P_h = 2T_{33}(\theta) - P_{cw}$
yes ↓ $\qquad$ if $P_h < 0$, then
$\qquad$ $P_h = 0$, $P_{cw} = 2T_{33}(\theta)$

$P_{od} = P_{cd} = 0$ $\quad P_{cw} < 0$ — yes
$P_h = 2T_{33}(\theta)$ $\quad$ no ↓

yes ← $P_{od} > P_{cd}$ → no

$P_h, P_{od}, P_{cd} = P_{cw} - P_{od}$ $\qquad\qquad$ $P_h, P_{cd}, P_{od} = P_{cw} - P_{cd}$
if $P_{cd} < 0$, then $\qquad\qquad$ if $P_{od} < 0$, then
$P_{cd} = 0$, $P_{od} = P_{cw}$ $\qquad\qquad$ $P_{od} = 0$, $P_{cd} = P_{cw}$

VV/HH power ratio : $10\log\left[\dfrac{T_{11}(\theta) + T_{22}(\theta) - 2\mathrm{Re}\,\{T_{12}(\theta)\}}{T_{11}(\theta) + T_{22}(\theta) + 2\mathrm{Re}\,\{T_{12}(\theta)\}}\right]$

-2 dB $\qquad\qquad$ 2 dB

$P_v = \frac{15}{8}P_v$	$P_v = 2P_v$	$P_v = \frac{15}{8}P_v$	$P_v = \frac{15}{16}P_v$
$S = T_{11}(\theta) - \frac{P_v}{2} - \frac{P_{od}}{2} - \frac{P_{cd}}{2}$	$S = T_{11}(\theta) - \frac{P_v}{2} - \frac{P_{od}}{2} - \frac{P_{cd}}{2}$	$S = T_{11}(\theta) - \frac{P_v}{2} - \frac{P_{od}}{2} - \frac{P_{cd}}{2}$	$S = T_{11}(\theta) - \frac{P_{od}}{2} - \frac{P_{cd}}{2}$
$D = T_{22}(\theta) - \frac{7}{30}P_v - \frac{P_h}{2}$	$D = T_{22}(\theta) - \frac{P_v}{4} - \frac{P_h}{2}$	$D = T_{22}(\theta) - \frac{7}{30}P_v - \frac{P_h}{2}$	$D = T_{22}(\theta) - \frac{7}{15}P_v - \frac{P_h}{2}$
$C - T_{12}(\theta) - \frac{1}{6}P_v$	$C = T_{12}(\theta)$	$C = T_{12}(\theta) + \frac{1}{6}P_v$	$C = T_{12}(\theta)$

yes ← $P_v + P_h + P_{od} + P_{cd} > TP$ — no → $C_0 = 2T_{11} + P_h - TP$

$P_s = P_d = 0$ $\qquad\qquad$ yes ↓ $\qquad$ no

$\qquad\qquad\qquad$ $C_0 > 0$ →

Surface scattering dominant $\qquad\qquad\qquad$ Double bounce dominant

$P_s = S + \frac{|C|^2}{S}$ $\qquad\qquad$ $P_d = D + \frac{|C|^2}{D}$
$P_d = D - \frac{|C|^2}{S}$ $\qquad\qquad$ $P_s = S - \frac{|C|^2}{D}$

For P_s and P_d, if

Decomposition powers	$P_s > 0, P_d > 0$	$P_s > 0, P_d < 0$	$P_s < 0, P_d > 0$
$P_h, P_{od}, P_{cd}, P_s = P_d = 0$	P_s, P_d, P_v	$P_v, P_h, P_{od}, P_{cd}, P_d = 0$	$P_s = 0, P_v, P_h, P_{od}, P_{cd}$
$P_v = TP - P_h - P_{od} - P_{cd}$	P_h, P_{od}, P_{cd}	$P_s = TP - P_v - P_h - P_{od} - P_{cd}$	$P_d = TP - P_v - P_h - P_{od} - P_{cd}$

FIGURE 8.13 Flow chart of six-component scattering power decomposition (6SD).

FIGURE 8.14   Decomposition image of San Francisco by 6SD.

First, we discriminate the *HV* contribution using the branch condition C_1. If the *HV* comes from natural vegetation, we go to the natural tree volume scattering. If the *HV* is caused by double-bounce scattering, we go to check the double-bounce scattering mainly in man-made structure by oriented and compound dipole scatterings. Then the same procedure determines if each scattering power is executed.

The final image is shown in Figure 8.14. We can see man-made structures are well-displayed as red, even in the triangular oriented urban area. The image as a whole looks improved.

8.7   7SD

When we checked the coherency matrix image in Figure 7.3, $\left|\text{Re}\left\{T_{23}\right\}\right|$ was rather strong in the oriented urban area. This element vanishes when rotation operation is applied. One idea is to use this element effectively in the decomposition without rotation operation. Since the term is mainly caused by oriented building blocks, the oriented surface model in Chapter 7 may apply to the non-rotated coherency matrix. Therefore, the model expansion can be written as

$$\text{Total power} = P_s + P_d + P_v + P_h + P_{cd} + P_{md} \tag{8.7.1}$$

$$\left\langle\left[T\right]\right\rangle = P_s\left\langle\left[T\right]\right\rangle_s + P_d\left\langle\left[T\right]\right\rangle_d + P_v\left\langle\left[T\right]\right\rangle_v + P_{md}\left[T\right]_{md} + P_h\left[T\right]_h + P_{od}\left[T\right]_{od} + P_{cd}\left[T\right]_{cd} \tag{8.7.2}$$

In addition to the existing six-components, we add one more model for Re$\{T_{23}\}$ to non-rotation matrix,

Pmd: Mixed dipole power with spacing, which is realizable in actual situation.

- For volume scattering with $\left|\sigma_{HH} - \sigma_{vv}\right| < 2$ dB, the expansion becomes as follows:

$$
\begin{bmatrix} T_{11} & T_{12} & T_{13} \\ T_{21} & T_{22} & T_{23} \\ T_{31} & T_{32} & T_{33} \end{bmatrix} = \frac{P_s}{1+|\beta|^2} \begin{bmatrix} 1 & \beta^* & 0 \\ \beta & |\beta|^2 & 0 \\ 0 & 0 & 0 \end{bmatrix} + \frac{P_d}{1+|\alpha|^2} \begin{bmatrix} |\alpha|^2 & \alpha & 0 \\ \alpha^* & 1 & 0 \\ 0 & 0 & 0 \end{bmatrix} + \frac{P_v}{4} \begin{bmatrix} 2 & 0 & 0 \\ 0 & 1 & 0 \\ 0 & 0 & 1 \end{bmatrix}
$$

$$
+ \frac{P_{md}}{2} \begin{bmatrix} 0 & 0 & 0 \\ 0 & 1 & \pm 1 \\ 0 & \pm 1 & 1 \end{bmatrix} + \frac{P_h}{2} \begin{bmatrix} 0 & 0 & 0 \\ 0 & 0 & \pm j \\ 0 & \mp j & 1 \end{bmatrix} \tag{8.7.3}
$$

$$
+ \frac{P_{od}}{2} \begin{bmatrix} 1 & 0 & \pm 1 \\ 0 & 0 & 0 \\ \pm 1 & 0 & 1 \end{bmatrix} + \frac{P_{cd}}{2} \begin{bmatrix} 1 & 0 & \pm j \\ 0 & 0 & 0 \\ \mp j & 0 & 1 \end{bmatrix}
$$

This expansion directly derives the following scattering powers:

$$
P_{md} = 2\left|\text{Re}\{T_{23}\}\right| \text{ Mixed dipole power} \tag{8.7.4}
$$

$$
P_h = 2\left|\text{Re}\{T_{23}\}\right| \text{ Helix power} \tag{8.7.5}
$$

$$
P_{od} = 2\left|\text{Re}\{T_{23}\}\right| \text{ Oriented dipole power} \tag{8.7.6}
$$

$$
P_{cd} = 2\left|\text{Re}\{T_{13}\}\right| \text{ Compound dipole power} \tag{8.7.7}
$$

$$
P_v = 2\left[2T_{23} - P_{md} - P_h - P_{od} - P_{cd}\right] \text{ Volume-scattering power} \tag{8.7.8}
$$

Then we have three equations with four unknowns $\left(P_s, P_d, \alpha, \beta\right)$,

$$
\begin{cases} \dfrac{P_s}{1+|\beta|^2} + \dfrac{P_d|\alpha|^2}{1+|\alpha|^2} = S \\[3mm] \dfrac{P_s|\beta|^2}{1+|\beta|^2} + \dfrac{P_d}{1+|\alpha|^2} = D, \\[3mm] \dfrac{P_s\beta^*}{1+|\beta|^2} + \dfrac{P_d\,\alpha}{1+|\alpha|^2} = C \end{cases} \quad \begin{cases} S = T_{11} - \dfrac{P_v}{2} - \dfrac{P_{od}}{2} - \dfrac{P_{cd}}{2} \\[3mm] D = T_{22} - \dfrac{P_v}{4} - \dfrac{P_{md}}{2} - \dfrac{P_h}{2} \\[3mm] C = T_{12} \end{cases} \tag{8.7.9}
$$

According to the magnitude imbalance of the *HH* and *VV* power, we choose an appropriate volume scattering matrix for the submatrix expansion as shown in equations (8.3.5–8.3.6). The results of three equations with unknowns are summarized in Table 8.5.

- For volume scattering caused by oriented dihedral scatter, the expansion becomes

$$
\begin{bmatrix} T_{11} & T_{12} & T_{13} \\ T_{21} & T_{22} & T_{23} \\ T_{31} & T_{32} & T_{33} \end{bmatrix} = \frac{P_s}{1+|\beta|^2}\begin{bmatrix} 1 & \beta^* & 0 \\ \beta & |\beta|^2 & 0 \\ 0 & 0 & 0 \end{bmatrix} + \frac{P_d}{1+|\alpha|^2}\begin{bmatrix} |\alpha|^2 & \alpha & 0 \\ \alpha^* & 1 & 0 \\ 0 & 0 & 0 \end{bmatrix} + \frac{P_v}{15}\begin{bmatrix} 0 & 0 & 0 \\ 0 & 7 & 0 \\ 0 & 0 & 8 \end{bmatrix}
$$

$$
+ \frac{P_{md}}{2}\begin{bmatrix} 0 & 0 & 0 \\ 0 & 1 & \pm1 \\ 0 & \pm1 & 1 \end{bmatrix} + \frac{P_h}{2}\begin{bmatrix} 0 & 0 & 0 \\ 0 & 0 & \pm j \\ 0 & \mp j & 1 \end{bmatrix} \tag{8.7.10}
$$

$$
+ \frac{P_{od}}{2}\begin{bmatrix} 1 & 0 & \pm1 \\ 0 & 0 & 0 \\ \pm1 & 0 & 1 \end{bmatrix} + \frac{P_{cd}}{2}\begin{bmatrix} 1 & 0 & \pm j \\ 0 & 0 & 0 \\ \mp j & 0 & 1 \end{bmatrix}
$$

This expansion yields,

$$
P_{md} = 2\left|\mathrm{Re}\left\{T_{23}\right\}\right|, \qquad P_h = 2\left|\mathrm{Im}\left\{T_{23}\right\}\right|, \qquad P_{cd} = 2\left|\mathrm{Re}\left\{T_{13}\right\}\right|,
$$

$$
P_{cd} = 2\left|\mathrm{Im}\left\{T_{13}\right\}\right|, \qquad P_v = \frac{15}{16}\left|2T_{23} - P_{md} - P_h - P_{od} - P_{cd}\right|,
$$

$$
\text{Three-equations:} \begin{cases} \dfrac{P_s}{1+|\beta|^2} + \dfrac{P_d|\alpha|^2}{1+|\alpha|^2} = S \\[2mm] \dfrac{P_s|\beta|^2}{1+|\beta|^2} + \dfrac{P_d}{1+|\alpha|^2} = D, \\[2mm] \dfrac{P_s\beta^*}{1+|\beta|^2} + \dfrac{P_d\,\alpha}{1+|\alpha|^2} = C \end{cases} \begin{cases} S = T_{11} - \dfrac{P_{od}}{2} - \dfrac{P_{cd}}{2} \\[2mm] D = T_{22} - \dfrac{7}{15}P_v - \dfrac{P_{md}}{2} - \dfrac{P_h}{2} \\[2mm] C = T_{12} \end{cases} \tag{8.7.11}
$$

The preceding expansions bring three equations with unknowns as in Table 8.5. The remaining powers P_s and P_d are given by the aid of the branch condition C_0 (8.2.4–8.2.6). In this case,

$$
C_0 = 2T_{11} - TP + P_1 + P_{md} \tag{8.7.12}
$$

Branch condition C_1 becomes,

$$
C_1 = T_{11} - T_{22} + \frac{7}{8}T_{33} + \frac{1}{16}\left(P_{md} + P_h\right) - \frac{15}{16}\left(P_{od} + P_{cd}\right) \tag{8.7.13}
$$

The flow chart of the seven-component scattering power decomposition is shown in Figure 8.15. The preceding procedure is directly reflected in the decomposition flow. Figure 8.16 shows a

TABLE 8.5

Helix Power, Volume-Scattering Power, Oriented Dipole Power, Compound Dipole Power, and Three Equations for 7SD Decomposition Scheme

	Vegetation			Dihedral
	$\sigma_{HH} - \sigma_{VV} > 2\,\text{dB}$	$|\sigma_{HH} - \sigma_{VV}| < 2\,\text{dB}$	$\sigma_{VV} - \sigma_{HH} > 2\,\text{dB}$	
Mixed dipole power	$P_{md} = 2\left|\text{Re}\{T_{23}\}\right|$	$P_{md} = 2\left|\text{Re}\{T_{23}\}\right|$	$P_{md} = 2\left|\text{Re}\{T_{23}\}\right|$	$P_{md} = 2\left|\text{Re}\{T_{23}\}\right|$
Helix power	$P_h = 2\left|\text{Im}\{T_{23}\}\right|$	$P_h = 2\left|\text{Im}\{T_{23}\}\right|$	$P_h = 2\left|\text{Im}\{T_{23}\}\right|$	$P_h = 2\left|\text{Im}\{T_{23}\}\right|$
Oriented dipole power	$P_{od} = 2\left|\text{Re}\{T_{13}\}\right|$	$P_{od} = 2\left|\text{Re}\{T_{13}\}\right|$	$P_{od} = 2\left|\text{Re}\{T_{13}\}\right|$	$P_{od} = 2\left|\text{Re}\{T_{13}\}\right|$
Compound dipole power	$P_{cd} = 2\left|\text{Im}\{T_{13}\}\right|$	$P_{cd} = 2\left|\text{Im}\{T_{13}\}\right|$	$P_{cd} = 2\left|\text{Im}\{T_{13}\}\right|$	$P_{cd} = 2\left|\text{Im}\{T_{13}\}\right|$
Volume-scattering power	$P_v = \frac{15}{8}\times\left[2T_{23} - P_{md} - P_h - P_{od} - P_{cd}\right]$	$P_v = 2\times\left[2T_{33} - P_{md} - P_h - P_{od} - P_{cd}\right]$	$P_v = \frac{15}{8}\times\left[2T_{33} - P_{md} - P_h - P_{od} - P_{cd}\right]$	$P_v = \frac{15}{16}\times\left[2T_{33} - P_{md} - P_h - P_{od} - P_{cd}\right]$
	$S = T_{11} - \frac{P_v}{2} - \frac{P_{od}}{2} - \frac{P_{cd}}{2}$	$S = T_{11} - \frac{P_v}{2} - \frac{P_{od}}{2} - \frac{P_{cd}}{2}$	$S = T_{11} - \frac{P_v}{2} - \frac{P_{od}}{2} - \frac{P_{cd}}{2}$	$S = T_{11} - \frac{P_{od}}{2} - \frac{P_{cd}}{2}$
	$D = T_{22} - \frac{7}{30}P_v - \frac{P_{md}}{2} - \frac{P_h}{2}$	$D = T_{22} - \frac{P_v}{4} - \frac{P_{md}}{2} - \frac{P_h}{2}$	$D = T_{22} - \frac{7}{30}P_v - \frac{P_{md}}{2} - \frac{P_h}{2}$	$D = T_{22} - \frac{7}{15}P_v - \frac{P_{md}}{2} - \frac{P_h}{2}$
	$C = T_{12} - \frac{1}{6}P_v$	$C = T_{12}$	$C = T_{12} + \frac{1}{6}P_v$	$C = T_{12}$

$$\frac{P_s}{1+|\beta|^2} + \frac{P_d|\alpha|^2}{1+|\alpha|^2} = S$$

$$\frac{P_s|\beta|^2}{1+|\beta|^2} + \frac{P_d}{1+|\alpha|^2} = D$$

$$\frac{P_s\beta^*}{1+|\beta|^2} + \frac{P_d\alpha}{1+|\alpha|^2} = C$$

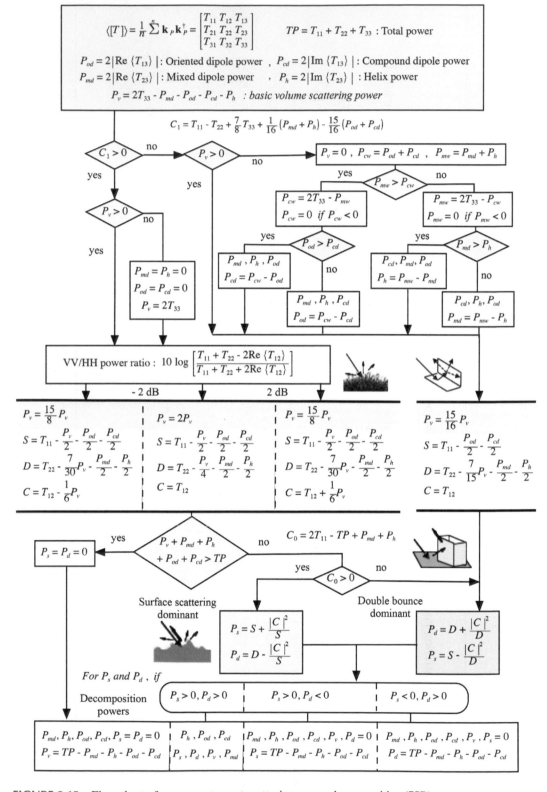

FIGURE 8.15 Flow chart of seven-component scattering power decomposition (7SD).

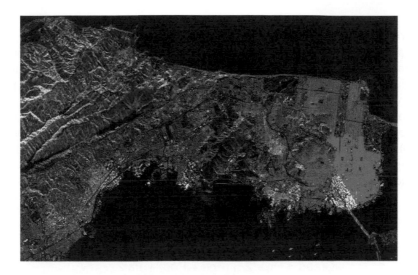

FIGURE 8.16   Decomposition image of San Francisco by 7SD.

decomposition image of San Francisco using the same data. The highly oriented urban area (triangular shaped in the lower corner) is now recognized as man-made structures with red colors. The area can be distinguished from vegetation area. P_{md} in this decomposition contributed to the double-bounce scattering.

8.8   COLOR-CODING

Color-coding is an important factor to create full-color images. If color-coding is arbitrarily chosen, the polarimetric color image will have tremendous variation. It will be difficult to evaluate. The important item is intuitive recognition for everybody.

RGB color-coding is frequently used for three-component decomposition as shown in Figure 8.17. Commonly used is P_s (blue), P_d (red), and P_v (green). Since this color-coding is widely used, it is better to follow it. If helix scattering power is added, we have four colors. We assign the power P_h to yellow (red/2 + green/2) as shown in Figure 8.18. Since the magnitude P_h is usually less than other

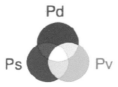

FIGURE 8.17   RGB color-coding.

Red = Pd + Ph/2
Green = Pv + Ph/2
Blue = Ps

FIGURE 8.18   Four-component color-coding and brightness assignment.

components, the yellow color does not affect color image so much. So we use the color-code for FDD, Y40, Y4R, S4R, and G4U.

For multiple scattering powers 6SD, we need to assign P_{od} and P_{cd} to different colors. There is no rule to assign colors. Two color composite images by assigning RGB color to six scattering powers can be considered; however, it is desirable to make one image using all scattering powers. After many trials, we assigned P_{od} and P_{cd} to orange as shown in Figure 8.19.

Another trial is to use intermediate colors such as pink, yellow, and light blue for corresponding coherency matrix elements and corresponding model powers referring to color combination (Figure 8.20). Since the combination of color-coding is almost infinite, we have made up a software that can composite any color arbitrarily as shown in Figure 8.21. This combination is used to create the San Francisco image in Figure 8.1.

Red = Pd + 3/5 (Pcd + Pod) + Ph/2
Green = Pv + 2/5 (Pcd + Pod) + Ph/2
Blue = Ps
Yellow = Ph (helix) = Ph/2 + Ph/2
Orange = 3/5 (Pcd + Pod) + 2/5 (Pcd + Pod)

FIGURE 8.19 6SD color assignment.

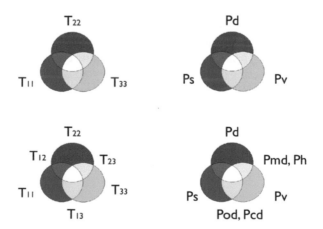

FIGURE 8.20 General idea for color-coding.

Color-code 2

	Red	Green	Blue	
Ps			1	
Pd	1	0.2		
Pv		1		
Ph	0.5	0.5		
Pod		0.5	0.5	
Pcd		0.5	0.5	

FIGURE 8.21 Color-composite with arbitrary ratio.

8.9 SUMMARY

As we have seen in this chapter, there are a number of decomposition methods. The methodology remains the same throughout from three- to seven-component decompositions. The model-based scattering power decomposition has the following advantages:

- Simple algorithm.
- Fast computation.
- Scattering powers are obtained. Power is a key radar parameter.
- Scattering powers are color-coded to composite full-color image.
- Color represents scattering mechanism.
- It is straightforward to interpret images by color for everybody.
- Time series change can be recognized by color change.
- Scattering powers can be further input to other applications such as classification and convolutional neural network (CNN).

Any decomposition method from three- to seven-component can be used according to analysis purpose (with or without rotation. See Figure 7.2).

APPENDIX

A8 COVARIANCE MATRIX FORMULATION AND BRANCH CONDITION

In the beginning of scattering power decomposition, the covariance matrix formulation was employed [10,11] because the diagonal elements are directly proportional to radar channel powers, and the off-diagonal terms represent cross-correlations between elements. Especially important is

C_{13} term, which has special physical property, such as $\mathrm{Re}\{C_{13}\} < 0$ for the double-bounce scattering and $\mathrm{Re}\{C_{13}\} > 0$ for the surface scattering. Therefore, we expand the scattering power decomposition in the covariance matrix form for the sake of deriving branch conditions.

  Covariance matrix formulation

$$\langle[C]\rangle = \begin{bmatrix} C_{11} & C_{21} & C_{13} \\ C_{21} & C_{22} & C_{23} \\ C_{31} & C_{32} & C_{33} \end{bmatrix} = \begin{vmatrix} \langle|S_{HH}|^2\rangle & \sqrt{2}\langle S_{HH}S_{HV}^*\rangle & \langle S_{HH}S_{VV}^*\rangle \\ \sqrt{2}\langle S_{HV}S_{HH}^*\rangle & 2\langle|S_{HV}|^2\rangle & \sqrt{2}\langle S_{HV}S_{VV}^*\rangle \\ \langle S_{VV}S_{HH}^*\rangle & \sqrt{2}\langle S_{VV}S_{HV}^*\rangle & \langle|S_{VV}|^2\rangle \end{vmatrix} \tag{A8.1}$$

Seven-component scattering power decomposition can be expanded as

$$\begin{bmatrix} C_{11} & C_{12} & C_{13} \\ C_{21} & C_{22} & C_{23} \\ C_{31} & C_{32} & C_{33} \end{bmatrix} = f_s\begin{bmatrix} |\beta|^2 & 0 & 0 \\ 0 & 0 & 0 \\ \beta^* & 0 & 1 \end{bmatrix} + f_d\begin{bmatrix} 1 & 0 & \alpha^* \\ 0 & 0 & 0 \\ \alpha & 0 & |\alpha|^2 \end{bmatrix} + \frac{P_v}{8}\begin{bmatrix} 3 & 0 & 1 \\ 0 & 2 & 0 \\ 1 & 0 & 3 \end{bmatrix}$$

$$+ \frac{P_{md}}{4}\begin{bmatrix} 1 & \pm\sqrt{2} & -1 \\ \pm\sqrt{2} & 2 & \mp\sqrt{2} \\ -1 & \mp\sqrt{2} & 1 \end{bmatrix} + \frac{P_h}{4}\begin{bmatrix} 1 & \pm j\sqrt{2} & -1 \\ \mp j\sqrt{2} & 2 & \pm j\sqrt{2} \\ -1 & \mp j\sqrt{2} & 1 \end{bmatrix} \tag{A8.2}$$

$$+ \frac{P_{od}}{4}\begin{bmatrix} 1 & \pm\sqrt{2} & 1 \\ \mp\sqrt{2} & 2 & \pm\sqrt{2} \\ 1 & \mp\sqrt{2} & 1 \end{bmatrix} + \frac{P_{cd}}{4}\begin{bmatrix} 1 & \pm j\sqrt{2} & 1 \\ \mp j\sqrt{2} & 2 & \pm j\sqrt{2} \\ 1 & \mp j\sqrt{2} & 1 \end{bmatrix}$$

C_{13} and C_{22} becomes

$$C_{13} = \langle S_{HH}S_{VV}^*\rangle = f_s\beta + f_d\alpha^* + \frac{1}{8}P_v - \frac{1}{4}\left(P_{md} + P_h\right) + \frac{1}{4}\left(P_{od} + P_{cd}\right)$$

$$C_{22} = 2\langle|S_{HV}|^2\rangle = \frac{1}{4}P_v + \frac{1}{2}\left(P_{md} + P_h\right) + \frac{1}{2}\left(P_{od} + P_{cd}\right)$$

Therefore: $2C_{13} - C_{22} = 2\langle S_{HH}S_{VV}^*\rangle - 2\langle|S_{HV}|^2\rangle = 2\left(f_s\beta + f_d\alpha^*\right) - \left(P_{md} + P_h\right)$

$$2\mathrm{Re}\{f_s\beta + f_d\alpha^*\} = 2\mathrm{Re}\{\langle S_{HH}S_{VV}^*\rangle\} - 2\langle|S_{HV}|^2\rangle + \left(P_{md} + P_h\right) \tag{A8.3}$$

$$= T_{11} - T_{22} - T_{33} + \left(P_{md} + P_h\right) = 2T_{11} - TP + P_h + P_{md} \quad (\text{coherency matrix expression})$$

- Branch condition C_0

  Since $\text{Re}\{\alpha\} < 0$ and $\text{Re}\{\beta\} > 0$ are assumed in the covariance matrix formulation [10], the sign of $\text{Re}\{f_s\beta + f_d\alpha^*\}$ determines which scattering mechanism is dominant, surface or double-bounce.

  Therefore, we put

$$C_0 = 2T_{11} - TP + P_h + P_{md} \ (\text{coherency matrix expression}) \tag{A8.4}$$

  and check the sign of C_0.

  $C_0 > 0 \Rightarrow$ surface scattering dominant

  $C_0 < 0 \Rightarrow$ double-bounce scattering dominant

  If P_{md} or P_h is not accounted for in the decomposition, discard it.

$$C_0 = 2T_{11} - TP + P_h \quad (\text{for 6SD, G4U, S4R, Y4R, Y40}) \tag{A8.5}$$

$$C_0 = 2T_{11} - TP \quad (\text{for FDD}) \tag{A8.6}$$

- Branch condition C_1

  This condition is used to decide which comes from the HV component, oriented dihedral surface or vegetation. Since the volume scattering model in the covariance matrix becomes

$$[T]^{dihedral} - \frac{1}{15}\begin{bmatrix} 0 & 0 & 0 \\ 0 & 7 & 0 \\ 0 & 0 & 8 \end{bmatrix} \Rightarrow [C]^{dihedral} = \frac{1}{30}\begin{bmatrix} 7 & 0 & -7 \\ 0 & 16 & 0 \\ -7 & 0 & 7 \end{bmatrix}$$

$$\begin{bmatrix} C_{11} & C_{12} & C_{13} \\ C_{21} & C_{22} & C_{23} \\ C_{31} & C_{32} & C_{33} \end{bmatrix} = f_s\begin{bmatrix} |\beta|^2 & 0 & 0 \\ 0 & 0 & 0 \\ \beta^* & 0 & 1 \end{bmatrix} + f_d\begin{bmatrix} 1 & 0 & \alpha^* \\ 0 & 0 & 0 \\ \alpha & 0 & |\alpha|^2 \end{bmatrix} + \frac{P_v}{30}\begin{bmatrix} 7 & 0 & -7 \\ 0 & 16 & 0 \\ -7 & 0 & 7 \end{bmatrix}$$

$$+ \frac{P_{md}}{4}\begin{bmatrix} 1 & \pm\sqrt{2} & -1 \\ \pm\sqrt{2} & 2 & \mp\sqrt{2} \\ -1 & \mp\sqrt{2} & 1 \end{bmatrix} + \frac{P_h}{4}\begin{bmatrix} 1 & \pm j\sqrt{2} & -1 \\ \mp j\sqrt{2} & 2 & \pm j\sqrt{2} \\ -1 & \mp j\sqrt{2} & 1 \end{bmatrix} \tag{A8.7}$$

$$+ \frac{P_{od}}{4}\begin{bmatrix} 1 & \pm\sqrt{2} & 1 \\ \mp\sqrt{2} & 2 & \pm\sqrt{2} \\ 1 & \mp\sqrt{2} & 1 \end{bmatrix} + \frac{P_{cd}}{4}\begin{bmatrix} 1 & \pm j\sqrt{2} & 1 \\ \mp j\sqrt{2} & 2 & \pm j\sqrt{2} \\ 1 & \mp j\sqrt{2} & 1 \end{bmatrix}$$

C_{13} and C_{22} becomes

$$C_{13} = \langle S_{HH}S_{VV}^* \rangle = f_s\beta + f_d\alpha. - \frac{7}{30}P_v - \frac{1}{4}\left(P_{md} + P_h\right) + \frac{1}{4}\left(P_{od} + P_{cd}\right)$$

$$C_{22} = 2\left\langle |S_{HV}|^2 \right\rangle = \frac{16}{30} P_v + \frac{1}{2}\left(P_{md} + P_h\right) + \frac{1}{2}\left(P_{od} + P_{cd}\right)$$

Therefore: $2C_{13} - C_{22} = 2\langle S_{HH}S_{VV}^* \rangle - 2\langle | S_{HV} |^2 \rangle = 2\left(f_s\beta + f_d\alpha^*\right) - \left(P_v + P_{md} + P_h\right)$

$$2\mathrm{Re}\left\{f_s\beta + f_d\alpha^*\right\} = 2\mathrm{Re}\left\{\langle S_{HH}S_{VV}^* \rangle\right\} - 2\langle | S_{HV} |^2 \rangle + \left(P_v + P_{md} + P_h\right)$$

$$= T_{11} - T_{22} - T_{33} + P_v + P_{md} + P_h = 2T_{11} - TP + P_v + P_h + P_{md}$$

$$\text{(coherency matrix expression)} \tag{A8.8}$$

Since $P_v = \frac{15}{16}\left[2T_{33} - P_{md} - P_h - P_{od} - P_{cd}\right]$ for the oriented dihedral, the expression of coherency matrix form can be further rewritten as

$$C_1 = T_{11} - T_{22} + \frac{7}{8}T_{33} + \frac{1}{16}\left(P_{md} + P_h\right) - \frac{15}{16}\left(P_{od} + P_{cd}\right) \text{ for 7SD} \tag{A8.9}$$

to check the sign of $\mathrm{Re}\left\{f_s\beta + f_d\alpha^*\right\}$.

$$C_1 > 0 \Rightarrow \text{vegetation scattering}$$

$$C_1 < 0 \Rightarrow \text{dihedral scattering}$$

If scattering models are not so many, we just discard equation (A8.9) and use the following:

$$c_1 = T_{11}(\theta) - T_{22}(\theta) + \frac{7}{8}T_{33}(\theta) + \frac{1}{16}P_h - \frac{15}{16}\left(P_{od} + P_{cd}\right) \text{ for 6SD} \tag{A8.10}$$

$$C_1 = T_{11}(\theta) - T_{22}(\theta) + \frac{7}{8}T_{33}(\theta) + \frac{1}{16}P_h \text{ for S4R, G4U} \tag{A8.11}$$

The degree of polarization can be used to discriminate vegetation and man-made targets adaptively [25].

REFERENCES

1. S. R. Cloude and E. Pottier, "A review of target decomposition theorems in radar polarimetry," *IEEE Trans. Geosci. Remote Sens.*, vol. 34, no. 2, pp. 498–518, 1996.
2. Y. Yamaguchi, *Radar Polarimetry from Basics to Application (in Japanese)*, IEICE, Tokyo, 2007.
3. J. S. Lee and E. Pottier, *Polarimetric Radar Imaging from Basics to Applications*, CRC Press, Boca Raton, FL, 2009.
4. S. R. Cloude, *Polarisation Applications in Remote Sensing*, Oxford University Press, Oxford, UK, 2009.
5. Y.-Q. Jin and F. Xu, *Theory and Approach for Polarimetric Scattering and Information Retrieval of SAR Remote Sensing (In Modern Chinese)*, Science Press, Beijing, 2008, ISBN978-7-03-022649-5.
6. J. van Zyl and Y. Kim, *Synthetic Aperture Radar Polarimetry*, Wiley, Hoboken, NJ, 2011.
7. A. Moreira et al., A tutorial on synthetic aperture radar, *IEEE GRSS Magazine*, pp. 6–43, March 2013.
8. S. W. Chen, X.-S. Wang, S.-P. Xiao, and M. Sato, *Target Scattering Mechanism in Polarimetric Synthetic Aperture Radar—Interpretation and Application*, Springer, Singapore, 2017.

9. Y. Cui, Y. Yamaguchi, H. Kobayashi, and J. Yang, "Filtering of polarimetric synthetic aperture radar images: a sequential approach," *Electronic Proceedings of IEEE-IGARSS 2012*, Munich, July 2012.

10. A. Freeman and S. Durden, "A three-component scattering model for polarimetric SAR data," *IEEE Trans. Geosci. Remote Sens.*, vol. 36, no. 3, pp. 963–973, May 1998.

11. Y. Yamaguchi, T. Moriyama, M. Ishido, and H. Yamada, "Four-component scattering model for polarimetric SAR image decomposition," *IEEE Trans. Geosci. Remote Sens.*, vol. 43, no. 8, pp. 1699–1706, August 2005.

12. F. Xu and Y. Q. Jin, "Deorientation theory of polarimetric scattering targets and application to terrain surface classification," *IEEE Trans. Geosci. Remote Sens.*, vol. 43, no. 10, pp. 2351–2364, October 2005.

13. L. Zhang, B. Zou, H. Cai, and Y. Zhang, "Multiple-component scattering model for polarimetric SAR image decomposition," *IEEE Geosci. Remote Sens. Lett.*, vol. 5, no. 4, pp. 603–607, October 2008.

14. W.-T. An, Y. Cui, and J. Yang, "Three-component model-based decomposition for polarimetric SAR data," *IEEE Trans. Geosci. Remote Sens.*, vol. 48, no. 6, pp. 2732–2739, June 2010.

15. J. S. Lee and T. Ainsworth, "The effect of orientation angle compensation on coherency matrix and polarimetric target decompositions," *Proc. of EUSAR 2010*, Germany, 2010, and *IEEE Trans. Geosci. Remote Sens.*, vol. 49, no. 1, pp. 53–64, January 2011.

16. Y. Yamaguchi, A. Sato, W.-M. Boerner, R. Sato, and H. Yamada, "Four-component scattering power decomposition with rotation of coherency matrix," *IEEE Trans. Geosci. Remote Sens.*, vol. 49, no. 6, pp. 2251–2258, June 2011.

17. W.-T. An, C. Xie, X. Yuan, Y. Cui, and J. Yang, "Four-component decomposition of polarimetric SAR images with deorientation," *IEEE Geosci. Remote Sens. Lett.*, vol. 8, no. 6, pp. 1090–1094, November 2011.

18. M. Arii, J. J. van Zyl, and Y. Kim, "Adaptive model-based decomposition of polarimetric SAR covariance matrices," *IEEE Trans. Geosci. Remote Sens.*, vol. 49, no. 3, pp. 1104–1113, March 2011.

19. A. Sato, Y. Yamaguchi, G. Singh, and S.-E. Park, "Four-component scattering power decomposition with extended volume scattering model," *IEEE Geosci. Remote Sens. Lett.*, vol. 9, no. 2, pp. 166–170, March 2012.

20. G. Singh, Y. Yamaguchi, and S.-E. Park, "General four-component scattering power decomposition with unitary transformation of coherency matrix," *IEEE Trans. Geosci. Remote Sens.*, vol. 51, no. 5, pp. 3014–3022, May 2013.

21. S. W. Chen, X. Wang, S. P. Xiao, and M. Sato, "General polarimetric model-based decomposition for coherency matrix," *IEEE Trans. Geosci. Remote Sens.*, vol. 52, no. 3, pp. 1843–1855, March 2014.

22. J. S. Lee, T. L. Ainsworth, and Y. Wang, "Generalized polarimetric model-based decompositions using incoherent scattering models," *IEEE Trans. Geosci. Remote Sens.*, vol. 52, pp. 2474–2491, 2014.

23. G. Singh and Y. Yamaguchi, "Model-based six-component scattering matrix power decomposition," *IEEE Trans. Geosci. Remote Sens.*, vol. 56, no. 10, pp. 5687–5704, October 2018.

24. G. Singh, R. Malik, S. Mohanty, V. S. Rathore, K. Yamada, M. Umemura, and Y. Yamaguchi, "Seven-component scattering power decomposition of POLSAR coherency matrix," *IEEE Trans. Geosci. Remote Sens.*, vol. 57, vol. 6, June 2019. doi:10.1109/TGRS.2019.2920762.

25. A. Bhattacharya, G. Singh, S. Manickman, and Y. Yamaguchi, "Adaptive general four-component scattering power decomposition with unitary transformation of coherency matrix," *IEEE Geosci. Remote Sens. Lett.*, vol. 12, no. 10, pp. 2110–2114, 2015.

9 Correlation and Similarity

The concept of correlation or coherence plays an important role in signal processing. The correlation coefficient is a key parameter for investigating object detection, classification, and identification. It has been used in radar signal analyses as well as checking radar channel hardware. The definition comes from the mathematical theory.

If there are two complex signals, $s_1(x)$ and $s_2(x)$, the Cauchy-Schwarz inequality holds,

$$\left| \int_a^b s_1(x) s_2^*(x) dx \right|^2 \leq \int_a^b |s_1(x)|^2 dx \int_a^b |s_2(x)|^2 dx \tag{9.0.1}$$

$$0 \leq \frac{\left| \int_a^b s_1(x) s_2^*(x) dx \right|}{\sqrt{\int_a^b |s_1(x)|^2 dx \int_a^b |s_2(x)|^2 dx}} \leq 1 \tag{9.0.2}$$

If we replace complex signals $s_1(x)$ and $s_2(x)$ to scattering matrix elements s_1 and s_2, we can define polarimetric correlation coefficient or polarimetric coherency as

$$\gamma = \frac{\left\langle s_1 s_2^* \right\rangle}{\sqrt{\left\langle s_1 s_1^* \right\rangle \left\langle s_2 s_2^* \right\rangle}} = \frac{\left\langle s_1 s_2^* \right\rangle}{\sqrt{\left\langle |s_1|^2 \right\rangle \left\langle |s_2|^2 \right\rangle}} \tag{9.0.3}$$

$$0 < |\gamma| < 1 \tag{9.0.4}$$

where $*$ denotes complex conjugation and $\langle \cdots \rangle$ shows ensemble average.

This becomes an indicator such that two signals are:

$\gamma \approx 1$: very close, very similar
$\gamma \approx 0$: not close to each other, different, or independent

This basic concept brings diverse applications. We may think of, what is the best polarization basis? Which polarization basis provides us with useful information for target detection or classification? In other words, what is the best straight line (basis) through the Poincaré sphere as shown in Figure 9.1.

In order to specify polarization basis, we can use the following notation:

$$\gamma_{XY-AB} = Cor(XY, AB) = \frac{\left\langle S_{XY} S_{AB}^* \right\rangle}{\sqrt{\left\langle S_{XY} S_{XY}^* \right\rangle \left\langle S_{AB} S_{AB}^* \right\rangle}} \tag{9.0.5}$$

where the subscripts X, Y, A, and B denotes polarization symbols, such as H, V, L, R, $\pm 45°$. Then we can define the polarimetric correlation precisely.

In this chapter, we deal with the utilization of the correlation coefficient with emphasis placed on detection or classification of objects.

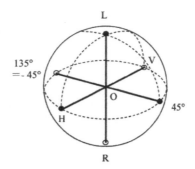

FIGURE 9.1   Poincaré sphere and polarization bases (straight lines).

9.1   COVARIANCE MATRIX AND CORRELATION COEFFICIENT

Using scattering matrices, the covariance matrix can be created and averaged in an imaging window. After ensemble averaging, the covariance matrix in the HV polarization basis is expressed by

$$\left\langle\left[C\left(HV\right)\right]\right\rangle = \begin{bmatrix} C_{11}C_{21}C_{13} \\ C_{21}C_{22}C_{23} \\ C_{31}C_{32}C_{33} \end{bmatrix} = \begin{bmatrix} \left\langle\left|S_{HH}\right|^2\right\rangle & \sqrt{2}\left\langle S_{HH}S_{HV}^*\right\rangle & \left\langle S_{HH}S_{VV}^*\right\rangle \\ \sqrt{2}\left\langle S_{HV}S_{HH}^*\right\rangle & 2\left\langle\left|S_{HV}\right|^2\right\rangle & \sqrt{2}\left\langle S_{HV}S_{VV}^*\right\rangle \\ \left\langle S_{VV}S_{HH}^*\right\rangle & \sqrt{2}\left\langle S_{VV}S_{HV}^*\right\rangle & \left\langle\left|S_{VV}\right|^2\right\rangle \end{bmatrix} \tag{9.1.1}$$

We normalize it by the H-pol channel power $\sigma_{HH} = C_{11} = \left\langle\left|S_{HH}\right|^2\right\rangle$ $\qquad$ (9.1.2)

and introduce power ratios $g = \dfrac{\left\langle\left|S_{VV}\right|^2\right\rangle}{\left\langle\left|S_{HH}\right|^2\right\rangle}$, and $e = \dfrac{\left\langle\left|S_{HV}\right|^2\right\rangle}{\left\langle\left|S_{HH}\right|^2\right\rangle}$ $\qquad$ (9.1.3)

Then the off-diagonal terms can be related to the correlation coefficients

$$\gamma_{HH-VV} = Cor\left(HH, VV\right) = \frac{\left\langle S_{HH}S_{VV}^*\right\rangle}{\sqrt{\left\langle\left|S_{HH}\right|^2\right\rangle\left\langle\left|S_{HV}\right|^2\right\rangle}} = \frac{C_{13}}{\sqrt{C_{11}C_{33}}} \tag{9.1.4}$$

$$\gamma_{HH-HV} = Cor\left(HH, HV\right) = \frac{\left\langle S_{HH}S_{HV}^*\right\rangle}{\sqrt{\left\langle\left|S_{HH}\right|^2\right\rangle\left\langle\left|S_{HV}\right|^2\right\rangle}} = \frac{C_{12}}{\sqrt{C_{11}C_{22}}} \tag{9.1.5}$$

$$\gamma_{VV-HV} = Cor\left(VV, HV\right) = \frac{\left\langle S_{VV}S_{HV}^*\right\rangle}{\sqrt{\left\langle\left|S_{VV}\right|^2\right\rangle\left\langle\left|S_{HV}\right|^2\right\rangle}} = \frac{C_{23}}{\sqrt{C_{22}C_{33}}} \tag{9.1.6}$$

Finally, the covariance matrix can be rewritten in terms of the correlation coefficient as

$$\left\langle\left[C\left(HV\right)\right]\right\rangle = \sigma_{HH} \begin{bmatrix} 1 & \sqrt{2e}\,\gamma_{HH-HV} & \sqrt{g}\gamma_{HH-VV} \\ \sqrt{2e}\,\gamma_{HH-HV}^* & 2e & \sqrt{2eg}\,\gamma_{HV-VV} \\ \sqrt{g}\,\gamma_{HH-VV}^* & \sqrt{2eg}\,\gamma_{HV-VV}^* & g \end{bmatrix} \tag{9.1.7}$$

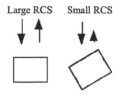

FIGURE 9.2   RCS difference by orientation.

For natural distributed areas such as forests and vegetation, the so-called "reflection symmetry condition" holds,

$$\langle S_{HH}S_{HV}^{*} \rangle \approx 0, \quad \langle S_{VV}S_{HV}^{*} \rangle \approx 0.$$

Under the reflection symmetry condition, the covariance matrix becomes a simple form with five parameters [1],

$$\left\langle \left[C(HV) \right] \right\rangle_{ref.sym.} = \sigma_{HH} \begin{bmatrix} 1 & 0 & \sqrt{g}\,\gamma_{HH-VV} \\ 0 & 2e & 0 \\ \sqrt{g}\,\gamma_{HH-VV}^{*} & 0 & g \end{bmatrix} \tag{9.1.8}$$

This expression means that five parameters (σ_{HH}, power ratio e and g, and the complex correlation coefficient γ_{HH-VV}) are sufficient for characterizing these area. The choice of correlation coefficient $0 < |\gamma_{HH-VV}| < 1$, or $-1 < \mathrm{Re}(\gamma_{HH-VV}) < 1$ is left for applications. Using these five parameters, it is possible to classify or identify object types that exhibit the reflection symmetry condition.

For non-reflection symmetric conditions ($\langle S_{HH}S_{HV}^{*} \rangle \neq 0, \langle S_{VV}S_{HV}^{*} \rangle \neq 0$), the covariance expression remains in the form of equation (9.1.7). This situation corresponds to complex urban areas where many objects are aligned partially orthogonal and partially oriented to radar illumination. If buildings are oriented as shown in Figure 9.2, small radar cross section (RCS) returns back to radar, compared to large RCS by orthogonal case [2]. The contribution ratio of the HV component becomes rather big in the oriented scattering case, which produces $\langle S_{HH}S_{HV}^{*} \rangle \neq 0, \langle S_{VV}S_{HV}^{*} \rangle \neq 0$. For practical applications, it is desirable to use a simple or similar scheme to classify complex man-made objects and natural objects. In order to expand the classification or detection capability to a general scattering case by PolSAR, it is necessary to check and find out a useful correlation coefficient in the various polarization basis.

9.2   CORRELATION COEFFICIENT IN THE CIRCULAR POLARIZATION BASIS

By the definition of the covariance matrix in the circular polarization basis, the correlation coefficient becomes [3]

$$Cor(LL,RR) = \gamma_{LL-RR} = |\gamma_{LL-RR}| \angle \varphi_{LL-RR}$$

$$= \frac{\left\langle 4|S_{HV}|^{2} - |S_{HH}-S_{VV}|^{2} \right\rangle - j4\,\mathrm{Re}\left\langle S_{HV}^{*}(S_{HH}-S_{VV}) \right\rangle}{\sqrt{\left\langle |S_{HH}-S_{VV}+j\,2S_{HV}|^{2} \right\rangle \left\langle |S_{HH}-S_{VV}-j2S_{HV}|^{2} \right\rangle}} \tag{9.2.1}$$

$$\varphi_{LL-RR} = \tan^{-1} \frac{4\,\mathrm{Re}\left\langle S_{HV}^{*}(S_{HH}-S_{VV}) \right\rangle}{\left\langle |S_{HH}-S_{VV}|^{2} - 4|S_{HV}|^{2} \right\rangle} \tag{9.2.2}$$

Since $Cor(LL,LR)$ and $Cor(LR,RR)$ are not so helpful for applications, we omit them. Under the reflection symmetry condition, the correlation coefficient γ_{LL-RR} in equation (9.2.1) becomes equation (9.2.3), which is directly related to the coherency matrix elements. For discrimination of the non-reflection symmetry case, we denote it for the reflection symmetry condition as

$$\gamma_{LL-RR}(0) = \frac{\left\langle 4|S_{HV}|^2 - |S_{HH} - S_{VV}|^2 \right\rangle}{\left\langle 4|S_{HV}|^2 + |S_{HH} - S_{VV}|^2 \right\rangle} = \frac{T_{33} - T_{22}}{T_{33} + T_{22}} \tag{9.2.3}$$

$$\varphi_{LL-RR}(0) = \pi, 0 \tag{9.2.4}$$

From this expression, we can understand that $\gamma_{LL-RR}(0)$ can be expressed in terms of coherency matrix elements and becomes real-valued, that is, it locates close to the real axis when drawn in a complex plane. Other objects with non-reflection symmetry locate away from the real axis [4]. We will check this property in the next section.

Under the non-reflection symmetry condition, the phase expression has an interesting feature,

$$\varphi_{LL-RR} = \tan^{-1} \frac{4\,\mathrm{Re}\left\langle S_{HV}^*(S_{HH} - S_{VV})\right\rangle}{\left\langle |S_{HH} - S_{VV}|^2 - 4|S_{HV}|^2\right\rangle} = \tan^{-1}\left(\frac{2\mathrm{Re}\{T_{23}\}}{T_{22} - T_{33}}\right) \tag{9.2.5}$$

If we rotate the circular polarization covariance matrix by θ, the cross-correlation between LL and RR becomes

$$\left\langle S_{LL}(\theta)S_{RR}^*(\theta)\right\rangle = \{S_{LL}S_{RR}^*\}e^{-j4\theta} \Rightarrow |\gamma_{LL-RR}|e^{j\varphi}e^{-j4\theta} \tag{9.2.6}$$

For this term to be real, $\varphi = 4\theta$ \hfill (9.2.7)

Therefore, the rotation angle must satisfy

$$\theta = 4\varphi = \frac{1}{4}\tan^{-1}\left(\frac{2\mathrm{Re}\{T_{23}\}}{T_{22} - T_{33}}\right) \tag{9.2.8}$$

which is the same as the rotation angle of coherency matrix (Y4R) and the basis of polarization orientation compensation [5,6].

9.3   CORRELATION COEFFICIENT IN THE 45°- AND 135°-ORIENTED LINEAR POLARIZATION BASIS

The 45°-oriented linear polarization basis is denoted here as the X–Y basis. By the definition, the correlation coefficient in the X–Y basis becomes,

$$\gamma_{XX-YY} = |\gamma_{XX-YY}|\angle\varphi_{XX-YY} = \frac{\left\langle |S_{HH} + S_{VV}|^2 - 4|S_{HV}|^2\right\rangle - j\,4\,\mathrm{Im}\left\langle S_{HV}^*(S_{HH} + S_{VV})\right\rangle}{\sqrt{\left\langle |S_{HH} - S_{VV} + 2S_{HV}|^2\right\rangle \left\langle |S_{HH} - S_{VV} - 2S_{HV}|^2\right\rangle}} \tag{9.3.1}$$

$$\varphi_{XX-YY} = \tan^{-1}\frac{4\,\mathrm{Im}\left\langle S_{HV}^*(S_{HH} + S_{VV})\right\rangle}{\left\langle 4|S_{HV}|^2 - |S_{HH} + S_{VV}|^2\right\rangle}. \tag{9.3.2}$$

Under the reflection symmetry condition, they become

$$\gamma_{XX-YY}\left(0\right) = \frac{\left\langle \left|S_{HH} + S_{VV}\right|^2 - 4\left|S_{HV}\right|\right\rangle}{\left\langle \left|S_{HH} + S_{VV}\right|^2 + 4\left|S_{HV}\right|^2\right\rangle} = \frac{T_{11} - T_{33}}{T_{11} + T_{33}} \tag{9.3.3}$$

$$\varphi_{XX-YY}\left(0\right) = 0 \tag{9.3.4}$$

$\gamma_{XX-YY}(0)$ can be written in terms of coherency matrix elements. However, equation (9.3.3) yields positive real value (close to 1) since $T_{11} \gg T_{33}$ for most objects. This will not serve classification effectively.

9.4   CORRELATION COEFFICIENT IN THE ARBITRARY POLARIZATION BASIS

The expression of correlation coefficient is dependent on the polarization basis. For general expression of correlation expression, we use scattering matrix elements in the arbitrary polarization basis (equation 4.2.7).

$$\gamma_{AA-BB} = \frac{\left\langle S_{AA}S_{BB}^{*}\right\rangle}{\sqrt{\left\langle S_{AA}S_{AA}^{*}\right\rangle\left\langle S_{BB}S_{BB}^{*}\right\rangle}} \tag{9.4.1}$$

$$S_{AA} = \frac{e^{j2\alpha}}{1 + \rho\rho^{*}}\left(S_{HH} + 2\rho\, S_{HV} + \rho^2 S_{VV}\right),\ S_{BB} = \frac{e^{-j2\alpha}}{1 + \rho\rho^{*}}\left(\rho^{*2}S_{HH} - 2\rho^{*}S_{HV} + S_{VV}\right) \tag{9.4.2}$$

$$\rho = \frac{\sin\tau\cos\varepsilon + j\cos\tau\sin\varepsilon}{\cos\tau\cos\varepsilon - i\sin\tau\sin\varepsilon},\quad \alpha = \tan^{-1}\left(\tan\tau\tan\varepsilon\right) \tag{9.4.3}$$

If $\alpha = 0$, $p = 0$, the correlation coefficient (equation 9.1.4) in the HV polarization basis is derived. If $\alpha = 0$, $\rho = j$, the expression (equation 9.2.1) in the circular polarization basis is obtained. In addition, $\alpha = 0$, $\rho = 1$ leads to (equation 9.3.1) in the XY polarization basis.

   If we choose special polarization states such as co-pol max, it is possible to define and evaluate the correlation coefficient on that specific polarization state. However, the value becomes scene dependent. The corresponding values will change in imaging window by window, which is not suitable for classification throughout one image.

9.5   CORRELATION COEFFICIENT IN THE PAULI BASIS

In the coherency matrix expression, the off-diagonal terms represent cross-correlations between Pauli vector elements. In a similar way to the covariance matrix, we can define the correlation coefficients using coherency matrix elements,

$$\gamma_{HH+VV,\,HH-VV} = Cor\left(HH + VV,\, HH - VV\right) = \frac{T_{12}}{\sqrt{T_{11}T_{22}}} \tag{9.5.1}$$

$$_{HH+VV,\,HV} = Cor\left(HH + VV,\, HV\right) = \frac{T_{13}}{\sqrt{T_{11}T_{33}}} \tag{9.5.2}$$

$$\gamma_{HH+VV,HV} = Cor\left(HH-VV, HV\right) = \frac{T_{23}}{\sqrt{T_{22}T_{33}}} \qquad (9.5.3)$$

Since coherency matrix elements are related to physical scattering mechanisms, it is easy to interpret the meaning of these values (T_{11}: surface scattering, T_{22}: double-bounce scattering, T_{33}: cross-polarization generation).

S. W. Chen investigated the rotation angle dependency of the correlation coefficient in the various polarization basis [7]. The resultant figure correlation coefficient as a function of angle looks like an antenna radiation pattern which is dependent on target.

9.6 COMPARISON AMONG VARIOUS CORRELATION COEFFICIENTS

9.6.1 Magnitude of Correlation Coefficient

A question comes out. Which correlation coefficient is better for the classification of objects? Is *HH–VV* information better than *LL–RR* information? How about the correlation *XX–YY*?

We have calculated some correlation coefficients using PiSAR-L data over Niigata City [8]. The window size was chosen as 7×7. The color composite and the correlation coefficient magnitude images are shown in Figure 9.3.

It is seen that $|\gamma_{HH-VV}|$ image and $|\gamma_{HH+VV,HH-VV}|$ image are similar to each other, and $|\gamma_{XX-YY}|$ and $|\gamma_{LL-RR}|$ images also look similar. $|\gamma_{HH-VV}|$ image appears a little bit noisy. The red color indicates a large value that can be seen in urban man-made structures for $|\gamma_{HH-VV}|$ and $|\gamma_{HH+VV,HH-VV}|$ images. On the other hand, the red color can be found mainly in water surface area of the Toyano Lagoon and Shinanogawa River for the $|\gamma_{XX-YY}|$ and $|\gamma_{LL-RR}|$ images. This means $|\gamma_{HH-VV}|$ or $|\gamma_{LL-RR}|$ is suitable for detecting a flat surface area. The bright red pattern is completely different from these two

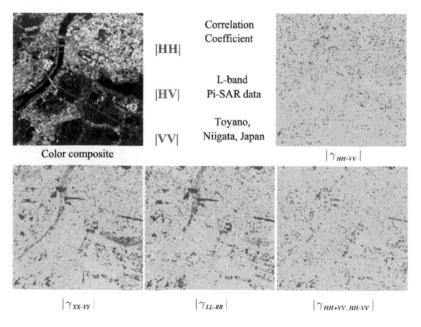

FIGURE 9.3 Correlation coefficient magnitude images.

groups. From Figure 9.3, we can understand that correlation coefficient becomes high when neighboring pixels have similar scattering characteristics regardless of the scatterer itself.

9.6.2 CORRELATION COEFFICIENT VERSUS WINDOW SIZE

Range resolution and azimuth resolution are dependent on the radar system. The PiSAR-L airborne radar system had approximately 3×3 m resolution on the ground, and the PiSAR-X2 system had a 30×30 cm resolution. If we take an imaging window with 10×10 pixels, the actual ground size becomes completely different from each other. The big window would contain various targets inside the window and blur the image.

For reliable calculation of the coefficient, we have checked the variation of correlation coefficient vs. window size such as 3×3, 5×5, 7×7, and 9×9. After checking the convergence by window size, 5×5 pixels are used in the calculation of correlation coefficient. The number of pixels more than 25 would ensure the statistical property.

9.6.3 CORRELATION COEFFICIENT VERSUS SCATTERER

If we take a 5×5 pixel window, what is the best correlation coefficient for the classification? We tried to find out the correlation coefficient response to various targets. Figure 9.4 shows the variation of these values for each target.

It is seen that correlation value is dependent on the target. The correlation coefficient in the circular polarization basis has the largest dynamic range. If the dynamic range is big, it is useful for discriminating the target according to the value. Therefore, the correlation coefficient in the circular polarization will play an important role for the classification of targets among others.

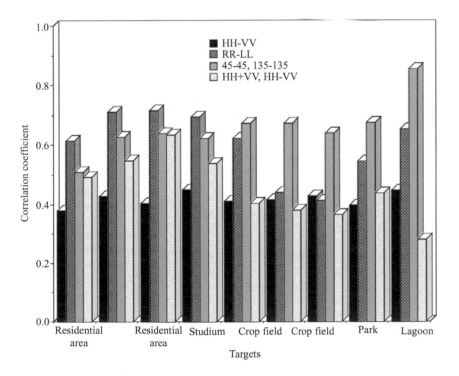

FIGURE 9.4 Correlation coefficient vs. various targets.

9.6.4 Phase Information of Correlation Coefficient

Phase information may become one of the important parameters for discriminating target. We calculated the X-band Pi-SAR data over Niigata University. The scene contains oriented buildings, residential houses, a playground, pine trees, a river, a water gate, sea, etc. Figure 9.5 shows a Pauli-based image and close-up sin(phase) images.

From this figure, we can see the $\sin \varphi_{LL-RR}$ image provides us with some interesting results. The phase φ_{LL-RR} among others seems to be related to object information.

$$\varphi_{LL-RR} = \tan^{-1}\left(\frac{2\mathrm{Re}\{T_{23}\}}{T_{22}-T_{33}}\right) \qquad (9.6.1)$$

The next step is to check the value distribution along a transect in Figure 9.5. The resultant value profile $\left(\sin \varphi_{LL-RR}\right)$ and validation image along path B are shown in Figure 9.6. The photos around the ground are displayed. It is very interesting to see the profile roughly corresponds to target distribution.

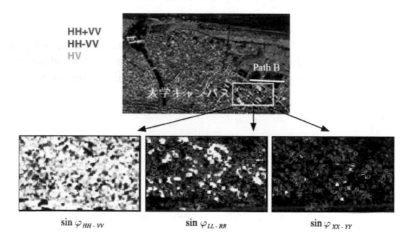

FIGURE 9.5 Phase images by correlation coefficients over the Niigata University area.

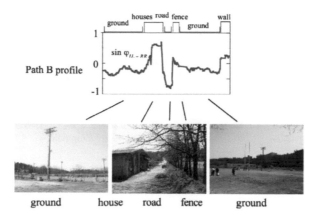

FIGURE 9.6 Phase information along path B of Niigata University.

9.7 APPLICATION OF CORRELATION COEFFICIENT IN THE CIRCULAR POLARIZATION

9.7.1 FREQUENCY DEPENDENCY

Figure 9.7 shows the correlation coefficient images in the circular polarization at the L-band and X-band by airborne PiSAR systems. Due to the difference in scattering phenomena by frequency band and the difference in radar resolution, we can see the difference in the images. The lower-half area in the image is a rice paddy field, and the upper area is residential areas and pine trees, next to the Sea of Japan. For the X-band wavelength, rice stems, and leaves have complicated structures. Hence, the scattering becomes random, yielding almost zero correlations. On the other hand, the correlation coefficient in the L-band in the paddy field is high, due to a similar scattering mechanism in the rice field. This characteristic appears as a big correlation value in the L-band, and small in the X-band.

9.7.2 DISTRIBUTION OF THE CORRELATION COEFFICIENT

We have picked up typical characteristic areas (patch A: sea, patch B: pine trees, patch C: urban area with non-orthogonal illumination, and patch D: urban area with orthogonal illumination) in Figure 9.9 and derived the correlation coefficient in the X-band as shown in Figure 9.8. From this figure, we can see,

1. $|\gamma_{LL-RR}|$ has big values for man-made structures, where the reflection symmetry condition does not hold.
2. $|\gamma_{LL-RR}|$ has small values for vegetation, where the reflection symmetry condition holds.

The correlation coefficient is frequency dependent; however, we may use these characteristics for the detection of man-made objects and vegetation.

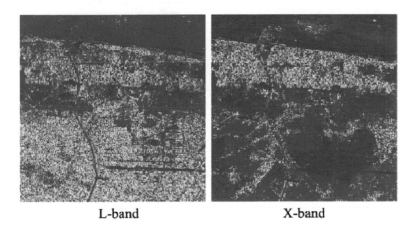

L-band   **X-band**

FIGURE 9.7 Correlation coefficient images in the circular polarization basis.

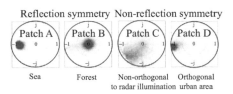

FIGURE 9.8 Distribution of the correlation coefficient in the circular polarization basis.

9.8 DETECTION OF SPECIFIC AREA USING CORRELATION COEFFICIENT IN THE CIRCULAR POLARIZATION BASIS

9.8.1 Oriented Urban Area with Non-Reflection Symmetry Condition

Under the reflection symmetry condition, the expression of circular polarization correlation coefficient becomes,

$$\gamma_{LL-RR}(0) = \frac{\left\langle 4|S_{HV}|^2 - |S_{HH} - S_{VV}|^2 \right\rangle}{\left\langle 4|S_{HV}|^2 + |S_{HH} - S_{VV}|^2 \right\rangle} = \frac{T_{33} - T_{22}}{T_{33} + T_{22}}$$

If we normalize the correlation coefficient by $\gamma_{LL-RR}(0)$, what kind of results are expected? We will check the value $\gamma'_{LL-RR} = \frac{|\gamma_{LL-RR}|}{\gamma_{LL-RR}(0)}$

We call this value a modified correlation coefficient. The value will be 1 for reflection symmetry condition $(\gamma'_{LL-RR} = 1)$, and it will be larger than 1 for a man-made object $(\gamma'_{LL-RR} > 1)$. Using this property, we calculated γ'_{LL-RR} using X-band PiSAR data in Figure 9.9 and obtained the typical values in Table 9.1.

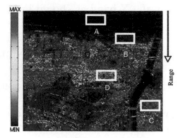

Total Power image (X-band)

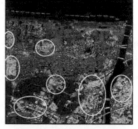

Scattering power decomposition (Y40)

Category	γ'_{LL-RR}
A: Sea	1.01
B: Pine tree	1.44
C: Non-orthogonal urban area	2.54
D: Orthognal urban area	1.12
Sandy seashore	1.10
Rice paddy field	1.11
Crop field	1.22

Modified correlation coefficient

FIGURE 9.9   Kobari, Niigata, Japan.

TABLE 9.1
ALOS2 Quad Pol Data Sets

Data	Data Number	Off-Nadir Angle	Observation Area
A	ALOS2044980740-150324	30.4°	San Francisco
B	ALOS2157350670-170422	30.4°	Los Angeles
C	ALOS2043450820-150313	30.4°	Barcelona
D	ALOS2035863740-150121	25°	Amazon
E	ALOS2072670740-150927	32.7°	Niigata
F	ALOS2066310860-150815	25°	Hokkaido

As can be seen in the preceding figures, oriented (non-orthogonal to radar illumination) urban/residential areas are not well-detected in the total power (TP) image. In the scattering power decomposition image, they appear as green areas due to the cross-polarized HV component generation. If we calculate the modified correlation coefficient, these non-orthogonal residential areas become easily detectable. They are indicated by yellow circles. On the other hand, orthogonal residential areas in the modified correlation coefficient image look dark. The average values for other typical areas are listed in Table 9.1, showing the detecting capability of man-made structures by the modified correlation coefficient [8,9].

9.8.2 Tree Detection with Reflection Symmetry Condition

For natural distributed objects such as trees, forests, and rice fields, the value of correlation coefficient $|\gamma_{LL-RR}|$ is very small. Conversely, $\frac{1}{|\gamma_{LL-RR}|}$ should be very big. If we make a $\frac{1}{|\gamma_{LL-RR}|}$ map, we may be able to retrieve the tree area very easily. Using this simple idea, we calculated $\frac{1}{|\gamma_{LL-RR}|}$ with some threshold to exclude noise as shown in Figure 9.10. It is clearly seen that the pine tree area exactly corresponds to an actual situation in Figure 9.11. This is a simple and effective way of using the reflection symmetry condition to find a natural vegetation area in the circular polarization correlation coefficient.

9.8.3 Combination of Power and Correlation Coefficients

Based on the distribution characteristics of $|\gamma_{LL-RR}|$ in Figure 9.8 and the TP, it is possible to create an algorithm for classifying objects in the PolSAR image. Figure 9.12 shows one example of classification using the correlation coefficient and power [8].

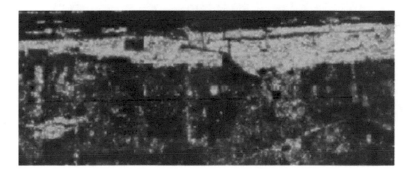

FIGURE 9.10 $\dfrac{1}{\gamma_{LL-RR}}$ image around Kobari, Niigata. Green corresponds to pine trees.

FIGURE 9.11   Aerial photo corresponding to Figure 9.10.

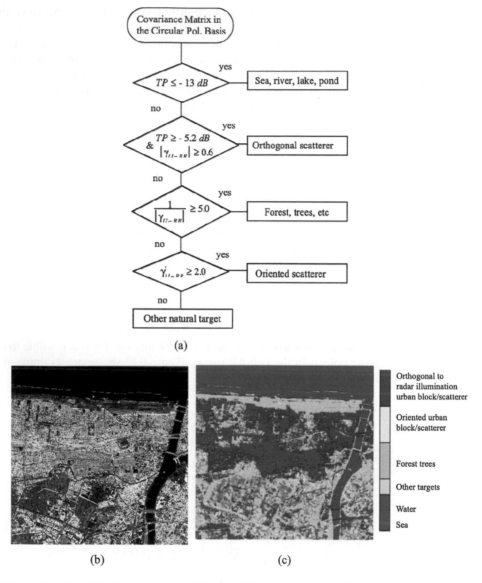

FIGURE 9.12 Classification result of Kobari, Niigata, using correlation coefficient and power: (a) classification algorithm, (b) scattering power decomposition image, and (c) classification result.

9.8.3.1 Water Surface Detection

Flooding causes devastating damage to our lives. If flooding occurs, the water surface covers tremendous areas including residential houses, crop fields, and rice paddy fields. The extent of flooding can be monitored by the water surface area.

For the detection of the water surface by PolSAR, it is difficult to use power information only, since the RCS from the water surface is too small and close to the noise level. The water surface looks dark, as seen in Figure 9.13 G4U image. In this situation, the use of correlation coefficients can be a candidate. Several polarization parameters are shown in Figure 9.13. Scattering mechanisms

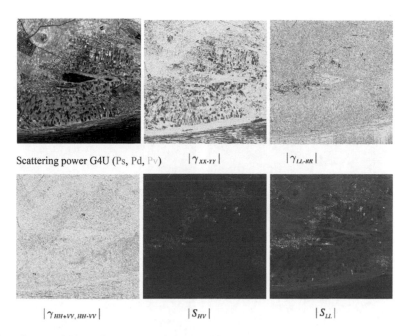

Scattering power G4U (Ps, Pd, Pv) $|\gamma_{xx\text{-}yy}|$ $|\gamma_{LL\text{-}RR}|$

$|\gamma_{HH+VV,\,HH\text{-}VV}|$ $|S_{HV}|$ $|S_{LL}|$

FIGURE 9.13   Some polarimetric parameter images of Sakata lagoon, Niigata.

can be seen easily in the scattering power G4U image. If we turn to the correlation images, $|\gamma_{LL-RR}|$ information looks like a good indicator for the water surface among others.

If a certain threshold level for the TP is set up to pick up the water surface, it becomes possible to extract a water covered (flooded) area. Figure 9.14 shows the extracted water surface using some correlation coefficients and the power information. The combination of power and correlation information plays a very important role for extracting the water surface. This will serve to create a flooded area map using PolSAR data.

9.8.3.2   Urban Area Detection

As the second example, ALOS2 quad pol data over Sapporo, Hokkaido, is used to detect urban areas with different orientation residential blocks. Since ALOS2 has rather fine resolution in the L-band, it is interesting to see how the correlation coefficient serves classification purposes. The data set used is ALOS2066310860-150815. First we checked the distribution of correlation coefficient in the circular polarization basis in the oriented urban area and forest, which are shown in Figure 9.15.

As seen in the previous analysis, the distribution is quite similar to those of the X-band data. The oriented urban area has spread distribution on the complex circle plane. On the other hand, the forest has a characteristic distribution centered at the origin. These are typical distributions for two targets. Using these distributions, a simple algorithm to distinguish or classify objects can be derived as shown in Figure 9.16. In the classification algorithm, TP is used to drive the sea and water surface with low RCS. Then according to the value of correlation coefficient, objects are classified into various targets. Sapporo City and its surrounding areas are classified as shown in Figure 9.17. The scattering power decomposition Y4R image (a) has a green (volume scattering) region in urban area due to highly oriented buildings and houses. These man-made structures are detected by the correlation coefficient and are overlaid on the scattering power image (b) colored by red. It is easy to recognize residential buildings and houses (red) around Sapporo City. A Google Earth image (c) is attached for comparison of (b).

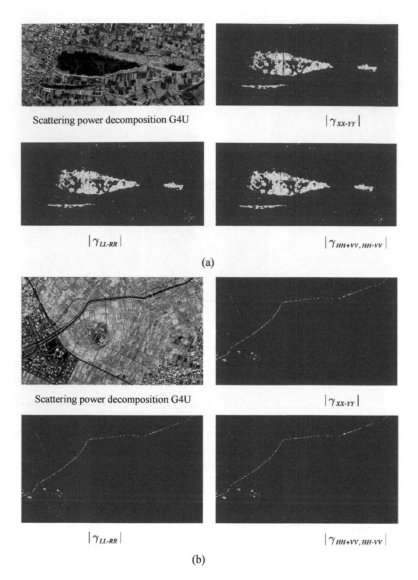

FIGURE 9.14 Extracted water surface by the correlation coefficient and the TP: (a) water surface detection, and (b) irrigation detection.

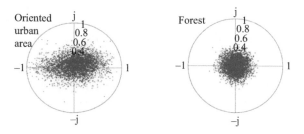

FIGURE 9.15 Distribution of the correlation coefficient γ_{LL-RR}.

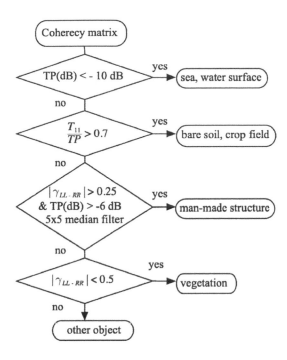

FIGURE 9.16 Simple classification of flow chart.

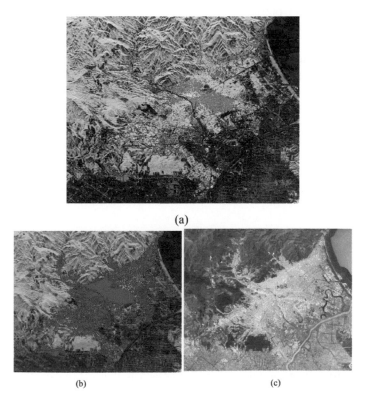

FIGURE 9.17 Detection of man-made object by the correlation coefficient superimposed on the scattering power decomposition image: (a) Y4R image of Sapporo, (b) detection result of a man-made structure, and (c) Google Earth optical image.

9.9   EXTENDED SIMILARITY AND ITS APPLICATIONS

There is a magnitude imbalance in the elements of the coherency matrix. Generally, the diagonal elements have large values, whereas off-diagonal elements are small. If we calculate the similarity parameter, the contribution of diagonal elements becomes dominant, and the contribution of off-diagonal elements becomes negligible. This means the polarimetric information contained in the off-diagonal term is lost in the inner product operation. The same is true for the correlation coefficient.

In this section, we propose a compensation scheme for solving the magnitude imbalance problem and illustrate a model-based target classification technique using the compensated polarimetric similarity parameter.

As it is well-known, the ensemble average coherency matrix is given as,

$$\langle [T] \rangle = \frac{1}{n} \sum_{p}^{n} \mathbf{k}_p \mathbf{k}_p^{\dagger} = \begin{bmatrix} T_{11} & T_{12} & T_{13} \\ T_{12}^* & T_{22} & T_{23} \\ T_{13}^* & T_{23}^* & T_{33} \end{bmatrix} \tag{9.9.1}$$

The scattering power decomposition is carried out in the following form:

$$\langle [T] \rangle = \sum_{i=1}^{n} P_i [T]_i \tag{9.9.2}$$

P_i: Model-based scattering power
$[T]_i$: Scattering model in the normalized coherency form

Now we try to use the similarity parameter (based on the correlation method) for the scattering mechanism retrieval. Since coherence or correlation is defined using two complex signals s_1 and s_2 as

$$\gamma = \frac{\langle s_1 s_2^* \rangle}{\sqrt{\langle |s_1|^2 \rangle \langle |s_2|^2 \rangle}} \tag{9.9.3}$$

Here we replace s_1 and s_2 by the vectorial form $\mathbf{T}_A$ and $\mathbf{T}_B$ of coherency matrix in equation (9.9.1). Then we have

$$\gamma = \frac{\mathbf{T}_A \cdot \mathbf{T}_B}{\|\mathbf{T}_A\| \|\mathbf{T}_B\|} \tag{9.9.4}$$

where $\cdot$ denotes the inner product and $\|\cdot\|$ denotes the Euclidean norm.

$$\mathbf{T} = \left[T_{11} T_{22} T_{33} \operatorname{Re}\{T_{12}\} \operatorname{Im}\{T_{12}\} \operatorname{Re}\{T_{13}\} \operatorname{Im}\{T_{13}\} \operatorname{Re}\{T_{23}\} \operatorname{Im}\{T_{23}\} \right]^t \tag{9.9.5}$$

where t denotes transpose. The form of equation (9.9.4) is an extension of the similarity parameter [10,11]. The problem in the conventional approach is that off-diagonal terms are too small compared to diagonal terms in the coherency form as shown in Figure 9.18.

This magnitude imbalance leads to neglect minor terms in the correlation. For example, diagonal terms only play dominant contributions in the inner product calculation, leaving off-diagonal terms negligible.

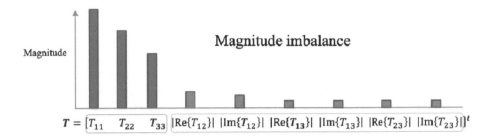

$$T = \begin{bmatrix} T_{11} & T_{22} & T_{33} & |\text{Re}\{T_{12}\}| & |\text{Im}\{T_{12}\}| & |\text{Re}\{T_{13}\}| & |\text{Im}\{T_{13}\}| & |\text{Re}\{T_{23}\}| & |\text{Im}\{T_{23}\}| \end{bmatrix}^t$$

FIGURE 9.18 Magnitude imbalance in the coherency matrix elements.

$$T_A \cdot T_B = T_{11}^A T_{11}^B + T_{22}^A T_{22}^B + T_{33}^A T_{33}^B + \ldots + \text{Im}\{T_{23}\}\,\text{Im}\,T_{23}^B$$
$$\approx T_{11}^A T_{11}^B + T_{22}^A T_{22}^B + T_{33}^A T_{33}^B \tag{9.9.6}$$

The off-diagonal terms do not affect the final value of the inner product. This is not desirable from the viewpoint of full utilization of polarimetric information. In addition, a subtraction may occur in equation (9.9.6) when the signs of the same element are opposite. For full and effective utilization of nine independent pieces of polarimetric information, we modify the magnitude imbalance as shown in Figure 9.19 and in the following vector form (9.9.7):

$$T \Rightarrow \begin{bmatrix} T_{11} \\ T_{22} \\ T_{33} \\ \text{Re}\{T_{12}\} \\ \text{Im}\{T_{12}\} \\ \text{Re}\{T_{13}\} \\ \text{Im}\{T_3\} \\ \text{Re}\{T_{23}\} \\ \text{Im}\{T_{23}\} \end{bmatrix} \Rightarrow \begin{bmatrix} T_{11} \\ a\,T_{22} \\ b\,T_{33} \\ c\,|\text{Re}\{T_{12}\}| \\ d\,|\text{Im}\{T_{12}\}| \\ e\,|\text{Re}\{T_{13}\}| \\ f\,|\text{Im}\{T_{13}\}| \\ g\,|\text{Re}\{T_{23}\}| \\ h\,|\text{Im}\{T_{23}\}| \end{bmatrix}, \quad T^{comp} = \begin{bmatrix} T_{11} \\ a\,T_{22} \\ b\,T_{33} \\ c\,|\text{Re}\{T_{12}\}| \\ d\,|\text{Im}\{T_{12}\}| \\ e\,|\text{Re}\{T_{13}\}| \\ f\,|\text{Im}\{T_{13}\}| \\ g\,|\text{Re}\{T_{23}\}| \\ h\,|\text{Im}\{T_{23}\}| \end{bmatrix} \tag{9.9.7}$$

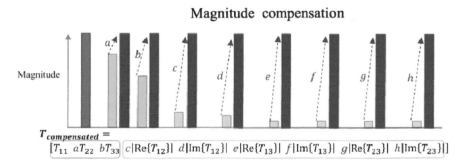

$$T_{compensated} = \begin{bmatrix} T_{11} & a T_{22} & b T_{33} & c|\text{Re}\{T_{12}\}| & d|\text{Im}\{T_{12}\}| & e|\text{Re}\{T_{13}\}| & f|\text{Im}\{T_{13}\}| & g|\text{Re}\{T_{23}\}| & h|\text{Im}\{T_{23}\}| \end{bmatrix}$$

FIGURE 9.19 Magnitude compensation of elements in the coherency matrix.

where $a, b, c,..., i$ are amplification coefficients. The absolute value is chosen so that the inner product of each element becomes additive. The basic idea is to make each element contribution equal in the inner product operation (equation 9.9.6) [12].

In order to determine the proper coefficients, we have examined the average absolute value of the coherency matrix elements using six data sets (level 1.1, single look complex) acquired by ALOS-2/PALSAR-2. The results of the relative magnitude of the entire images are shown in Table 9.2. The ID, off-nadir angle, and observation area of data $A, B, C, D, E,$ and F are listed in Table 9.1.

By checking the values listed in Table 9.2, and by consideration of almost equal-contribution of each element, the amplification coefficients of the magnitude-compensated vector $\mathbf{T}^{comp}$ are chosen as follows:

$$a = \frac{4}{3}, b = 4, c = 5, d = e = f = g = h = 10 \qquad (9.9.8)$$

We use the extended vector form of equation (9.9.7) with (9.9.8) for scattering mechanism retrieval, similar to the expression of scattering power decomposition in equation (9.9.2).

$$\mathbf{T}^{comp} = \sum_{i=1}^{n} \gamma_i^{comp} \mathbf{T}_i^{comp} \qquad (9.9.9)$$

where $\mathbf{T}^{comp}$ denotes the observed data, and $\mathbf{T}_i^{comp}$ denotes the theoretical scattering model. Compensated similarity parameter γ_i^{comp} is defined as

$$\gamma_i^{comp} = \frac{\mathbf{T}^{comp} \cdot \mathbf{T}_i^{comp}}{\left\| \mathbf{T}^{comp} \right\| \left\| \mathbf{T}_i^{comp} \right\|} \qquad (9.9.10)$$

where $\cdot$ denotes the inner product and $\|\cdot\|$ denotes the Euclidean norm.

To retrieve scattering mechanisms in equation (9.9.9) effectively, we select the largest gamma among the values of equation (9.9.10). Then we subtract the term from equation (9.9.9) with the corresponding $\mathbf{T}_i^{max}$, resulting in

$$\mathbf{T}^{comp} - \gamma_{max} \mathbf{T}_i^{max} = \sum_{i=2}^{n} \gamma_i \mathbf{T}_i^{comp} \qquad (9.9.11)$$

We repeat the procedure up to n, where n is the number of scattering models. In this way, we can expand $\mathbf{T}_{comp}$ with full utilization of polarimetric information and retrieve the scattering mechanism in a simple manner.

TABLE 9.2
Relative Magnitude Relation of the Coherency Matrix Elements

Data	T_{11}	T_{22}	T_{33}	$\left\|\mathrm{Re}\{T_{12}\}\right\|$	$\left\|\mathrm{Im}\{T_{12}\}\right\|$	$\left\|\mathrm{Re}\{T_{13}\}\right\|$	$\left\|\mathrm{Im}\{T_{13}\}\right\|$	$\left\|\mathrm{Re}\{T_{23}\}\right\|$	$\left\|\mathrm{Im}\{T_{23}\}\right\|$
A	1.0	0.75	0.25	0.23	0.14	0.07	0.07	0.08	0.05
B	1.0	0.86	0.26	0.22	0.15	0.08	0.07	0.12	0.05
C	1.0	0.86	0.26	0.25	0.14	0.09	0.07	0.10	0.04
D	1.0	0.68	0.47	0.15	0.11	0.08	0.08	0.07	0.07
E	1.0	0.66	0.23	0.17	0.18	0.07	0.07	0.07	0.04
F	1.0	0.69	0.24	0.22	0.12	0.06	0.06	0.07	0.04
Ave.	1.0	0.75	0.29	0.21	0.14	0.08	0.07	0.09	0.05

9.9.1 SELECTION OF SCATTERING MECHANISM

It is important to select scattering mechanism $\mathbf{T}_i^{comp}$. This can be based on the theoretical and experimental data as summarized in Chapter 6. Here we have selected four mechanisms as an example.

1. Surface scattering $\left\langle [T] \right\rangle_{surface} = \begin{bmatrix} 1 & \beta^* & 0 \\ \beta & |\beta|^2 & 0 \\ 0 & 0 & 0 \end{bmatrix}$ with $\beta = 0.1+0.1j$ as representative value.

2. Double-bounce scattering $\left\langle [T] \right\rangle_{double} = \begin{bmatrix} |\alpha|^2 & \alpha & 0 \\ \alpha^* & 1 & 0 \\ 0 & 0 & 0 \end{bmatrix}$ with $\alpha = 0.1+0.1j$ as representative value.

3. Volume scattering $\left\langle [T] \right\rangle_{vol} = \begin{bmatrix} 1 & 0 & 0 \\ 0 & 1/2 & 0 \\ 0 & 0 & 1/2 \end{bmatrix}$

4. 22.5°-oriented dihedral corner reflector scattering

The coherency matrix of the dihedral corner reflector rotated θ around the radar line of sight becomes

$$[T]_{dihedral}^{\theta} = \begin{bmatrix} 0 & 0 & 0 \\ 0 & \cos^2\theta & -\dfrac{\sin 4\theta}{2} \\ 0 & -\dfrac{\sin 4\theta}{2} & \sin^2\theta \end{bmatrix}$$

Considering the ensemble averaged data, we adopt a probability distribution function with its peak at 22.5° as

$$p(\theta) = \frac{1}{2}\cos\left(\theta - \frac{\pi}{8}\right) \qquad -\frac{\pi}{2} < \theta - \frac{\pi}{8} < \frac{\pi}{2}$$

The theoretical ensemble-averaged matrix for 22.5°-oriented dihedral scattering can be calculated by

$$\left\langle [T] \right\rangle_{dihedral}^{22.5} = \int_{-\frac{\pi}{2}+\frac{\pi}{8}}^{\frac{\pi}{2}+\frac{\pi}{8}} [T]_{dihedral}^{\theta} p(\theta)\, d\theta = \begin{bmatrix} 0 & 0 & 0 \\ 0 & 1 & \dfrac{1}{15} \\ 0 & \dfrac{1}{15} & 1 \end{bmatrix}$$

9.9.2 CLASSIFICATION RESULTS

This method is applied to ALOS2/PALSAR2 data over San Francisco, USA, acquired on March 24, 2015. The Google Earth image of the same area is shown in Figure 9.20a. The window size for the

(a)

(b)

(c)

FIGURE 9.20 Classification results and Google Earth image: (a) Google Earth optical image; (b) classification result before the compensation using four scattering models (red: double-bounce scattering, green: volume scattering, blue: surface scattering, magenta: 22.5°-oriented dihedral scattering); and (c) classification result after the compensation using four scattering models (red: double-bounce scattering, green: volume scattering, blue: surface scattering, magenta: 22.5°-oriented dihedral scattering).

TABLE 9.3

Classification Rates on a Typical Area

Patch	Surface		Double Bounce		Volume		Oriented Dihedral	
	Before [%]	After [%]	Before	After	Before	After	Before	After
A	2.7	6.4	0	0.2	97.3	89.1	0	4.3
B	0.2	18.7	28.4	75.4	71.4	5.4	0	0.5
C	2.4	11.7	4	8.3	91.8	36.6	1.8	43.4

ensemble average is 6 in the range direction and 12 in the azimuth direction. In all images, the radar illumination direction is from top to bottom. The classification result before compensating the contribution degree of each coherency matrix element is shown in Figure 9.20b, and the classification result after the compensation is shown in Figure 9.20c. Patches A, B, and C are vegetation areas, urban building areas orthogonal to radar illumination, and urban buildings areas oriented with respect to radar illumination, respectively.

In order to evaluate the classification accuracy quantitatively, the classification rates of Patches A, B, and C are shown in Table 9.3.

For Patch B (orthogonal urban area), 28.4% of pixels are classified as the double-bounce scattering, and 71.4% are classified as the volume scattering before the compensation scheme. After the compensation, 75.4% are classified into the double-bounce scattering, and 5.4% are classified into the volume scattering. It is also noticed that the compensation scheme is effective for detecting oriented urban areas. The classification rate of 22.5°-oriented dihedral scattering increases from 1.8% to 43.4% by the compensation in oriented urban areas. The classification accuracy in vegetation areas is still high, as much as 89.1%, although in a decreasing tendency. These results are due to the increase of the contribution of T_{12} and T_{23} in orthogonal scattering model and 22.5°-oriented dihedral scattering model.

From the classification results, it is confirmed that the implementation of the compensation scheme is very effective for classification using polarimetric similarity, especially for detecting urban areas. The idea of similarity can be extended to other applications using 100% utilization of polarimetric information.

9.10 SUMMARY

The correlation coefficient is a very important radar parameter. Since the value is completely independent from power, it seems better to combine correlation and power information to apply for detection, classification, and identification purposes. Among correlation coefficients in various polarization bases, the correlation in the circular polarization basis performs the best up to now. It has interesting features such as polarization orientation angle and reflection/non-reflection ratio.

In the calculation process, we often face higher-order scattering terms that do not contribute to the final value of correlation coefficient or similarity parameter. A compensation scheme to small elements by amplification is shown for the equipolarimetric contribution purpose. This may enrich the full utilization of polarimetric information.

REFERENCES

1. J. A. Kong, A. A. Swartz, H. A. Yueh, L. M. Novak, and R. T. Shin, "Identification of terrain cover using the optimal polarimetric classifier," *J. Electromag. Waves*, vol. 2, no. 2, pp. 171–194, 1988.
2. T. Moriyama, Y. Yamaguchi, S. Uratsuka, T. Umehara, H. Maeno, M. Satake, A. Nadai, and K. Nakamura, "A study on polarimetric correlation coefficient for feature extraction of polarimetric SAR data," *IEICE Trans. Commun.*, vol. E88-6, no. 6, pp. 235–2361, 2005.

3. M. Murase, Y. Yamaguchi, and H. Yamada, "Polarimetric correlation coefficient applied to tree classification," *IEICE Trans. Commun.*, vol. E84-C, no. 12, pp. 1835–1840, 2001.

4. K. Kimura, Y. Yamaguchi, and H. Yamada, "Circular polarization correlation coefficient for detection of non-natural targets aligned not parallel to SAR flight path in the X-band POLSAR image analysis," *IEICE Trans. Commun.*, vol. E87-B, no. 10, pp. 3050–3056, 2004.

5. J. S. Lee and T. Ainsworth, "The effect of orientation angle compensation on coherency matrix and polarimetric target decompositions," *Proceedings of EUSAR 2010*, Germany, 2010; *IEEE Trans. Geosci. Remote Sens.*, vol. 49, no. 1, pp. 53–64, 2011.

6. W.-T. An, C. Xie, X. Yuan, Y. Cui, and J. Yang, "Four-component decomposition of polarimetric SAR images with deorientation," *IEEE Geosci. Remote Sens. Lett.*, vol. 8, no. 6, pp. 1090–1094, 2011.

7. S. W. Chen, X.-S. Wang, S.-P. Xiao, and M. Sato, *Target Scattering Mechanism in Polarimetric Synthetic Aperture Radar – Interpretation and Application*, Springer, Singapore, 2017. ISBN: 978-981-10-7268-0.

8. Y. Yamaguchi, Y. Yamamoto, J. Yang, W.-M. Boerner, and H. Yamada, "Classification of terrain by implementing the correlation coefficient in the circular polarization basis using X-band POLSAR data," *IEICE Trans. Commun.*, vol. E91-B, no. 1, pp. 297–301, 2008.

9. T. L. Ainsworth, D. L. Schuler, and J. S. Lee, "Polarimetric SAR characterization of man-made structures in urban areas using normalized circular-pol correlation coefficients," *Remote Sens. Environ.*, vol. 112, pp. 2876–2885, 2008.

10. J. Yang, Y. N. Peng, and S. M. Lin, "Similarity between two scattering matrices," *Electron. Lett.*, vol. 37, no. 3, pp. 193–194, 2001. doi:10.1049/el:20010104.

11. Q. Chen, Y. M. Jiang, L. J. Zhao, and G. Y. Kuang, "Polarimteric scattering similarity between a random scatterer and a canonical scatterer," *IEEE Geosci. Remote Sens. Lett.*, vol. 7, no. 4, pp. 866–869, 2010. doi:10.1109/LGRS.2010.2053912.

12. M. Umemura, Y. Yamaguchi, and H. Yamada, "Model-based target classification using polarimetric similarity with coherency matrix elements," *IEICE Commun. Express*, vol. 8, no. 3, pp. 73–80, 2019. doi:10.1587/comex.2018XBL0152.

10 Polarimetric Synthetic Aperture Radar

Synthetic aperture radar (SAR) is a high-resolution two-dimensional imaging system and is useful for monitoring the earth's surface [1–13]. In this chapter, we see how a high-resolution radar image is created by SAR. Once the imaging principle is understood, it serves to interpret SAR images from a technical aspect such as resolutions in the range and in the azimuth direction. In addition, the SAR principles can be further extended to fully polarimetric SAR (PolSAR), interferometric SAR (InSAR), PolInSAR, tomographic SAR (Tomo-SAR), and holographic SAR (Holo-SAR) [13].

The range resolution of SAR depends on the bandwidth of the transmitting signal. Table 10.1 shows the name of frequency bands used in radar and the frequency allotted to satellite applications. Given the bandwidth, the maximum attainable resolutions for satellite SAR are also shown in the list. The SAR system operative at higher frequencies above the S-band can achieve very high resolution (less than 1 m). Especially, the X-band and Ku-band SAR will be very high-resolution radar systems in the future.

TABLE 10.1

Frequency Band Name and Allocation to Satellite Application

Name	Frequency (GHz)	Allotted to Satellite	Bandwidth (MHz)	Maximum Resolution
P	0.25–0.5	0.432–0.438	6	25 m
L	1–2	1.125–1.3	85	176 cm
S	2–3.75	3.1–3.3	200	75 cm
C	3.75–7.5	5.25–5.57	320	47 cm
X	7.5–12	9.3–9.9	600	25 cm
Ku	12–17.6	13.25–13.75	500	30 cm
K	17.6–26.5	24.05–24.25	200	75 cm
Ka	25–40	34.5–36	500	30 cm

Compared to single polarization SAR, fully polarimetric or quad PolSAR, which is denoted here as PolSAR, has lot of information in its scattering matrix data. The relative scattering matrix (10.1.1) is rewritten to remind us that there are five independent parameters assuming $S_{VH} = S_{HV}$. There are three amplitudes and two phases.

$$[S]_{relative} = \begin{bmatrix} |S_{HH}| & |S_{HV}| \angle \left(\phi^{HV} - \phi^{HH} \right) \\ |S_{VH}| \angle \left(\phi^{VH} - \phi^{HH} \right) & |S_{VV}| \angle \left(\phi^{VV} - \phi^{HH} \right) \end{bmatrix} \tag{10.1.1}$$

If the scattering matrices are averaged in a 3×3 covariance or coherency matrix for incoherent scatterings, the number of independent parameters becomes as much as nine, compared to one and three parameters by single polarization or dual-polarization SAR, respectively (see Chapter 4).

With the advent of the PolSAR system, not only airborne systems but also spaceborne systems have become available in recent years. The first fully polarimetric SAR system in space was demonstrated by SIR-C/X-SAR (NASA JPL, USA) onboard the Space Shuttle in 1994, which operated at the X/C/L-band. Then phased array L-band SAR (PALSAR) on the Advanced Land Observing Satellite (ALOS) was launched in 2006 (JAXA, Japan) and continued experimental quad pol data acquisition until 2011. It brought us more than 270,000 polarimetric scenes over the world. RadarSAT-2 is a follow-on mission of the Canadian satellite RadarSAT-1 and also carried the C-band quad PolSAR system (2007). TerraSAR-X (2007) and TanDEM-X (DLR, Germany, 2010) had cooperative flight and acquired some X-band quad pol data. Followed by ALOS, the second-generation ALOS-2 was launched in 2014 by JAXA. The third-generation ALOS-4 with fully polarimetric data acquisition mode will be available in 2021. Figure 10.1 shows some spaceborne PolSAR systems.

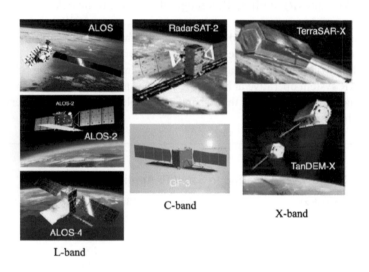

FIGURE 10.1   PolSAR system in space.

10.1   SYNTHETIC APERTURE RADAR (SAR)

As we have seen in Chapter 3, radar measures target range by the time delay of the pulse and detect target by the reflected wave information [amplitude or radar cross section (RCS)]. There are roughly two types in radar operations and its signal processing. The most frequently used one is pulse radar in the time domain in which a very short pulse is used for high resolution in the range direction. The other is in the frequency domain where a continuous wave is used for target detection using such a network analyzer system or a frequency modulated continuous wave (FMCW) radar. Although the hardware constitution is different, the fundamental principle is the same because the time signal and the frequency signal are connected through the Fourier transform. They are convertible to each other.

Table 10.2 shows the comparison of radar types. Pulse radar is suitable for far-range sensing and is used mostly in various applications. Step frequency radar (mainly the network analyzer system) and FMCW radar are suitable for short-range sensing. Usually the network analyzer system has the largest dynamic range for receiving signals; however, it takes time to complete measurement. It is often employed for research purposes. The FMCW radar operates in the radio frequency (RF), and the signal processing is carried out in the intermediate frequency (IF). It is suited for cost-effective short-range sensing. However, the dynamic range may be restricted since the noise floor

TABLE 10.2
Comparison of Radar Operation Types

	Pulse	Step Frequency	FMCW
Operation domain	Time	Frequency	Frequency
Range resolution	Bandwidth	Bandwidth	Bandwidth
Range accuracy	Number of FFT points	Number of FFT points	Number of FFT points
Signal domain	RF	RF	IF
Hardware	Complex	Simple	Simple
Near range	△	○	○
Far range	○	△	△
SAR processing	○	○	○
Polarimetry	○	○	○

is rather high. The range resolution is the same for all radar types and is decided by the signal bandwidth. Signal processing for the synthetic aperture technique as well as polarimetric analysis is available to all radar systems.

10.1.1 SAR PRINCIPLE

SAR is a kind of high-resolution microwave imaging radar and is an extension of radar function. It generates two-dimensional images comprising of the range axis by the pulse compression technique, and the cross-range (azimuth) axis by the synthetic aperture technique. The final image becomes a high-resolution and two-dimensional distribution of reflected signals from the radar scene.

As shown in Figure 10.2a, the radar on a platform moves along a line trajectory with emitting microwave pulses to the side-looking direction. The side-looking direction is the radar range direction, and the cross-range direction is orthogonal to the range and is parallel to the fight path. The cross-range direction is also called as azimuth direction.

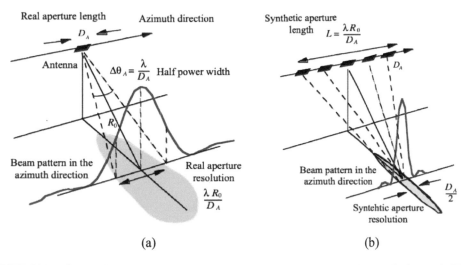

(a)   (b)

FIGURE 10.2 Conceptional view of radar resolution: (a) real aperture radar resolution and (b) synthetic aperture radar resolution.

If a single antenna emits a pulse, the antenna beam spreads proportionally to range R_0. The beam width on the ground becomes $\lambda R_0 / D_A$ where λ is the wavelength and D_A is the real aperture length as shown in Figure 10.2a. This width $\lambda R_0 / D_A$ is the real aperture resolution on the ground. The resolution becomes worse with increasing R_0. It is proportional to R_0.

Synthetic aperture comes from an artificial antenna array configuration. It is known that large aperture antenna has a narrow beam. By arranging many antennas in a straight line along the trajectory, a hypothetical array antenna can be achieved. As shown in Figure 10.2b, a large array antenna of length L can be realized by antennas with repetitive pulse emission and reception at each point determined by pulse repetition frequency making use of constant movement. At each point along the aperture length L, the radar signal is recorded. This artificial array can have a very high resolution by signal processing so the beam is always focused on the target. The beam width $D_A / 2$ can be achieved by this synthetic aperture. This value is called azimuth resolution of SAR. This is a significant feature that the resolution is independent of R_0. Therefore, the airborne radar and spaceborne radar can have the same azimuth resolution regardless of its flight height.

Some radar terminologies with respect to directions and angles for radar observation are illustrated in Figure 10.3. Radar swath is defined as the width of the radar footprint length on the ground. If the radar beam direction changes, the incidence angle changes and hence the swath width changes. This beam scanning in the range direction is used for the ScanSAR mode in which a large swath area mapping is desired. Figure 10.4 shows the radar beam of the squint direction (offset from the side looking) and corresponding angles used in SAR observations.

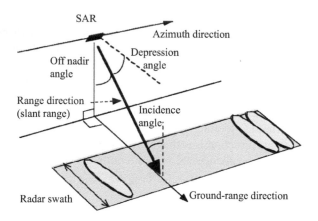

FIGURE 10.3   Directions and angles in SAR image formulation.

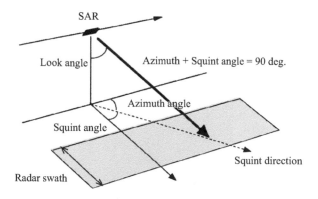

FIGURE 10.4   Directions and angles in SAR image formulation (cont.).

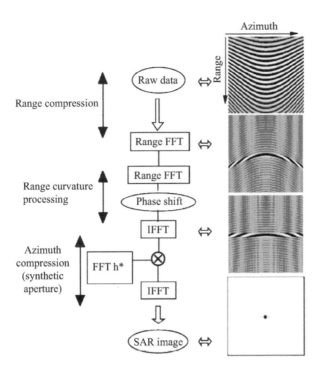

FIGURE 10.5   SAR image processing flow.

Here we see the range compression and synthetic aperture processing in the azimuth direction in more detail. In general, SAR processing indicates both range compression and azimuth compression. Fourier transformation and its execution algorithm, fast Fourier transform (FFT), are frequently used for the signal processing. Figure 10.5 shows the flow chart of SAR processing. Here we take an example of FMCW SAR processing, which has been used in our lab.

10.1.2   RANGE COMPRESSION

The range resolution of the radar is dependent on the pulse width. If a short pulse is used, the spatial resolution increases. For high resolution in the range direction, it is desirable to use a pulse width as short as possible. Suppose we measure objects in the 1.5-m range, it is necessary to design a radar system, which can discriminate less than a 10-ns time resolution. However, the generation of very short pulse demands an ultra-broadband radar system. A 1-ns pulse needs 1 GHz bandwidth. In addition, the huge transmitting power is needed to emit short pulses repeatedly, which is not suitable for satellite or aircraft platform. There is a limitation on the broadband radar system and the power generation for airborne or spaceborne SAR.

A pulse compression technique was devised to overcome these problems by using relatively long pulses with fewer powers. This can be achieved by a linear frequency modulated (FM) signal as shown in Figure 10.6. A linear FM signal has a special property in which the instantaneous frequency increases linearly with time. This linear FM signal is sometimes called a chirp signal because it looks like birds singing that varies from a lower to higher tone.

Figure 10.7 explains how to measure time delay or radar range precisely. Since the pulse radar and FMCW radar utilizes the same linear FM signal, the FMCW radar example is shown to explain the principle.

Suppose the target is located at distance R_1 (near) and R_2 (far) as shown in Figure 10.7. Once the linear FM signal is transmitted from the antenna, it is reflected by target at R_1 and returns to radar

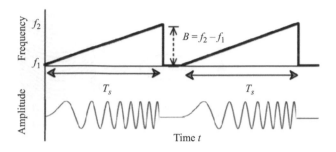

FIGURE 10.6   Linear FM signal.

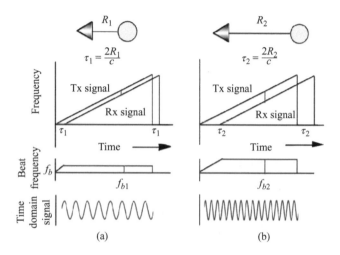

FIGURE 10.7   Principle of FMCW radar.

with time delay τ_1. Since the speed is constant for any frequency, the relation between the time and the instantaneous frequency becomes as shown in the triangular part of Figure 10.7a. The reflected wave delays time by τ_1 and shifts to the right direction on the time axis by τ_1. On the other hand, the corresponding time delay τ_2 by the target at R_2 (far) becomes larger than τ_1 as shown in Figure 10.7b. The time delay is proportional to the range distance.

An important factor here is the frequency difference of transmitting and receiving signals, which is called a beat frequency. If we denote this beat frequency as f_b, there is a linear relation between time delay τ as,

$$\tau = \frac{2R}{c}\sqrt{\varepsilon_r} \propto K\, f_b \quad \text{K : coefficient} \tag{10.1.2}$$

where ε_r is the relative dielectric constant in the propagation medium. The time domain beat signals are depicted in the lower portion of Figure 10.7. For a target in the near range R_1, the beat frequency is low. If the target is in far range R_2, the beat frequency becomes high. The beat frequency is constant throughout the pulse width.

Therefore, if we measure the beat frequency of beat signal, the time delay and hence the range distance can be determined by equation (10.1.2) Since frequency can be measured precisely by the Fourier transform rather than time itself, we can obtain a precise distance. This is the conceptional ranging principle of FMCW radar.

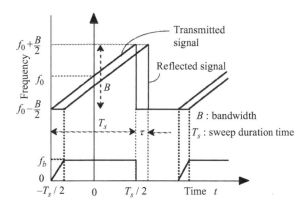

FIGURE 10.8 Relation among linear FM transmitting and receiving signals and its beat frequency with respect to time.

In addition, it is easy to handle the beat signal for circuit design because the beat frequency falls in the intermediate frequency range (around several kHz to MHz). Once the IF circuit of FMCW system is developed, it can apply for L-band, X-band, and/or Ku-band as far as the beat signal falls in the same frequency range. This is another merit of the FMCW radar system.

Now referring to Figure 10.8, a pulse compression procedure can be explained in equations. Let the center frequency of transmitting signal as f_0, and pulse duration time as T_s. The transmitting signal is swept by bandwidth B during the duration time $T_s(-T_s/2 < t < T_s/2)$. The instantaneous frequency becomes,

$$f(t) = f_0 + \frac{B}{T_s}t = f_0 + Mt \tag{10.1.3}$$

The phase $\psi(t)$ of the signal can be derived from $f(t) = \frac{1}{2\pi}\frac{d\psi(t)}{dt}$

$$\psi(t) = 2\pi\left(f_0 t + \frac{M}{2}t^2\right)$$

Therefore, the FM transmitting signal $S_{tx}(t)$ during T_s can be written as

$$S_{tx}(t) = A\cos\left[2\pi\left(f_0 t + \frac{M}{2}t^2\right)\right] \tag{10.1.4}$$

where A is the amplitude, f_0 is the center frequency, t is the time, $M = B/T_s$ is the frequency modulation rate, B is the swept frequency bandwidth, and T_s is the sweep duration time.

Let g represent the reflection coefficient of a point target located at distance R from the antenna.

$$g = g(x_0, z_0) \quad (x_0, z_0) : \text{target coordinate} \tag{10.1.5}$$

The receiving signal $S_{rx}(t)$ delays time by τ, so that time factor changes from t to $(t - \tau)$.

$$S_{rx}(t) = gA'\cos\left[2\pi\left\{f_0(t-\tau) + \frac{M}{2}(t-\tau)^2\right\}\right] \tag{10.1.6}$$

where A' is an amplitude factor affected by propagation loss.

By combining receiving signal equations (10.1.6) and reference signal (10.1.4) in a nonlinear mixer, we can pick up the following low frequency signal ($0 < f \ll f_0$). This operation is corresponding to using matched filtering in pulse radar.

$$S_b(t) = gAA' \cos\left[2\pi\left(f_0\tau + M\tau t\right)\right]$$ (10.1.7)

$M\tau$ in the cosine function represents a frequency that is caused by the difference of two signals. This frequency is called beat frequency f_b and can be represented as

$$M\tau = f_b = \frac{2B}{cT_s}R$$ (10.1.8)

Since the target range R is proportional to f_b, we need to find f_b instead of finding the time delay.

The most popular method to determine frequency is to use the Fourier transform. The Fourier transform of (10.1.7) in the sweep (duration) time ($-T_s/2 < t < T_s/2$) yields

$$S_b(\omega) = \frac{gAA'}{2}e^{j\omega_0\tau}T_s\frac{\sin\left[(\omega-\omega_b)\frac{T_s}{2}\right]}{(\omega-\omega_b)\frac{T_s}{2}} + \frac{gAA'}{2}e^{-j\omega_0\tau}T_s\frac{\sin\left[(\omega+\omega_b)\frac{T_s}{2}\right]}{(\omega+\omega_b)\frac{T_s}{2}}$$

The beat spectrum $S_b(f)$ with positive frequency component becomes,

$$S_b(f) = CgT_s\,\exp\left(j\frac{4\pi R}{\lambda_0}\right)\frac{\sin\left[\pi(f-f_b)T_s\right]}{\pi(f-f_b)T_s}$$ (10.1.9)

$$\left|S_b(f)\right|^2 = \left|CgT_s\right|^2 Sinc^2\left[\pi(f-f_b)T_s\right]$$

where $C = \frac{AA'}{2}$ is a constant.

The beat frequency f_b can be obtained from the maximum value of (10.1.9). The same equation can be obtained in pulse radar system by using the matched filtering of receiving signal [12]. Then a complex signal is created by the Hilbert transform, that is, by mixing transmitting signal (I-component) and a 90° phase-shifted transmitting signal (Q-component).

$$E(t) = gA''T_s\,\exp\left(j\frac{4\pi R}{\lambda_0}\right)Sinc\left[\pi B(t-\tau)\right]$$ (10.1.10)

Equation (10.1.10) is written as a function of time, while (10.1.9) is a function of frequency. The meaning is the same. Equations (10.1.9) and (10.1.10) consist of an exponential phase function and $Sinc$ function, which relates to the time delay. If the magnitude is considered, the $Sinc$ function plays the most important role. If the interval of the first null points is regarded as a pulse width in the envelope of $Sinc$ function, the width becomes $2/B$ as shown in Figure 10.9. This value is equal to $\frac{2}{BT_s}$ times of the original pulse T_s. The compression rate is $\frac{2}{BT_s}$. For example, if we take $T_s = 5$ ms, $B = 200$ MHz, then the compression rate is 1/500,000. So the range resolution increases as much as 500,000 times from that of the original T_s. This process is called pulse compression or range compression. The range compression for FMCW radar is performed by Fourier transform of time domain beat signal.

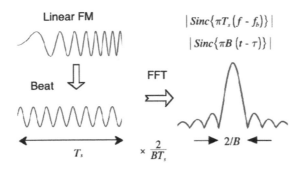

FIGURE 10.9 Pulse compression.

10.1.3 RANGE RESOLUTION ΔR

Range resolution is one of the radar parameters that represents radar performance. It is defined as the minimum distance ΔL that can separate two targets aligned in the range direction. Let ΔL denote the target interval as shown in Figure 10.10 [9]. According to the range resolution ΔR, overlapping of the beams between two targets can occur: (a) separable $(\Delta L > \Delta R)$, (b) limit of separation $(\Delta L = \Delta R)$, and (c) unable to separate $(\Delta L < \Delta R)$. The range resolution can be determined in the case (b), where two targets are aligned closest to each other with respect to ΔR. It is a separable limit.

This overlapping is caused by the main lobes of the Fourier transform. The range resolution ΔR is defined as the half power width as shown in Figures 10.11 and 10.12. Half power means 3 dB smaller than the maximum power. If we check the *Sinc* function, we have the following:

$$Sinc^2(\pi Bt) = \left(\frac{\sin x}{x}\right)^2 = \frac{1}{2} \Rightarrow \frac{\sin x}{x} = \frac{1}{\sqrt{2}} \Rightarrow x = 1.39156$$

$$2x = 2\pi Bt = 2.78 \approx \pi \quad \Rightarrow \quad T = \frac{2.78}{2\pi B} = 0.88\frac{1}{2B} \approx \frac{1}{2B} \tag{10.1.11}$$

$$\Delta R = cT = 0.88\frac{c}{2B} \approx \frac{c}{2B}$$

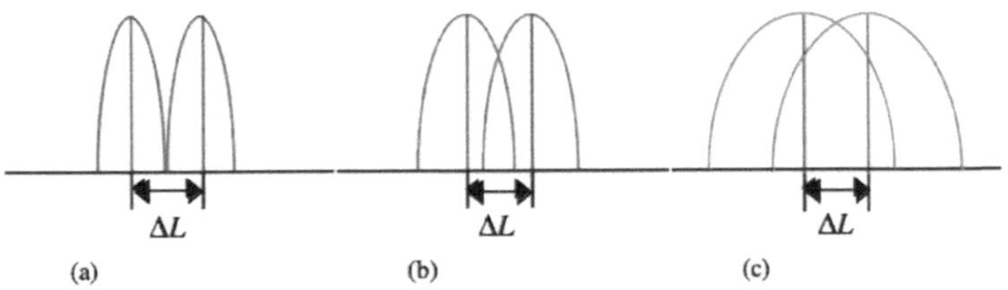

(a)   (b)   (c)

FIGURE 10.10 Relation of target interval ΔL and range resolution ΔR: (a) $\Delta L > \Delta R$, (b) $\Delta L = \Delta R$, and (c) $\Delta L < \Delta R$.

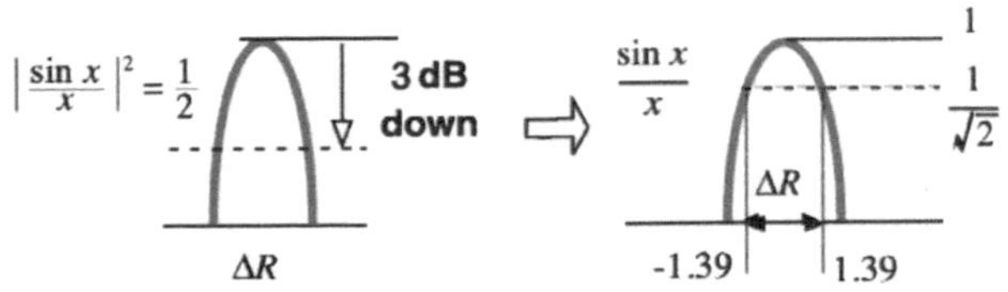

FIGURE 10.11 Beam width and range resolution.

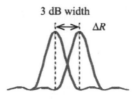

FIGURE 10.12   Definition of range resolution.

As can be seen in (10.1.11), the range resolution ΔR is inversely proportional to the bandwidth B. The wider the bandwidth, the finer the resolution. This is a common characteristic to all radar systems. The coefficient 0.88 in (10.1.11) is usually regarded as 1, so the range resolution is often described by $c/2B$, where c is the speed of light.

For the ground-range resolution in SAR observation, we have the following relation with respect to the incidence angle θ referring to Figure 10.13:

$$\Delta R_g = \frac{\Delta R}{\sin \theta} = \frac{c}{2B \sin \theta} \tag{10.1.12}$$

This equation indicates that the ground-range resolution ΔR_g becomes coarse for small incidence angles and becomes finer for large incidence angles. These phenomena are reflected in the SAR image that at near range is expanded and at far range has a fine image.

10.1.4   RANGE ACCURACY

FFT is often used in the calculation of beat spectrum. The discrete interval of spectrum (corresponding to range accuracy) can be written as

$$\Delta R_{acc} = \Delta R \frac{f_s T_s}{N} \tag{10.1.13}$$

where N is the number of the FFT point, and f_s is the data sampling frequency. If N is increased, the range accuracy will increase, but the envelope of spectrum does not change. Therefore, range accuracy and range resolution are totally different, although they look similar.

10.1.5   RANGE PROFILE AFTER COMPRESSION

If we have two targets in the range direction, the beat spectrum of the FMCW radar becomes as shown in Figure 10.14. There are two main lobes corresponding to the positions of target 1 and 2. The positions, f_{b1} and f_{b2}, in the horizontal axis after FFT show the target distance from the radar. Therefore, this beat spectrum is identical with the range profile. We can recognize main lobes as well as side lobes, which are not desirable. Big side lobes sometimes mask desired small main lobes. In order to suppress side lobes, various window functions such as Kaiser, Hanning, and Hamming are employed in FFT execution (see Appendix A10.1).

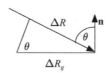

FIGURE 10.13   Ground-range resolution.

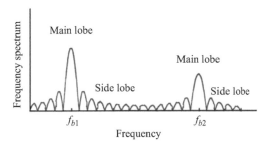

FIGURE 10.14   Beat spectrum of FMCW radar.

10.1.6   AZIMUTH COMPRESSION TECHNIQUE = SYNTHETIC APERTURE PROCESSING (SAR PROCESSING)

Next we consider synthetic aperture processing in a two-dimensional plane as shown in Figure 10.15. The plane consists of the azimuth axis in the x-direction and the range axis in the z-direction.

We rewrite the Equations (10.1.9) into the form of Figure 10.15. If a point target is in the Fresnel region of antenna, the distance R from the antenna to the point target can be written as

$$R = \sqrt{z_0^2 + \left(x - x_0\right)^2} \approx z_0 + \frac{\left(x - x_0\right)^2}{2\,z_0} \tag{10.1.14}$$

Therefore,

$$f_0\tau = f_0 \frac{2R}{c} = \frac{2}{\lambda_0}\left\{ z_0 + \frac{\left(x - x_0\right)^2}{2\,z_0} \right\} \tag{10.1.15}$$

where $\lambda_0 = c/f_0$ is the wavelength at the center frequency.

If an assumption $R \approx z_0$ holds, the beat frequency can also be approximated as

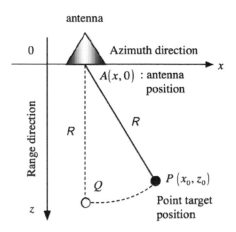

FIGURE 10.15   Position of antenna and point target.

$$f_b \approx \frac{2B}{cT_s} z_0 \tag{10.1.16}$$

As a result, the next approximation holds in the *Sinc* function:

$$\pi T_s \left(f - f_b \right) = \pi B \left(t - \tau \right) \approx \frac{2\pi B}{c} \left(R - z_0 \right) = \pi \frac{\left(R - z_0 \right)}{\Delta R} \tag{10.1.17}$$

$$\frac{\sin \left[\pi \left(f - f_b \right) T_s \right]}{\pi \left(f - f_b \right) T_s} = \frac{\sin \left[\pi \left(R - z_0 \right) / \Delta R \right]}{\pi \left(R - z_0 \right) / \Delta R} = Sinc \left[\pi \frac{\left(R - z_0 \right)}{\Delta R} \right] \tag{10.1.18}$$

Equations (10.1.9) and (10.1.10) become a function of space only. We rename the beat spectrum as $U(x,z)$;

$$U(x,z) = Bg(x_0,z_0) \exp \left[j \frac{4\pi}{\lambda_0} \left\{ z_0 + \frac{\left(x - x_0 \right)^2}{2 z_0} \right\} \right] Sinc \left[\pi \frac{\left(R - z_0 \right)}{\Delta R} \right] \tag{10.1.19}$$

This function $U(x,z)$ consists of the following three functions:

Range function $\qquad\qquad f\left(R - z_0 \right) = Sinc \left[\pi \frac{\left(R - z_0 \right)}{\Delta R} \right] \tag{10.1.20}$

Propagation function $\qquad h\left(x - x_0, z_0 \right) = \exp \left[j \frac{4\pi}{\lambda_0} \left\{ z_0 + \frac{\left(x - x_0 \right)^2}{2 z_0} \right\} \right] \tag{10.1.21}$

Object function $\qquad\qquad g = g\left(x_0, z_0 \right) \tag{10.1.22}$

Now the antenna is scanned along the *x*-axis. At each point of the antenna position, the radar records the beat frequency data as shown in Figure 10.16a. Figure 10.16a shows the beat frequency in the time domain.

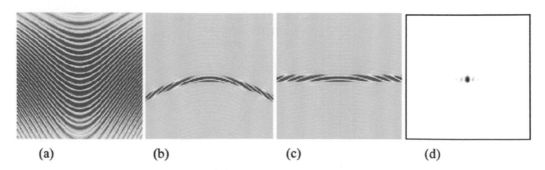

    (a)   (b)   (c)   (d)

FIGURE 10.16 Example of SAR image generation of point target: (a) beat signal, (b) Re{beat spectra}, (c) range migration, and (d) SAR image.

If the Fourier transform is applied to the beat signal, we obtain the beat spectrum (10.1.9) or
(10.1.19) as shown in Figure 10.16b. Figure 10.16b shows the real part of (10.1.9). There is one peak
in the range direction. The peak trajectory looks like a parabola. Range function has a maximum
value when $R = z_0$. The problem comes from the same range R. The value of the range function
from point P is the same as the one from point Q because of the same range R. Radar cannot distinguish the location of P and Q. Received data is saved as the data from point Q beneath the antenna.
Therefore, a curved trajectory is obtained after antenna scanning in the azimuth direction as shown
in Figure 10.16b. When the antenna comes on the zenith of point P, the distance to the point target
is the shortest so that the peak locates on the trajectory summit.

If we proceed to the Fourier transform to this curved data in the azimuth direction, the final
image collapses. Since the echo from the point target spreads in the range direction, range migration
processing is carried out to arrange the curved data to straight line data. Range migration arranges
the data in the $z = z_0$ line.

$$f(R - z_0) \Rightarrow f(z - z_0) = Sinc\left[\pi \frac{(z - z_0)}{\Delta R}\right] \tag{10.1.23}$$

There are various migration methods [2–4]. In this text, a phase shift using FFT (Appendix A10.2)
was employed. The beat spectrum $U(x, z)$ after range migration can be written as

$$U(x, z) = A'' g(x_0, z_0) f(z - z_0) h(x - x_0, z_0) \tag{10.1.24}$$

where A'' is a coefficient. When a distributed target is observed, the beat spectrum $U(x, z)$ can be
expressed in the integral form,

$$U(x, z) = A'' \int_0^\infty \int_{-\infty}^\infty g(x_0, z_0) f(z - z_0) h(x - x_0, z_0) dx_0 dz_0 \tag{10.1.25}$$

Now at around $z \approx z_0$, we have $f(z - z_0) = 1$, and (10.1.25) reduces to

$$U(x, z_0) = A'' \int_{-\infty}^\infty g(x_0, z_0) h(x - x_0, z_0) dx_0 \tag{10.1.26}$$

This form is the same as Fresnel-Kirchhoff diffraction integral. It can be regarded as one of the
Fresnel holograms. Object function can be obtained by multiplication of an inverse propagation
function $h^*(x - x_0, z_0)$,

$$g(x_0, z_0) = A'' \int_{-L/2}^{L/2} U(x, z_0) h^*(x - x_0, z_0) dx \tag{10.1.27}$$

where L is the antenna scan width and * denotes complex conjugation.

This equation represents the SAR processing principle directly. We obtain beat spectrum data
$U(x, z)$ at each point of scan width. Then the inverse propagation function is multiplied, that is, it
is a back-propagation method. Each back-propagated wave is summed up to create the image of
the object. The integration is carried out along the antenna scan, that is, along with the equivalent
synthetic aperture. Since the object function is obtained by the synthetic aperture data, the method
is called synthetic aperture processing. Actual data processing can be executed by

$$g(x_0, z_0) = FT^{-1}\left[FT(U) \cdot FT(h^*)\right] \tag{10.1.28}$$

where FT denotes the Fourier transform, and FT^{-1} denotes the inverse Fourier transform. The calculation is usually carried out by an FFT algorithm. It is recommended to add zero data (zero-padding) to the data sets of U or h, in order to suppress the artifact.

Figure 10.16 shows one example of a SAR image generation of a point target located at the center of the image: (a) is a two-dimensional image of beat frequency in the time domain. If the Fourier transform is performed in the vertical direction, (b) is created. (b) is the range compressed image of (a) and is corresponding to the real part of equation (10.1.9). According to the location of the antenna and target, the curved trajectory is created in (b). By range migration, the curve is rectified to a straight line (c). Then the Fourier transform is performed along the horizontal direction of (c), yielding (d) SAR image of the point target. (d) is a response of point target and hence is called the point spread function (PSF). This image (d) is the final SAR image we are dealing with. The same images are depicted in Figure 10.5 where the image processing procedure is shown together.

10.1.7 Two-Dimensional SAR

By scanning antenna on a two-dimensional plane, we can conduct a three-dimensional measurement. Using the result of a 1D scanning case, we extend it to 2D (x and y axes) scanning.

If the target is located in the Fresnel zone of the transmitting antenna, the distance between the antenna and target can be written as

$$R = \sqrt{z_0^2 + (x-x_0)^2 + (y-y_0)^2} \approx z_0 + \frac{(x-x_0)^2 + (y-y_0)^2}{2z_0} \tag{10.1.29}$$

Object function can be written as

$$g = g(x_0, y_0, z_0)$$

where x_0, y_0 is the coordinate of object, z_0 is the object range.

Other factors are the same as the 1D case. The observed beat signal can be written as

$$U(x,y,z) = \int_0^\infty \int_{-\infty}^\infty \int_{-\infty}^\infty g(x_0, z_0) f(z-z_0) h(x-x_0, y-y_0, z_0) dx_0 dy_0 dz_0 \tag{10.1.30}$$

where:

$$f(z-z_0) = Sinc\left[\frac{(z-z_0)}{\Delta R}\right],$$

$$h(x-x_0, y-y_0, z_0) = \exp\left[j\frac{4\pi}{\lambda_0}\left\{z_0 + \frac{(x-x_0)^2 + (y-y_0)^2}{2z_0}\right\}\right] \tag{10.1.31}$$

At around $z \approx z_0$, we have $f(z-z_0) = 1$, so that

$$U(x,z_0) = \int_{-\infty}^\infty \int_{-\infty}^\infty g(x_0, y_0, z_0) h(x-x_0, y-y_0, z_0) dx_0 dy_0 \tag{10.1.32}$$

This form is again one of the Fresnel holograms. Therefore, its inverse Fresnel transform can be applicable

$$g\left(x_0, y_0, z_0\right) = \int_{-L_{y/2}}^{L_{y/2}} \int_{-L_{x/2}}^{L_{x/2}} U\left(x, y, z_0\right) h^*\left(x - x_0, y - y_0, z_0\right) dxdy \qquad (10.1.33)$$

using the back-propagation function $h^*\left(x - x_0, y - y_0, z_0\right)$. L_x, L_y represents the antenna scan widths in the x and y directions, respectively. * denotes complex conjugation. Equation (10.1.33) establishes 2D synthetic aperture processing. We can extract object function $g = g\left(x_0, y_0, z_0\right)$ by this signal processing.

There are two ways to make a 2D image as shown in Figure 10.17. The left-hand side corresponds to image creation by airborne or spaceborne SAR, whereas the right-hand side corresponds to slice image by ground-penetrating radar (GPR) or experimental radar in laboratories. In remote sensing, the default meaning of SAR processing is assumed to be the case of the left-hand side. In this text, both imaging methods are used for illustrating examples of PolSAR data.

10.1.8 Example of Two-Dimensional SAR Image

Real aperture and synthetic aperture FMCW radar were applied to obtain images of airplane models (Christmas gift) in laboratory measurement. Figure 10.18 shows the difference between these images. The shape of the airplane is clearly imaged by SAR. The imaging algorithm is based on Equation (10.1.33).

10.1.8.1 On Synthetic Aperture Width

According to the inverse Fresnel transformation (10.1.27) and (10.1.33), the resolution may be dependent on the antenna scan width, that is, synthetic aperture length. It is assumed in the equation the antenna has an omnidirectional property. Therefore, the scan length should be large enough. Suppose the origin of measurement system is placed on the target and looks toward the antenna scan length. If the length seen from the origin is wide, the higher-order phase term in the propagation function is captured in the signal processing, resulting in high-resolution imaging. If the length from the origin is limited, the higher-order phase term in the propagation function is eliminated, resulting in a low-resolution image.

This situation is depicted in Figure 10.19. The same aperture length is used to compare the image due to the difference of a higher-order phase term in the propagation function. Even if the same

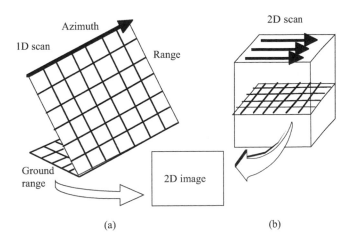

FIGURE 10.17   2D image generation by scanning antenna: (a) 1D scan and (b) 2D scan.

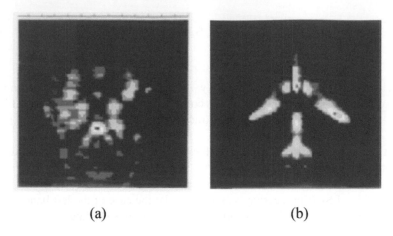

FIGURE 10.18   Difference between real and synthetic aperture images: (a) real aperture image and (b) synthetic aperture image.

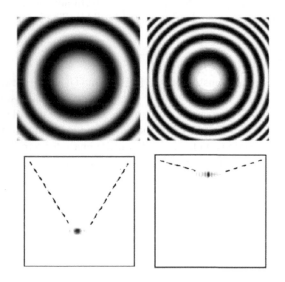

FIGURE 10.19   Difference of image by synthetic aperture length.

aperture is used, the higher-order phase information is different. Since a deep target has less phase information, the corresponding image becomes a blur. On the other hand, since a shallow target can have higher terms, the image becomes vivid.

Therefore, synthetic aperture length should be large enough taking into account the antenna beam width. In order to obtain the same resolution, the angle spanned by the target should be the same. The far target needs a long synthetic aperture length. The earth observation satellite needs a very long synthetic aperture length.

An actual antenna has its directivity, that is, its beam pattern is not uniform. The beam shape is usually in the *Sinc* function, Gaussian function, etc. The Fresnel transformation assumes the omni-direction. In order to match the transformation formula, some antenna pattern modifications are needed to adjust the actual beam pattern to the omni pattern. The antenna beam pattern modification is another issue.

If we use a very small antenna with an omnidirectional pattern, then the synthetic aperture radar works well by definition. However, there arises a problem of power dissipation. Since the power density from small antenna becomes very small, the sensitivity of the far target becomes worse. If the propagation medium is lossy, no signal can be detected. Energy is dissipated. In such a case, we need to use a high-gain antenna, which causes a small synthetic aperture length.

10.2 POLARIMETRIC FMCW RADAR

Polarimetric measurement can be performed by the combination of transmitting and receiving antenna polarization in FMCW SAR. Since (10.1.27) and (10.1.33) are object functions representing object shape, we can regard them as an element of scattering matrix.

$$[S(HV)] = \begin{bmatrix} S_{HH} & S_{HV} \\ S_{VH} & S_{VV} \end{bmatrix} = \begin{bmatrix} g_{HH} & g_{HV} \\ g_{VH} & g_{VV} \end{bmatrix} \tag{10.2.1}$$

Figure 10.20 shows the procedure for scattering matrix acquisition. At first, H-pol is transmitted and both H- and V-pol components are simultaneously received. Then V-pol is transmitted, and both H- and V-pol components are simultaneously received. This procedure is repeated to obtain raw data of scattering matrix.

Then the SAR processing is applied to the polarimetric raw data of each channel and yields a fully polarimetric SAR image as shown in Figure 10.21. The fully polarimetric image consists of pixels with scattering matrix data.

Figure 10.22 shows one example of polarimetric FMCW SAR images obtained by 2D antenna scanning (Figure 10.17). The center frequency was 16 GHz, and the object was an airplane model of 30 cm. The amplitude images $|S_{HH}|$ and $|S_{VV}|$ look similar, and the $|S_{HV}|$ image looks different. Although the magnitude of $|S_{HV}|$ is small compared to $|S_{HH}|$ and $|S_{VV}|$, the differences come from the scattering nature of the object. Furthermore, the phases $\varphi_{HH}, \varphi_{HV}, \varphi_{VV}$ have different values. Therefore, scattering matrix brings various information not only by amplitude but also by phase.

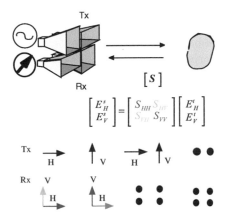

FIGURE 10.20 PolSAR data acquisition.

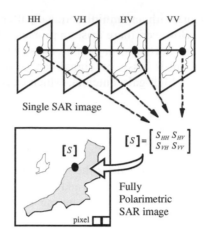

FIGURE 10.21 PolSAR image.

In addition, the total power image is shown as Span [S], the sum of scattering matrix elements squared. It has the maximum S/N ratio. Also, the phase $\frac{\varphi_{HH}-\varphi_{VV}}{2}$ is known to represent a curved surface, and the image is shown for reference.

All images show that the combination of polarimetric measurement by FMCW SAR performs as polarimetric FMCW SAR. Therefore, Pol-FMCW SAR seems like a cost-effective instrument for a short-range sensing system.

It should be noted that the radar reflection signal is a result of swept frequency bandwidth. In the polarimetry principle, the reflected wave is assumed to be a monochromatic wave. On the other hand, we need to have a certain bandwidth in order to achieve range resolution. It is assumed the beat signal is constant in the bandwidth B. If B is very wide, the beat signal amplitude may not be constant within the bandwidth and becomes an amplitude modulated wave. In such a case, we may divide the signal into narrow bandwidths where the beat signal is constant.

10.2.1 FMCW HARDWARE

The hardware constitution of FMCW radar is rather simple and easy to construct. Figure 10.23 shows an example of polarimetric FMCW radar. Basic components are: sweep oscillator, directional coupler or power divider, switch, antennas, and mixer, operative at microwave frequency. IF portions are filter, amplifier, A/D converter, PC controller, etc. Once the IF portion is created, it can be used at any frequency band. Actually, the system below is operated at L-, C-, X-, and Ku-band.

An important issue for polarimetric radar design is the antenna system in the operative frequency band. A highly polarimetric antenna, that is, high polarization purity, is the most important factor as a polarimetric radar in the operative band. For this purpose, we used four standard horn antennas in a compact arrangement as shown in Figure 10.24. Although the phase center by HV and HH, and the center by HV and VV, are different, the effect can be corrected by a polarimetric calibration [14].

This system was used to target classification in anechoic chamber [15]. A metallic sphere of radius 10 cm, dihedral corner reflector (10 cm), and wire target were imaged by FMCW PolSAR independently. Once the scattering matrix is obtained, it is possible to carry out a coherent scattering decomposition based on the coherency matrix [15]. Measured coherency

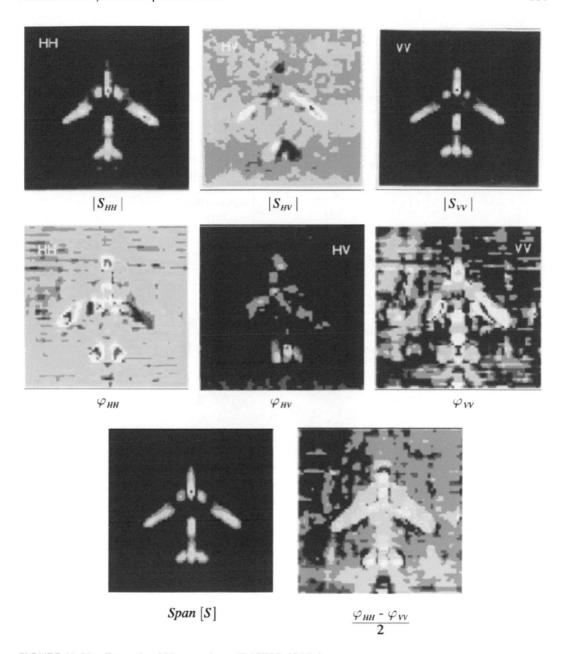

FIGURE 10.22   Example of 2D scanning a FMCW PolSAR image.

matrix $\langle[T]\rangle$ is expanded as a sum of the surface-scattering power P_s, double-bounce scattering power P_d, wire-scattering power P_w (without integration of oriented dipole), and helix-scattering power P_c,

$$\langle[T]\rangle = P_s[T]_s + P_d[T]_d + P_w[T]_w + P_c[T]_c \qquad (10.2.2)$$

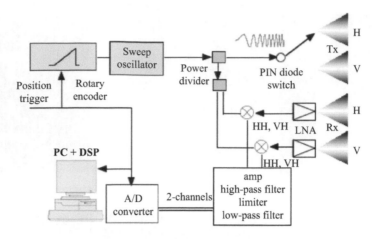

FIGURE 10.23   Polarimetric FMCW radar block diagram.

FIGURE 10.24   Polarimetric radar antennas.

From this expansion, expansion powers can be directly obtained as

$$P_c = \left| 2 \operatorname{Im} \{T_{23}\} \right|$$

$$P_w = \sqrt{\left(\operatorname{Re}\{T_{12}\}\right)^2 + \left(2\operatorname{Re}\{T_{13}\}\right)^2}$$

$$P_s = T_{11} - P_w / 2$$

$$P_d = T_{22} + T_{33} - P_c - P_w / 2 \tag{10.2.3}$$

The decomposition result together with the decomposition ratio is shown in Figure 10.25. It is clearly seen that the system performs as a fully polarimetric radar.

Using the same Pol-FMCW system, three targets aligned in the range direction were measured. These targets were a tree, a metallic sphere, and a dihedral shown in Figure 10.26a. The detection result (b) shows the capability of classifying objects.

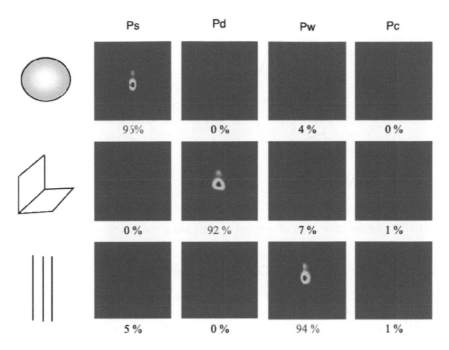

FIGURE 10.25 Decomposition result of sphere, dihedral, and wire target by FMCW PolSAR. Decomposition ratio of each power is also depicted.

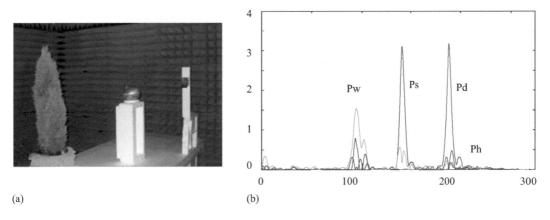

(a)   (b)

FIGURE 10.26 Measurement of three targets aligned in the range direction by Pol-FMCW radar: (a) photos of three targets in anechoic chamber and (b) detection and decomposition result.

The orientation angle θ of the wire target can be obtained by

$$\theta = \frac{1}{2}\cos^{-1}\left(\frac{\mathrm{Re}\{T_{12}\}}{P_w}\right) \tag{10.2.4}$$

Using this equation, it was possible to find the orientation rather accurately. The detection result is shown in Figure 10.27.

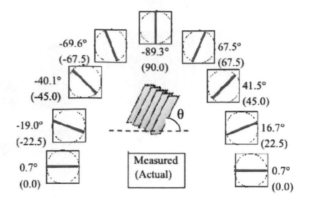

FIGURE 10.27   Radar detection result of orientation angle and actual setting angle.

10.2.2   EQUIVALENT SENSITIVITY TIME CONTROL TECHNIQUE

The receiving power of radar decreases with increasing distance as r^{-4}, as shown in the radar Equation (3.1.10). Therefore, the power from a far target is very small. In order to increase detection range, a sensitivity time control (STC) technique is used in the pulse radar system, which amplifies the receiving signal according to the arrival time so that late arriving signal is enhanced. There was no such technique for FMCW radar. We turned back to the FMCW principle and found an equivalent STC technique for FMCW radar system [16]. This method enhances the far target.

The time domain beat signal can be written as

$$S_b(t) = gAA' \exp\left[j2\pi\left(f_0\tau + f_b t\right)\right] \tag{10.2.5}$$

In order to obtain beat spectrum, we make use of the Fourier transform.

$$FT\left[S_b(t)\right] = S_b(f) \tag{10.2.6}$$

According to the property of the Fourier transform, we know

$$FT\left[\frac{\partial^n S_b(t)}{\partial t_n}\right] = \left(j2\pi f_b\right)^n S_b(f) \tag{10.2.7}$$

This means the differentiation of the beat spectrum makes the original signal $\left(j2\pi f_b\right)^n$ times bigger. Since f_b is proportional to range, the target can be amplified according to its range. Therefore, the propagation loss is compensated as shown in Figure 10.28. This equation (10.2.7) establishes the equivalent STC technique for FMCW radar, which is different from the time domain STC.

The advantage of this method is just multiplication of $\left(j2\pi f_b\right)^n$ after the Fourier transform of beat signal. It can be executed in a computer. If a circuit is needed, it is feasible by adding just a simple differential circuit in the IF band.

By multiplication of $\left(j2\pi f_b\right)^n$, signals in far region are amplified, not only the desired echo but also the noise is amplified. The S/N ratio does not improve; however, the combination of polarimetric data will serve detection improvement which follows.

This method can be applied not only in the free space, but also in lossy media such as snow or soil. GPR always suffers from severe attenuation in the soil medium. This equivalent STC can

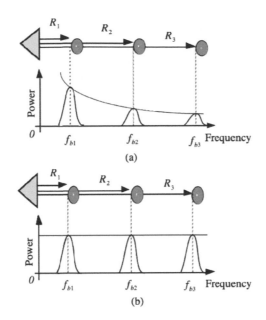

FIGURE 10.28   Amplitude compensation: (a) before compensation and (b) after compensation.

improve detection performance. Figure 10.29 shows an example of GPR detection. A metallic plate was buried in sandy underground at 120 cm deep. Polarimetric FMCW radar was scanned over the surface, and scattering matrices were obtained. The left-hand side of Figure 10.29 is the original co-pol null image for the surface point. We can see the target echo around 120 cm deep. Since there are so many clutter echoes in the original image, then the middle image was created by an equivalent STC method with first-order differentiation. The near-target echo reduces, and the far target echo is enhanced. A metallic plate can be clearly identified in the middle image. The right-most image is created by the second-order differentiation. The far target (metallic plate) is highly enhanced.

The same characteristics are found in the X-pol null images for the surface. Since surface clutter is always severe in GPR detection, it is useful to make a co-pol null or X-pol null image of surface and apply equivalent STCs to detect objects buried in the underground.

Another example is snow radar application for detection of objects buried in accumulated snow-pack [18]. Figure 10.30 shows the measurement situation conducted in Yamakoshi Village, Niigata, Japan, in 1997. Two targets were placed orthogonal to each other as shown in the figure. Target 1 is a metallic pipe, and target 2 is a plate. They were buried at the depth of 60 and 110 cm, respectively. L-band FMCW radar antennas were scanned over the surface, which produced a 3D polarimetric data set.

Figure 10.31 shows polarimetric channel images of (a) *HH*, (b) *HV*, and (c) *VV*. It is seen that the *HV* image does not provide big echo from these targets. Due to the attenuation of waves, the detection of these targets is difficult in the *HV* image. On the other hand, clutter masks desired echoes from targets in the *HH* and *VV* images.

By selecting the co-pol null polarization state of target 1, the receiving power is recalculated to show (d) co-pol null image, where target 1 is eliminated and target 2 is clearly picked up without clutter. In this way, once the scattering matrix is obtained, it is possible to make a less-cluttered image.

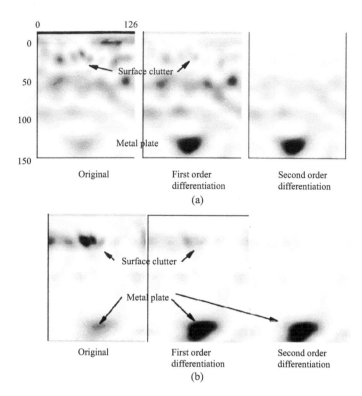

FIGURE 10.29 Detection result of a metallic object at 120 cm deep in the underground and the effect of equivalent STC technique for FMCW radar: (a) co-pol null polarimetric images and (b) X-pol null polarimetric images.

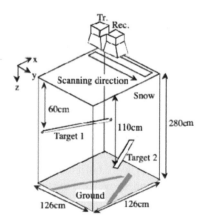

FIGURE 10.30 Measurement situation.

10.2.3 REAL-TIME POLARIMETRIC FMCW RADAR

The signal processing of FMCW radar is mainly based on the Fourier transform. The Fourier transform can be executed very fast by an FFT algorithm using digital signal processing board. In order to obtain a range profile, one execution of FFT is sufficient. Therefore, FMCW radar is suitable

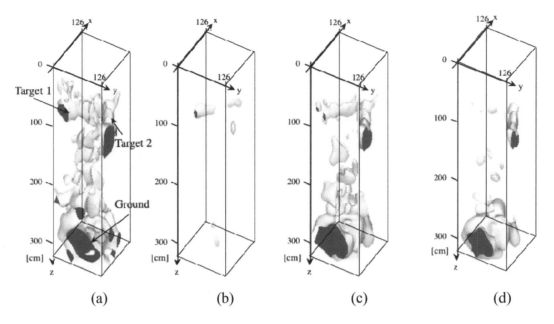

(a)   (b)   (c)   (d)

FIGURE 10.31 3D polarimetric detection result of orthogonal targets (target 1: pipe, target 2: plate): (a) HH, (b) HV, (c) VV, and (d) co-pol null of target 1.

for real-time operation such as car application [17]. In 2006, we have made up a polarimetric and real-time FMCW radar as shown in Figure 10.32. Scattering matrices of a single snapshot can be obtained in less than 20 ms. A polarimetric range profile example is shown to display as a real-time system.

Using the real-time data, SAR processing can be carried out immediately as shown in Figure 10.33.

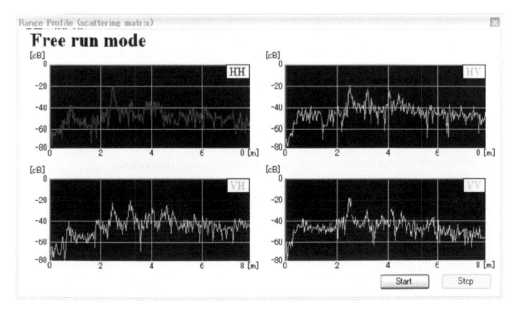

FIGURE 10.32 Range profile example of real-time polarimetric FMCW radar.

Azimuth ⟶   Azimuth ⟶

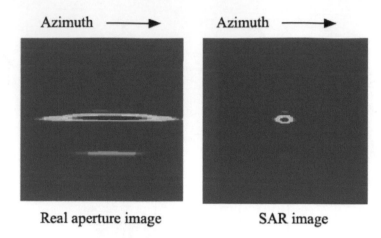

Real aperture image   SAR image

FIGURE 10.33 Example of real-time Pol-FMCW image.

10.3 POLARIMETRIC HOLO-SAR

Spaceborne SAR and airborne SAR move along a straight pass to obtain a 2D SAR image. There are various flight configurations for SAR as shown in Figure 10.34. If the flight path and trajectory become circular, it is called a circular-SAR (C-SAR). C-SAR observes a target from 360° view angles. Therefore, the target structure can be reconstructed using the angular data. If this C-SAR observation is repeated at different heights along the vertical direction, it is called Holo-SAR. Then the Holo-SAR can provide a 3D image viewed from 360° angles.

If polarimetric measurement is applied to Holo-SAR, the system becomes polarimetric Holo-SAR. In order to check the 3D polarimetric imaging capability, we conducted the following experiments in an anechoic chamber.

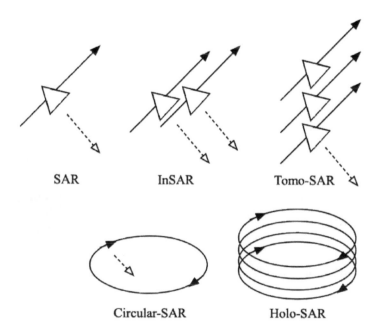

SAR   InSAR   Tomo-SAR

Circular-SAR   Holo-SAR

FIGURE 10.34 SAR and trajectory.

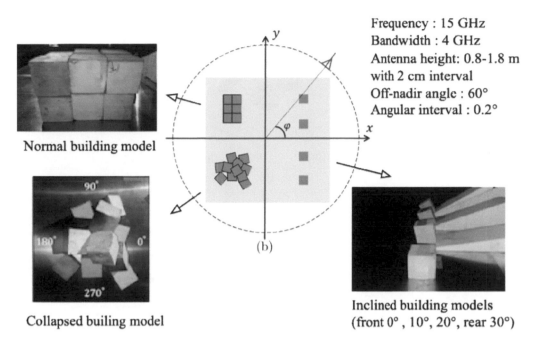

Frequency : 15 GHz
Bandwidth : 4 GHz
Antenna height: 0.8-1.8 m
with 2 cm interval
Off-nadir angle : 60°
Angular interval : 0.2°

Normal building model

Collapsed builing model

Inclined building models
(front 0° , 10°, 20°, rear 30°)

FIGURE 10.35   Experimental setup for polarimetric Holo-SAR.

The first one is a mixture of concrete blocks as shown in Figure 10.35. The blocks are modeled as a normal building, collapsed building by an earthquake, and inclined buildings at 0, 10, 20, and 30° from the vertical direction. These blocks are imaged by a network analyzer-based polarimetric radar system on a turntable in the anechoic chamber. A C-SAR measurement was repeated 50 times to scan a total length of 1 m along the vertical direction in 2-cm increments.

After the Holo-SAR signal processing, polarimetric scattering decomposition is applied to the 3D data sets. The final images are shown in Figure 10.36. RGB color-coding is used to display the scattering mechanisms. Red (double-bounce scattering) is strong around $z = 0$ where the metallic ground plane and vertical concrete wall form right-angle structures. On the top of the normal building, the color exhibits green (volume scattering), which comes from concrete surface. This situation is the same as the scattering of an oblique flat surface, which induces the cross-polarized component, yielding the volume-scattering power. On the other hand, blue (surface scattering) is significant if the oblique angle of the surface is more than 20°. This situation happens in the collapsed building scenario. These angular characteristics are well detected on the rooftop of inclined buildings in Figure 10.36b. For small oblique angles less than 10°, the main scattering power is P_v (green). As the oblique angle becomes larger than 10°, the scattering tends to exhibit P_s (blue). The difference in the scattering mechanism in the vertical direction is well detected in the side-view image as shown in Figure 10.36c.

The second example is conifer and broad-leaf trees. Two trees were measured in the same way. Not only a concrete-metal object but also vegetation can be reconstructed as shown in Figure 10.37. It is seen that conifer tree produces the cross-polarized HV component more compared to the broad-leaf tree. This causes the volume scattering (green) to be dominant for the conifer tree. On the other hand, the surface scattering (blue) is rather significant in the broad-leaf tree. The decomposition powers are retrieved from rectangular boxes in the side-view image of Figure 10.37, and the power ratio is listed in Table 10.3. These values are typical for these tree species at Ku- and X-band [19,20].

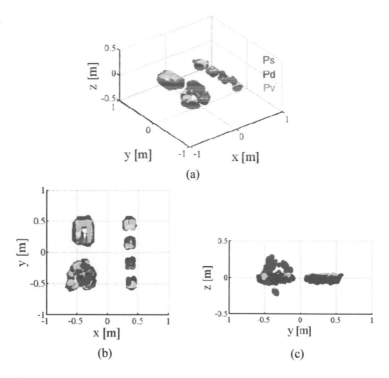

FIGURE 10.36 Polarimetric decomposition image of Holo-SAR (concrete): (a) 3D view of decomposition image, (b) top view, and (c) side view.

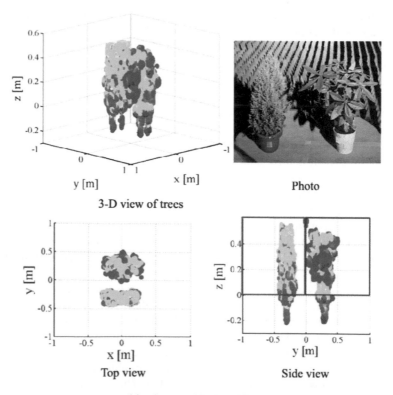

FIGURE 10.37 Polarimetric decomposition image of Holo-SAR (conifer and broad-leaf trees).

TABLE 10.3

Decomposition Power Ratio (%) of Conifer and Broad-Leaf Tree

	P_s	P_d	P_v	P_c
Conifer	22.5	5.4	71.8	0.3
Broad leaf	48.3	4.9	44.7	2.1

Therefore, we can confirm polarimetric Holo-SAR is capable of retrieving a 3D scattering mechanism of an object from all circumference directions.

10.4 SUMMARY

In this chapter, synthetic aperture processing is explained with illustrations. The range resolution is determined by the bandwidth of the transmitting signal. This point holds to pulse, step frequency, and FMCW radar systems. Since the frequency allocation to spaceborne radar is decided (Table 10.1), high resolution less than 1 m on the ground can be achieved only above S-band.

In SAR processing, Fourier transform is frequently used to generate high-resolution 2D images, not only in the range compression but also in the azimuth direction as well as for range migration processing. Since PolSAR is considered a multiple of single-polarization SAR, it requires computation four times more, in addition to polarimetric calibration. After obtaining a scattering matrix, some application examples are shown: decomposition, target angle estimation, equivalent STC for FMCW radar and its GPR application, 3D imaging of buried objects in snowpack with polarization state filtering, real-time polarimetric FMCW radar, and polarimetric Holo-SAR imaging viewed from 360°. All of the polarimetric results are satisfactory and promising.

APPENDIX

A10.1 WINDOW FUNCTIONS

A window function is defined as a mathematical function that is nonzero-valued inside of a chosen interval, normally symmetric around the middle of the interval, and usually tapering away from the middle. Mathematically, when another function is multiplied by a window function, the product is also nonzero inside the interval. The simplest form of a window function is a gate function, which has a unit magnitude inside the interval $\left(-T/2 < t < T/2\right)$ and zero-valued outside. Figure A10.1 shows the gate function in the time domain and its Fourier transform in the frequency domain. If the time interval is taken small, the frequency response spreads, and the frequency characteristics of the system become important to preserve the signal.

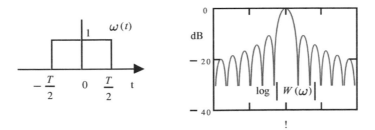

FIGURE A10.1 Rectangular window and its frequency characteristics.

When the Fourier transform is performed, we use a window function. In the actual signal processing, we only have signals of finite length. In order to suppress unwanted frequency responses, an appropriate window function for finite time interval should be selected.

There are several kinds of window functions. Each window has special characteristics. For example, a rectangular window (gate function) yields a *Sinc* function after the Fourier transform. The width of the main lobe is the smallest among other functions. This sharp beam is suitable for making a fine resolution images. The width of the main lobe is also called half-power width. The first side-lobe peak is −13 dB, which is a rather big value. If this value is larger than the echo of other targets, this side lobe masks the main beam of other desired targets. For example, the echo from the land area is big compared to those from the sea surface in radar sensing, and the side-lobe image of land sometimes appears on the sea surface. In order to overcome this undesired situation, various window functions are devised to suppress the side-lobe level. Representative window functions are listed in Table A10.1. The side-lobe levels are suppressed by these functions; however, the main lobe width increases instead.

If we would like to reduce side-lobe levels, then the main beam increases, and vice versa. This trade-off is caused by the property of the Fourier transform. Since the ultimate main lobe width is $c/2B$ in radar, the window function provides us the choice of side-lobe level reduction at the sacrifice of the resolution. The window function modifies the amplitude of the data but does not affect phase information of the data in signal processing.

The main lobe width corresponds to the resolution. A rectangular window or gate function has the sharpest resolution. We take it as the basis and compare other window resolutions with respect to the basis. According to Table A10.1, the width becomes two times for Hanning and Hamming windows and more for Kaiser windows. The Hamming window is preferred because the Hamming window has a lower side-lobe level compared to the Hanning window, although they have the same main lobe width. By adjusting the parameter β, the Kaiser window [12] can be adjusted to any resolution.

- **Kaiser window**

$$W(x) = \frac{I_0\left(\beta\sqrt{1-x^2}\right)}{I_0(\beta)}, \quad -1 < x < 1 \tag{A10.1.1}$$

where $I_0(\bullet)$ is the modified Bessel function and x is a normalized variable.

It is easy to adjust the trade-off relation by changing parameter β in the Kaiser window. Figure A10.2 shows the Kaiser window as a function of β. $\beta = 0$ makes the Kaiser window to the gate function, $\beta = 2.5$ has the side-lobe level as −20 dB, although resolution width increases 20%. This is

TABLE A10.1

Representative Window Function

Window Function	Equation	First Side-Lobe Level	Main Lobe Width
Rectangular	$W(x) = 1, \quad -1 < x < 1$	−13	1
Hanning	$W(x) = 0.5 + 0.5\cos\left(\frac{\pi}{2}x\right)$	−32	2
Hamming	$W(x) = 0.54 + 0.46\cos\left(\frac{\pi}{2}x\right)$	−41	2
Kaiser	$W(x) = \dfrac{I_0\left(\beta\sqrt{1-x^2}\right)}{I_0(\beta)}$	$-46\ (\beta = 2\pi)$	$\sqrt{5}\ (\beta = 2\pi)$

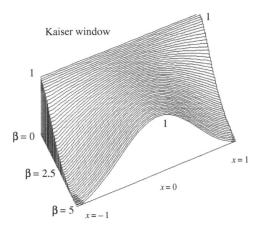

FIGURE A10.2 Kaiser window.

frequently used in SAR processing. $\beta = 5$ has side-lobe level as low as 37 dB, but the main lobe width increases 50%. $\beta = 2\pi$ derives -46 dB side-lobe level.

A10.2 RANGE MIGRATION PROCESSING

Suppose we measure a point target by radar as shown in Figure 10.15. The antenna scans over a target. The range-compressed trajectory becomes curved one as shown in the upper left side of Figure A10.3. The image axes consist of the azimuth direction and the range direction. This curve is caused by the distance between the antenna and the point target.

If we apply the Fourier transform to this curved trajectory image without range migration, then the right-hand side image comes out. Since the Fourier transform is carried out horizontally line by line, the transformed image does not show focusing at all but rather blurring the surroundings of the target. This degradation comes from the data arrangement in the compressed image. In order to arrange data in a straight line, range migration processing is needed.

The range migration arranges data from a curved one to a straight one as shown in Figure A10.3. After the range migration, the Fourier transform yields the focused image of the point target (PSF) as shown in the figure.

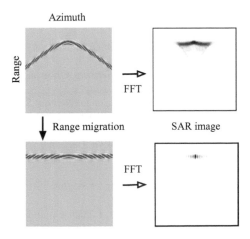

FIGURE A10.3 SAR image with/without range migration (left: range FFT, right: SAR image).

There are several methods proposed in range migration. Here, a simple method of using the Fourier transform is explained. Fourier transform has the following phase-shift property:

$$f(t) \Leftrightarrow F(\omega) = \int_{-\infty}^{\infty} f(t) e^{-j\omega t} dt \tag{A10.2.1}$$

$$f(t-a) \Leftrightarrow F(\omega) e^{j\omega a} = \int_{-\infty}^{\infty} f(t) e^{-j\omega(t-a)} dt \tag{A10.2.2}$$

Equation (A10.2.2) shows that if the position t moves by a, the phase shift in the frequency domain becomes $e^{j\omega a}$. If we can know the position shift a, then the corresponding phase shift will serve to rearrange the data location.

In the range-compressed image, we have a *Sinc* function. We would like to arrange the data location change as

$$Sinc\left[\pi \frac{(R-z_0)}{\Delta R}\right] \text{to } Sinc\left[\pi \frac{(z-z_0)}{\Delta R}\right].$$

Moving the space variable $\dfrac{(R-z)}{\Delta R}$ is equivalent to the phase shift of $\exp\left\{j2\pi \dfrac{(R-z)}{\Delta R} \dfrac{m}{M}\right\}$.

We first execute the Fourier transform of the data in the range direction. Then after multiplying the phase shift to the frequency domain data, the frequency data is again inversely Fourier transformed. This procedure is repeated along the azimuth direction. Then the compressed peak aligns in the horizontal line at $z = z_0$. In this way, the phase-shift property can be used for range migration.

REFERENCES

1. Y. Furuhama, K. Okamoto, and H. Masuko, *Microwave Remote Sensing by Satellite*, IEICE, Tokyo 1986.
2. D. R. Wehner, *High Resolution Radar*, Artech House, Boston, 1987.
3. J. P. Fitch, *Synthetic Aperture Radar*, Springer-Verlag, Germany, 1988.
4. M. I. Skolnik ed., *Radar Handbook*, 2nd ed., McGraw-Hill, 1990.
5. D. L. Mensa, *High Resolution Radar Cross-Section Imaging*, Artech House, Boston, 1991.
6. N. C. Currie ed., *Radar Reflectivity Measurement: Techniques and Applications*, Artech House, Boston, 1989.
7. F. T. Ulaby and C. Elachi, *Radar Polarimetry for Geoscience Applications*, Artech House, Boston, 1990.
8. K. Okamoto eds., *Global Environmental Remote Sensing*, Ohmsha Press, Tokyo, 2001. Wave Summit Course.
9. K. Ouchi, *Fundamentals of Synthetic Aperture Radar for Remote Sensing (in Japanese)*, Tokyo Denki University Press, Tokyo, 2004. ISBN: 978-4-501-32710-1; 2nd ed., 2009.
10. Y. Yamaguchi, *Radar Polarimetry from Basics to Applications: Radar Remote Sensing Using Polarimetric Information* (in Japanese), IEICE, Tokyo, December 2007. ISBN: 978-4-88554-227-7.
11. M. Takagi and H. Shimoda, eds., *New Handbook on Image Analysis*, Tokyo University Press, Tokyo, 2004.
12. I. G. Cumming and F. H. Wong, *Digital Processing of Synthetic Aperture Radar Data*, Artech House, Boston, 2005.
13. A. Moreira, P. Prats-Iraola, M. Younis, G. Krieger, I. Hajnsek, and K. Papathanassiou, "A Tutorial on synthetic aperture radar," *IEEE Geosci. Remote Sens. Mag.*, vol. 1, no. 1, pp. 6–43, 2013.

14. M. Nakamura, Y. Yamaguchi, and H. Yamada, "Real-time and full polarimetric FM-CW radar and its applications to the classification of targets," *IEEE Trans. Instrum. Meas.*, vol. 47, no. 2, pp. 572–577, 1999.
15. J. Nakamura, K. Aoyama, M. Ikarashi, Y. Yamaguchi, and H. Yamada, "Coherent decomposition of fully polarimetric FMCW radar data," *IEICE Trans. Commun.*, vol. E91-B, no. 7, pp. 2374–2379, 2008.
16. H. Kasahara, T. Moriyama, Y. Yamaguchi, and H. Yamada, "On an equivalent sensitivity time control circuit for FMCW radar," *Trans. IEICE B-II*, vol. J79-B-II, no. 9, pp. 583–588, 1996.
17. Y. Yamaguchi, M. Sengoku, and S. Motooka, "Using a van-mounted FM-CW radar to detect corner-reflector road-boundary markers," *IEEE Trans. Instrum. Meas.*, vol. 45, no. 4, pp. 793–799, 1996.
18. T. Moriyama, Y. Yamaguchi, and H. Yamada, "Three-dimensional fully polarimetric imaging in snow-pack by a synthetic aperture FM-CW radar," *IEICE Trans. Commun.*, vol. E83-B, no. 9, pp. 1963–1968, 2000.
19. Y. Yamaguchi, Y. Minetani, M. Umemura, and H. Yamada, "Experimental validation of conifer and broad-leaf tree classification using high resolution PolSAR data above X-band," *IEICE Trans. Commun.*, vol. E102-B, no. 7, pp. 1345–1350, 2019. doi:10.1587/transcom.2018EBP3288.
20. H. Shimoda, Y. Yamaguchi, and H. Yamada, "Experimental study on polarimetric-HoloSAR," *IEICE Commun. Exp. (ComEX)*, vol. 8, no. 4, pp. 122–128, 2019. doi:doi:10.1587/comex.2018XBL0153.

11 Material Constant

Radar measures target-scattering information. The target has its own material constants (dielectric constant ε or relative dielectric constant ε_r, equivalently permittivity ε for the electrical property, permeability μ for the magnetic property, and electric conductivity σ). Scattering behavior depends on the material constants $(\varepsilon, \mu, \sigma)$. Backscattering coefficient depends on both the material constant and the shape of the object. Even if the shape is the same, the backscattering coefficient becomes different for different material constants. Usually metallic objects reflect the largest signal,

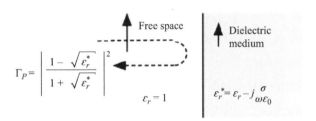

FIGURE 11.1   Power reflection by dielectric medium.

while dielectric materials reflect small signals.

The essential point of scattering is the mutual interaction between the radar wave and the object material. The boundary condition of the electromagnetic wave, that is, the continuity of the tangential electric field and magnetic field, plays the decisive role in scattering phenomena. After applying the boundary condition to the air-dielectric interface as shown in Figure 11.1, one can obtain the power reflection coefficient [1],

$$\Gamma_p = \left| \frac{\eta_1 - \eta_0}{\eta_1 + \eta_0} \right|^2 = \left| \frac{1 - \sqrt{\varepsilon_r^*}}{1 + \sqrt{\varepsilon_r^*}} \right|^2 \qquad (11.1)$$

where $\eta_0 = \sqrt{\frac{\mu_0}{\varepsilon_0}}$ and $\eta_1 = \sqrt{\frac{\mu_0}{\varepsilon_0 \varepsilon_r^*}}$ are intrinsic impedances of the free space and the medium, respectively. The complex relative dielectric constant (or relative permittivity) is given as

$$\varepsilon_r^* = \varepsilon_r - j \frac{\sigma}{\omega \varepsilon_0} \qquad (11.2)$$

The power reflection coefficient in equation (11.1) applies for most of the dielectric materials for which we can assume the permeability is equal to that of free space $(\mu = \mu_0)$. From this equation, we can see the effect of complex permittivity on the reflection power or the scattered power.

There are various materials, such as air, water, ice, soil, trees, and rocks. Their electric properties are completely different from each other. The object constitutes are usually a mixture of different

TABLE 11.1

Complex Permittivity of Medium

Debye model	$\varepsilon_r(f,T) = \varepsilon_r(\infty,T) + \dfrac{\varepsilon_r(0,T) - \varepsilon_r(\infty,T)}{1 + j 2\pi f \tau(T)}$			
	$\varepsilon_r(\infty, T)$	$\varepsilon_r(0, T)$	$2\pi f \tau(T)$	$f[GHz]$, $T[°C]$
Water		$77.66 + 103.3\,\theta$ $\theta = \dfrac{300}{273.5+T} - 1$		$-20 < T < 60$
Pure water	4.9	$88.045 - 0.4147\,T$ $+6.295 \times 10^{-4}\,T^2$	$1.11 \times 10^{-10} - 3.82 \times 10^{-12}\,T$ $+6.938 \times 10^{-14}\,T^2$	$f < 100$

Extended Debye Model

Sea water	$\varepsilon_{sw}^* = \varepsilon_{sw}' - j\,\varepsilon_{sw}'' = 4.9 + \dfrac{\varepsilon_{w0}(T) - 4.9}{1 + j\,2\pi f \tau_w(T)} - j\,\dfrac{\sigma}{2\pi\varepsilon_0 f}$	$\sigma = 0.18\,C^{0.93}\left[1 + 0.02(T-20)\right]$ $C = 3.254$ [0/00] (representative)
Ice	$\varepsilon_{ice}^* = \varepsilon_{ice}' - j\,\varepsilon_{ice}'' = 3.15 - j\,10^{-4}\left[\dfrac{\alpha(\theta)}{f} + \beta(\theta)f\right]$ $\alpha(\theta) = (50.4 + 62\theta)\exp\{-22.1\theta\}$ $\beta(\theta) = \dfrac{0.502 - 0.131\theta}{1+\theta} + 0.00542\left(\dfrac{1+\theta}{\theta+0.0073}\right)^2$	$\varepsilon_{ice}' = 3.15$ for fresh ice $\theta = \dfrac{300}{273.15+T} - 1$
Soil	$\varepsilon_{soil}^{0.65} = 1 + 0.655\rho + m_v^\beta\left(\varepsilon_{fw}^{0.65} - 1\right)$	$\rho\left[g/cm^3\right]$ density, m_v (%): volume soil moisture content $\beta = 1.0(sand) \sim 1.17(clay)$
Rock	$\varepsilon_r^* = \varepsilon_r' - j\varepsilon_r'' = \varepsilon_r'\left(1.0 - j\tan\delta\right)$	$2.4 < \varepsilon_r' < 9.6$ $\tan\delta = 0.01 \sim 0.1$
Dry snow	$\varepsilon_{ds} = \begin{cases} 1.0 + 1.9\,\rho_s & \text{for } \rho_s < 0.5 \\ 0.51 + 2.88\,\rho_s & \text{for } \rho_s > 0.5 \end{cases}$	$3 \le f \le 37$ GHz $\rho_s\left[g/cm^3\right]$: snow density
Wet snow	$\varepsilon_{ws}' = 1.0 + 1.83\,\rho_s + 0.02\,m_v^{1.015} + \dfrac{0.073\,m_v^{1.31}}{1 + (f/f_0)^2}$ $\varepsilon_{ws}'' = \dfrac{0.073\,(f/f_0)^2\,m_v^{1.31}}{1 + (f/f_0)^2}$	$3 \le f \le 37$ GHz $f_0 = 9.07$ GHz m_v(%): volume soil moisture content

small particles, just as snow is a mixture of ice, air, and water. The electric parameters as a whole are characterized by the dielectric property based on measurements and the Debye model. In this chapter, the medium is regarded as homogenous and isotropic. We summarize what is known about the complex permittivity in the form of equation (11.2). In the following, empirical equations of complex permittivity are shown and summarized in Table 11.1.

11.1 COMPLEX PERMITTIVITY [2]

The general form of the complex permittivity by the **Debye model** is given as

$$\varepsilon(f) = \varepsilon' - j\varepsilon'' = \varepsilon_\infty + \frac{\varepsilon_0 - \varepsilon_\infty}{1 + j\, 2\pi f \tau(T)} \tag{11.3}$$

where f is frequency, ε_0 is relative permittivity at zero frequency, ε_∞ is relative permittivity at infinite frequency, $\tau(T)$ is relaxation time, and $T[°C]$ is temperature.

11.1.1 WATER AND SEA WATER [3]

Static water: $\quad \varepsilon_0(T) = 77.66 + 103.3 \left(\dfrac{3.00}{2735 + T} - 1 \right)$ $\qquad\qquad$ (11.4)

$\qquad\qquad\qquad$ where $T[°C]$ is the temperature effective for $-20°C < T < 60°C$

Pure water: $\quad \varepsilon_{fw}^* = \varepsilon_{fw}' - j\varepsilon_{fw}'' = 4.9 + \dfrac{\varepsilon_{w0}(T) - 4.9}{1 + j2\pi f \tau_w(T)}$ $\qquad\qquad$ (11.5)

$\qquad\qquad\qquad$ where $2\pi f \tau_w(T) = 1.11 \times 10^{-10} - 3.82 \times 10^{-12}T + 6.938 \times 10^{-14}T^2$

$$\varepsilon_{w0}(T) = 88.045 - 0.4147T + 6.295 \times 10^{-4}T^2$$

Conductivity of Sea Water

$$\sigma = 0.18C^{0.93}\left[1 + 0.02(T - 20) \right] \tag{11.6}$$

$\qquad\qquad$ effective for frequency below 1 GHz.
$\qquad\qquad$ Salinity $C = 3.254\%$ (typical),

$\qquad\qquad \sigma = 5\, S/m$ at $T = 20°C$: typical conductivity value

Sea Water

$$\varepsilon_{sw}^* = \varepsilon_{sw}' j\varepsilon_{sw}'' = 4.9 + \frac{\varepsilon_{w0}(T) - 4.9}{1 + j2\pi f \tau_w(T)} - j\frac{\sigma}{2\pi\varepsilon_0 f} \tag{11.7}$$

There is no difference between sea water and pure water at frequencies above 20 GHz.

11.1.2 ICE

For frequencies from 1 MHz to 1 GHz, the following equation for ice is derived [4].

$$\varepsilon_{ice}^* = \varepsilon_{ice}' - j_{ice}'' = 3.15 - j\,10^{-4}\left[\frac{\alpha(\theta)}{f} + \beta(\theta)f \right] \tag{11.8}$$

$$\alpha(\theta) = (50.4 + 62\theta)e^{-22.1\theta}$$

$$\beta(\theta) = \frac{0.502 - 0.131\theta}{1+\theta} + 0.00542\left(\frac{1+\theta}{\theta + 0.0073}\right)^2$$

$$\theta = \frac{300}{273.15 + T} - 1$$

$$f[\text{GHz}], T[°C]$$

11.1.3 Snow

Snow consists of ice particles, an air gap, and water, and hence is a mixture of these materials. The dielectric constant depends on the ratio of these materials. Water inclusion basically determines the complex permittivity. The water inclusion can be represented by $m_v(\%)$ in volume. Dry snow can be roughly classified as $m_v < 3\%$, whereas wet snow can be classified as $m_v > 3\%$. The following approximation has been derived by the Debye model and experimental results.

$$\textbf{Dry snow}: \varepsilon_{ds} = \begin{cases} 1.0 + 1.9\rho_s & \text{for } \rho_s < 0.5\left[\text{g/cm}^3\right] \\ 0.51 + 2.88\rho_s & \text{for } \rho_s > 0.5\left[\text{g/cm}^3\right] \end{cases} \tag{11.9}$$

$$\textbf{Wet snow}: \varepsilon_{ws}^* = \varepsilon_{ws}' - j\varepsilon_{ws}''$$

$$\varepsilon_{ws}' = 1.0 + 1.83\,\rho_s + 0.02\,m_v^{1.015} + \frac{0.073\,m_v^{1.31}}{1 + (f/f_0)^2} \tag{11.10}$$

$$\varepsilon_{ws}'' = \frac{0.073(f/f_0)^2\,m_v^{1.31}}{1 + (f/f_0)^2} \tag{11.11}$$

$$\text{Frequency range: } 3 \le f \le 37\text{ GHz}, f_0 = 9.07\text{ GHz}$$

$$\rho_s\left[\text{g/cm}^3\right]: \text{density}, 0.09 \le \rho_s \le 0.38\left[\text{g/cm}^3\right]$$

$$m_v[\%]: \text{water inclusion in volume } 0 \le m_v \le 15\%\text{ max}$$

In general, accumulated dry snow has a small real part of permittivity less than 1.5, and the imaginary part can be ignored. For accumulated wet snow, the maximum permittivity is less than 3 [5]. The imaginary part depends on the water inclusion ratio m_v.

11.1.4 Soil

Soil, rock, clay, sand, etc. are mixtures of dielectric materials by grains, air, and water. The dielectric property depends on the constitutional ratio of these materials. Based on numerous data sets, and using the Debye model, some empirical equations are presented [6].

$$\varepsilon_{soil}^{0.65} = 1 + 0.655\rho + m_v^{\beta}\left(\varepsilon_{fw}^{0.65} - 1\right) \tag{11.12}$$

where $\beta = 1.0(sand) \sim 1.17(clay)$

$$m_v[\%]: \text{water inclusion in volume}(0.01 < m_v < \text{saturation})$$

$$\rho\left[\text{g/cm}^3\right]: \text{density}$$

$$\varepsilon_{fw}^*: \text{complex permittivity of water}$$

It is [8] reported that the real part of permittivity is accurately calculated from this equation.

11.1.5 Rocks

Champbell and Ulrichs [7] derived an empirical equation for the frequency range 450 MHz–35 GHz,

$$\varepsilon_r^* = \varepsilon_r' - j\varepsilon_r'' = \varepsilon_r'(1.0 - j\tan\delta) \tag{11.13}$$

$$2.4 < \varepsilon_r' < 9.6: \text{no frequency dependency}$$

$$\tan\delta = 0.01 \sim 0.1$$

11.1.6 Volcanic Ashes

Oguchi et al. [9] reported the measurement results of volcanic ashes that were erupted from Asama Mountain. For the frequency range 2–12 GHz,

$$\text{volcanic ashes} \quad \varepsilon_r = 3 \sim 3.5, \quad \tan\delta = 0.06 \tag{11.14}$$

$$\text{stone} \quad \varepsilon_r = 5.3 \sim 5.4, \quad \tan\delta = 0.1$$

11.1.7 Flame (Plasma)

Flame is one of plasma in which electrons and ion particles are in a separated condition. For ionosphere propagation, the electromagnetic wave receives Faraday rotation and attenuation depending on frequency. For big fire, or ionosphere, the equivalent permittivity can be determined by the electron density N.

$$\varepsilon_r^* = 1 - \frac{\omega_p^2}{\omega_c^2 + \omega^2} - j\frac{\omega_c}{\omega}\frac{\omega_p^2}{\omega_c^2 + \omega^2} \tag{11.15}$$

where ω_c is the collision frequency of electrons,

$$\omega_p = \sqrt{\frac{Ne^2}{m\varepsilon_0}} : \text{Plasma frequency}$$

$$m = 9.109\times10^{-31}\left[\text{kg}\right], \quad e = -1.602\times10^{-19}\left[\text{C}\right], \quad \varepsilon_0 = 8.854\times10^{-12}\left[\text{F/m}\right]$$

Electromagnetic waves can propagate through plasma above the plasma frequency ω_p but cannot propagate below ω_p. For example, the plasma frequency becomes $f_p \approx 2.84$ MHz for electron density $N = 10^{11}/\text{m}^3$ in the ionosphere. Centimeter and millimeter propagation in explosions is reported in [10].

A good reference can be found in the *Recommendation of ITU-R* p. 527–5, "Electrical characteristics of the surface of the Earth" [2]. It shows frequency characteristics of the permittivity of sea water, wet soil, water, medium dry soil, dry soil, and ice.

11.2 POLARIZATION TRANSFORMATION BY FLAT GROUND SURFACE AND PERMITTIVITY OF THE GROUND

Suppose a linearly polarized wave is incident on the ground surface as shown in Figure 11.2. The reflected wave becomes, in general, an elliptically polarized wave. These phenomena can be regarded as polarization transformation by the ground surface. The reflected wave bears the information on the ground medium, that is, the permittivity of the ground. If we measure the polarization state of the reflected signal by bistatic measurement (Figure 11.3), it may be possible to retrieve the permittivity.

The scattering phenomenon in Figure 11.2 can be described by

$$
\begin{bmatrix} E_h^r \\ E_h^r \end{bmatrix} = \begin{bmatrix} R_h & 0 \\ 0 & R_v \end{bmatrix} \begin{bmatrix} E_h^i \\ E_h^i \end{bmatrix}
\tag{11.16}
$$

where R_h and R_v are Fresnel reflection coefficients for the horizontally and vertically polarized wave, respectively.

$$
R_h = \frac{\cos\theta - \sqrt{n^2 - \sin^2\theta}}{\cos\theta + \sqrt{n^2 - \sin^2\theta}}, \; R_v = \frac{n^2\cos\theta - \sqrt{n^2 - \sin^2\theta}}{n^2\cos\theta + \sqrt{n^2 - \sin^2\theta}}
\tag{11.17}
$$

θ: incidence angle

$$
n^2 = \varepsilon/\varepsilon_0 = \varepsilon_r^* = \varepsilon_r - j\frac{\sigma}{\omega\varepsilon_0}
$$

The polarization ratio of reflected signal can be written as

$$
\rho^r = \frac{E_v^r}{E_h^r} = \frac{R_v E_v^i}{R_h E_h^i} = \frac{R_v}{R_h}\rho^i
\tag{11.18}
$$

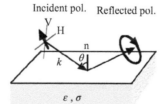

FIGURE 11.2 Polarization state transformation by reflection.

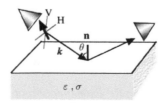

FIGURE 11.3 Bistatic measurement of permittivity by polarization ratio.

If we choose a linearly polarized wave for the incidence wave, the polarization ratio becomes

$$\rho^i = \frac{E_v^i}{E_h^i} = \tan \tau$$

If 45°-oriented linear polarization is chosen $(\tau = 45°)$, we have very simple value $\rho^i = 1$.

Therefore, the polarization ratio of the reflected signal becomes

$$\rho^r = \frac{R_v}{R_h} = \frac{\varepsilon_r^* \cos \theta - \sqrt{\varepsilon_r^* - \sin^2 \theta}}{\varepsilon_r^* \cos \theta + \sqrt{\varepsilon_r^* - \sin^2 \theta}} \frac{\cos \theta + \sqrt{\varepsilon_r^* - \sin^2 \theta}}{\cos \theta - \sqrt{\varepsilon_r^* - \sin^2 \theta}} \quad (11.19)$$

As seen in this equation, the polarization ratio is a function of incidence angle and permittivity. The relation of the polarization ratio magnitude and the relative permittivity is shown in Figure 11.4. The relative permittivity of the ground is chosen in the range 3–50 depending on water content. The incidence angle around 70° provides a wider dynamic range for the polarization ratio.

One of the measurement configurations is shown in Figure 11.5. A 45°-oriented horn antenna (T_x) is used for the transmitter, and horizontal (H) and vertical (V) horn antennas are used for the receiver. Since this system measures the ratio of received signals, it is easy to calibrate the measurement system by a metallic plate as shown in Figure 11.5. Since the metal plate is a perfect conductor, we have $\rho^r = 1$. If the plate has a concrete (dielectric material) surface, we always have $|R_v| < |R_h|$

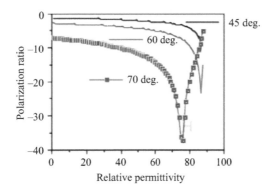

FIGURE 11.4 Polarization ratio and relative permittivity.

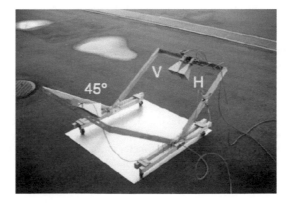

FIGURE 11.5 Bistatic measurement system.

so that $\left|\rho^r\right| < 1$. The measured polarization ratio ρ^r is dependent on the material constant and it becomes minimum at the Brewster angle. It was possible to classify the wet and dry road surface very easily by the simple system in Figure 11.5.

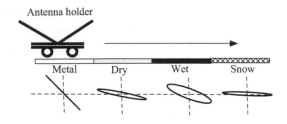

FIGURE 11.6   Road surface condition vs. polarization.

The measured polarization states corresponding to road surface conditions are shown in Figure 11.6.

11.3   REMARKS

The backscattering coefficient depends on both the material constant and the shape of object. The object has its own material constants (dielectric constant ε, permeability μ, and electric conductivity σ). Since there are various objects for radar measurement, it is worth checking the values of typical land objects. This chapter summarized the relative dielectric constant of typical land surface objects.

REFERENCES

1. F. T. Ulaby, R. K. Moore, and A. K. Fung, *Microwave Remote Sensing: Active and Passive*, vol. 1, pp. 78–82, and vol. 3, Appendix E, Artech House, Boston, 1986.
2. Recommendation of ITU-R, "Electrical characteristics of the surface of the Earth," p. 527–5, September 2019. https://www.itu.int/rec/R-REC-P.527/en.
3. H. J. Liebe, G. A. Hufford, and T. Manabe, "A model for the complex permittivity of water at frequencies below 1 THz," *Int. J. Infrared Millim. Waves*, vol. 12, no. 7, pp. 659–675, 1991.
4. G. Hufford, "A model for the complex permittivity of ice at frequencies below 1 THz," *Int. J. Infrared Millim. Waves*, vol. 12, no. 7, pp. 677–682, 1991.
5. T. Abe, Y. Yamaguchi, and M. Sengoku, "Experimental study of microwave transmission in snowpack," *IEEE Trans. Geosci. Remote Sens.*, vol. 28, no. 5, pp. 915–921, 1990.
6. M. C. Dobson, F. T. Ulaby, M. Hallikainen, and M. E. Rayes, "Microwave dielectric behavior of wet soil-part II: Four component dielectric mixing models," *IEEE Trans. Geosci. Remote Sens.*, GE-23, pp. 35–46, 1985.
7. C. K. Champbell and J. Ulrichs, "Electrical properties of rocks and their significance for lunar radar observations," *J. Geophys. Res.*, 74, pp. 5867–5881, 1969.
8. V. Mironov, Y. Kerr, J.-P. Wigneron, L. Kosolapova, and F. Demontoux, "Temperature- and texture-dependent dielectric model for moist soils at 1.4 GHz," *IEEE Geosci. Remote Sens. Lett.*, vol. 10, no. 3, pp. 419–423, May 2013.
9. T. Oguchi, M. Udagawa, N. Nanba, M. Maki, and Y. Ishimine, "Measurement of the dielectric constant of volcanic ash erupted from Asama volcano," *URSI-F Japan*, no. 515, June 2007.
10. G. P. Kulemin and V. B. Razskazovsky, "Centimeter and millimeter-wave radio signals attenuation in explosions," *IEEE Trans. Antenn. Propag.*, vol. 45, no. 4, pp. 740–743, 1997.

12 Interpretation of PolSAR Images

Due to global warming in recent years, we have experienced unusual weather conditions, such as localized heavy pouring rains, which cause flooding, landslides, and tornados, which unexpectedly occur and cause devastating damage in a local area. Not only small-scale events but also very large-scale typhoons, hurricanes, and cyclones have generated frequently in recent years. These events are attracting attention as global environmental problems. The IPCC's report on climate change warns the recent global warming based on the scientific data [1]. Actually, natural disasters are increasing year by year all around the world. Polarimetric SAR (PolSAR) will play an important role in coping with these problems by collecting accurate data. Especially, PolSAR is useful for the following categories (Figure 12.1).

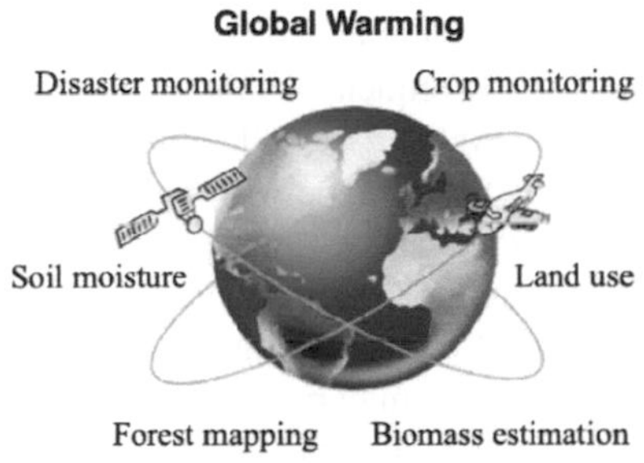

Global Warming

Disaster monitoring   Crop monitoring

Soil moisture   Land use

Forest mapping   Biomass estimation

FIGURE 12.1   Polarimetric SAR applications to monitor global warming.

Disaster monitoring: Earthquakes, landslides, tsunamis, flooding, typhoons, hurricanes, cyclones, tornados, severe and localized heavy rain, and volcano eruptions are typical natural disasters. They have occurred more frequently in recent years all over the world. In addition, localized heavy rainfall is increasing, which attacks specific areas and causes landslides and flooding. In rainy weather conditions, the optical sensor cannot serve its purpose because of opaque cloud cover. Since microwaves can penetrate clouds, SAR is expected to play an important role in monitoring such adverse situations. Similarly for volcano activity monitoring, the penetration capability of the radar signal plays a crucial role in the observation.

Biomass estimation: Carbon dioxide (CO_2) is a main source of greenhouse gas in global warming. Since vegetation absorbs CO_2 and mitigates global warming, it is really important to keep and maintain huge vegetation areas, as in the rainforest in the Amazon, Southeast Asia, and Central Africa. The amount of absorption depends on the amount of biomass. Accurate estimation of biomass therefore is a very important task worldwide. It is known that the *HV* polarimetric response is well-correlated to vegetation volume and that the information on the *HV* component can serve biomass estimation.

Forest monitoring: Biomass estimation is similar to forest monitoring. Due to illegal forest-cut or expansion of human activity in forests, deforestation areas have expanded year by year. Rainforests in the Amazon, Indonesia and Malaysia, for example, are severely facing deforestation problems. After forest-cut, the land is exposed to open field or used for palm oil plantation. The amount of CO_2 absorption there decreases significantly; instead, forest-cut increases exposure of methane gas from the bare ground, which further accelerates global warming. The role of healthy forests is the key for maintaining environmental issues [2].

Soil moisture: By soil moisture information, it becomes possible to assess information suitable for crop growing, extent of flooding area, dry area extension toward the desert, grass land monitoring, etc. Radar remote sensing is expected especially for a vast area where direct measurement is difficult.

Crop monitoring: In order to support a huge population in developing countries, crop monitoring is crucial for its production. Crops include rice, wheat, beans, corns, and various vegetation types. They appear as volume scattering in polarimetric radar sensing. Finer resolution for each crop monitoring is needed.

Land use monitoring: Urbanization with increasing populations is expanding in developing countries. This is accompanied with deforestation, and sometimes with transition from forest to farmland and to urban areas. The land use changes drastically in these areas. PolSAR is quite effective for monitoring these changes. It is expected to play a very important role.

Remote sensing itself does not provide direct solutions to these problems; however, it serves to provide valuable and accurate information on the events through the measurement. Radar remote sensing makes use of interaction of electromagnetic waves and objects under measurement. Scattering is a complicated phenomenon caused by frequency, bandwidth, polarization, object shape, material, and object size with respect to wavelength. After all, observation is based on the scattered wave from an object at a far distance. What can be done with radar remote sensing is to acquire the scattering phenomena and its change. Other issues such as expectations or forecasts are beyond the scope of scattering. It is important to understand what can be done with scattering and what cannot be done. In addition, the applicability of radar sensing to certain applications should not be overestimated. Otherwise, too many expectations may be misleading.

In this chapter, characteristics of the SAR image and the interpretation are reviewed to understand PolSAR images correctly [3–8]. There are so many targets for monitoring. If we take the case-study approach, the total number becomes huge, and sometimes the case study does not serve other similar events. To avoid many case studies, we try to recall the scattering mechanisms in Chapter 5 and investigate the scenario.

1. *Surface-scattering object*: Rough surface, bare soil, sea surface, volcano ashes, snow, ice, etc.
2. *Double-bounce scatter*: Right-angle structure, man-made object, urban area, tall vegetation, etc.
3. *Volume-scattering object*: Cross-polarization generator, tree, forest, crops, vegetation, oriented urban area, etc.

The basic point is the interaction of electromagnetic waves and objects under measurement. Material constants (permittivity and conductivity) are also an important factors to the scattering and interaction. Understanding wave interaction with objects is an essential part for radar data interpretation.

12.1 SAR IMAGE INTERPRETATION

We first review the fundamentals of radar remote sensing. Figure 12.2 shows the definition of angles and parameters [3,4].

The off-nadir angle is defined as the angle between main beam direction and the zenith direction and is equivalent to the incidence angle for flat-surface observation in low altitudes. However, for satellite observation, the off-nadir angle is different from the incidence angle as shown in the

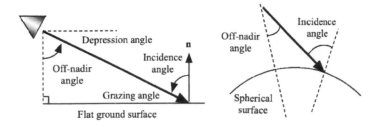

FIGURE 12.2   Off-nadir angle and incidence angle in radar observation.

right figure, due to the spherical shape of the earth's surface. The term "off-nadir angle" is a simple term for the radar operation side. But the definition of the incidence angle is different from that of off-nadir angle. It is defined as the angle spanned by the direction of wave incidence and the normal direction of the surface. It is an important parameter of the radar data analysis. This difference comes from the location of the coordinate origin. Therefore, both angles are used for their purpose. A typical difference can be seen in the following observation.

If a smooth, round mountain is illuminated by a radar beam as shown in Figure 12.3, the scattered waves go in different directions according to its location. Even if the off-nadir angle is the same, the incidence angle at each point becomes different. The incidence angle on the surface facing to the radar becomes small, whereas the angle on the surface away from the radar becomes large. This situation explains the difference between the off-nadir angle and incidence angle.

For data analysis, we often draw the backscattering magnitude radar cross section (RCS) as a function of incidence angle [3]. Typical RCS characteristics derived from both theoretical and experimental data are shown in Figure 12.4. It has the maximum value at 0° incidence angle and

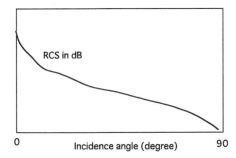

FIGURE 12.3   Incidence angle variation on a mountain surface.

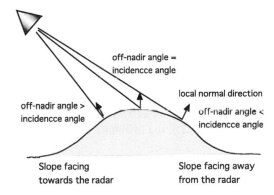

FIGURE 12.4   RCS vs. incidence angle.

FIGURE 12.5   Izu Oshima Island image by PiSAR-L.

gradually decreases with increasing incidence angle. The curve shape depends on the roughness of the surface and wavelength. If we take one example of PiSAR-L data over Izu Oshima Island, the radar image becomes as shown in Figure 12.5. The illumination direction is from the top to down. The mountain slope facing toward the radar is bright, whereas the mountain slope away from the radar looks dark. The RCS characteristics are well-represented in the SAR image and serve us to interpret 3D mapping. In addition, a flat surface "airport runaway" in the upper-right corner is a completely dark line due to mirror reflection.

SAR is a side-looking radar. It transmits pulses toward the oblique direction (slant range direction) as shown in Figure 12.6, where the wave front and its range resolution are depicted. The intersection point of the wave front and a flat ground surface becomes the sampling point of scattering. These sampling points are indicated in colored dots. Since level 1.1 data arranges each sampling on the straight line as shown in Figure 12.6, the near-range pixel has a compressed low-resolution image, and the far-range pixel has a finer resolution image.

If applied to a mountain as shown in Figure 12.7, the situation of the upper figure is called "foreshortening." As can be seen in Figure 12.6, the final data arrangement in level 1.1 shows that pixel

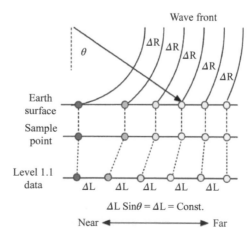

FIGURE 12.6   Data arrangement of side-looking radar system.

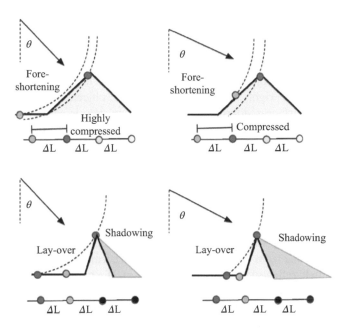

FIGURE 12.7   Foreshortening and layover.

interval between the first orange point and the red orange point at the top of the mountain is compressed. The information between the first pixel and the second pixel is compressed in the interval ΔL. This situation corresponds to foreshortening. When the off-nadir angle becomes smaller, the compression rate becomes higher (left image), hence leading to a steeper slope mountain image creation. If the off-nadir angle is moderate, the foreshortening effect is mitigated. This phenomenon always happens in the mountain area in a single-look complex image.

If the mountain is very steep and high, a layover effect appears in the radar image as shown in lower Figure 12.7. The top of the mountain (red dot) locates close to the radar in front of other geographical closer points (orange dots). The layover effect always appears in tall building or tower mapping.

Interesting images for our visual perception are shown in Figure 12.8. What is the left figure?

The same image is displayed on the right, but the radar illumination is different. We can easily recognize it as a volcano summit on the right. However, it is difficult for us to understand the left image because we are not accustomed to seeing images with the illumination from the bottom. We are familiar with the situation where light is coming from the top. This is our intuitive perception. In the radar image, the illumination direction corresponds to the range direction, and the cross range corresponds to the azimuth direction. It is better to write the range or illumination direction in the radar image.

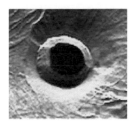

FIGURE 12.8   Volcano summit seen by PiSAR-L2.

12.1.1 Scattering Scenario

So many objects exist on the ground. Each object has its own shape, orientation, size, and material, and the scattering mechanism is dependent on the frequency. Figure 12.9 shows one example of a scattering scenario over each object. The corresponding scattering power is displayed beneath the scattering point. From left to right, we can see some typical objects under consideration. A rough sea surface reflects the surface-scattering power. If the sea state is calm and flat, the strength of the reflected signal becomes so small, close to zero, or equivalently the minimum detectable level to the radar which is equivalent to noise level, which results in black color in the radar image. If a tree stem stands on a flat surface, a strong double-bounce may occur. This situation happens when flooding water occupies crop fields. An inundated area sometimes reflects a strong double-bounce scattering signal caused by trees or pillows. In addition to this scattering, volume scattering will be mixed together in the flooding case. If the situation is represented by RGB color-coding, the area tends to exhibit an orange (mixed) color. For residential areas, the right-angle structure created by the water surface and building walls cause very strong reflections. The RCS value for this inundated case is stronger than the one from the normal dry case. For normal weather conditions, the bare soil surface such as crop field reflects surface-scattering power. If a periodic planting row is orthogonal to radar illumination and satisfies the Bragg-condition, it reflects strong Bragg surface scattering. These phenomena are frequently seen in rice paddies and crop fields. The cross-polarization, HV component, is generated at the edge of buildings, and at oriented surfaces, trees, vegetation, forests, etc. The magnitude is smaller than those of HH and VV components; however, it contributes to volume scattering in a complex scattering scenario.

Figure 12.10 shows how three major scattering powers change with respect to the incidence angle. In most cases, the off-nadir angle of a typical SAR system ranges from 20 to 60°.

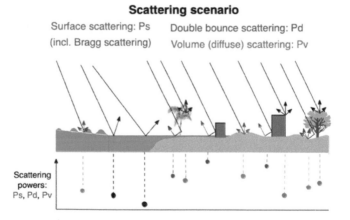

Scattering scenario

Surface scattering: Ps   Double bounce scattering: Pd
(incl. Bragg scattering)   Volume (diffuse) scattering: Pv

Scattering
powers:
Ps, Pd, Pv

FIGURE 12.9 Example of scattering mechanisms on the ground.

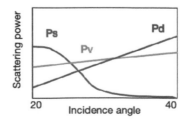

FIGURE 12.10 Incidence angle characteristics of scattering powers.

The surface-scattering power drops off at around 20 to 28° according to the surface roughness. The double-bounce scattering power and the volume-scattering power increase with incidence angles.

At around the 20° incidence angle, the surface scattering is strong. The corresponding surface-scattering power P_s is usually larger than P_d or P_v. This situation can be seen on the sea surface and bare soil surface such as crop fields or landslide areas. The surface scattering fades away at incidence angles more than 30°. For sea surfaces and sea-state monitoring, it is better to use the incidence angle or off-nadir angle smaller than 25°. Otherwise, the surface looks completely dark in the SAR image. Landslide areas, on the other hand, have rough surfaces and exhibit a big value in wider angle ranges, which is suitable for polarimetric radar sensing.

The volume-scattering power increases slightly with increasing incidence angle as seen in Figure 12.11. Since the volume scattering mainly comes from trees, forests, and vegetation, the reason of increase can be understood easily by a vegetation propagation model in Figure 12.11. If the propagation medium is assumed to be a homogeneous forest, then the cross-polarized HV component is generated proportional to the path length. The propagation path L1 for a small incidence angle is shorter than L2 for a large incidence angle case. This path length difference contributes to the difference in volume-scattering characteristics as a function of incidence angle in Figure 12.10.

The reason why the magnitude of the double-bounce scattering increases with increasing incidence angle can be explained in Figure 12.12. It is due to an increase of effective scattering area in the double-bounce scattering. S2 becomes bigger than S1 as shown in the figure. Roughly speaking, the double-bounce scattering at the 40° incidence angle is much larger than that of the 20° case.

These incidence characteristics can be seen in the following ALOS decomposition images of Istanbul, Turkey. The color-coding is red (double-bounce scattering power P_d), green (volume-scattering power P_v), and blue (surface-scattering power P_s). Blue can be seen in land for the 21.5° off-nadir angle; however, it vanishes in the 35° image. Instead, more green can be seen in the 35° image. Red areas slightly increase with increasing incidence angle. The 25° image has a moderate color combination, which indicates that various scattering appears in the image. We can see black stripe (road construction) and bridge construction in the upper land. Therefore, the best off-nadir angle for polarimetric radar seems to be 23–25°, which reflects all kinds of scattering in the decomposition image.

The incidence angle effects are well-recognized in Figure 12.13. We can confirm the scattering magnitudes of three scattering mechanisms following the characteristics in Figure 12.10. Therefore, when we see PolSAR images, we have to pay attention to the incidence angle or off-nadir angle of the scene.

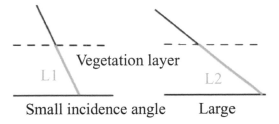

FIGURE 12.11 Incidence angle and propagation path effective length L1 and L2 contributing volume scattering.

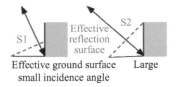

FIGURE 12.12 Incidence angle and double-bounce scattering area S1 and S2.

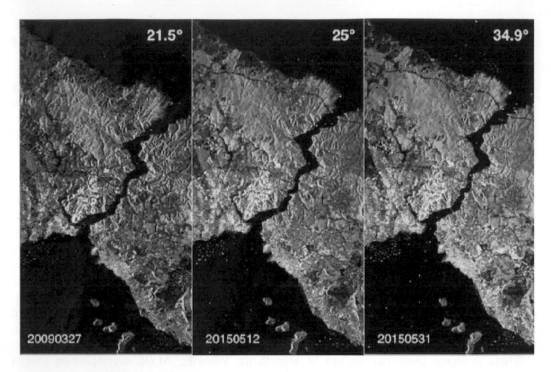

FIGURE 12.13 Scattering-power decomposition image of Istanbul, Turkey, as a function of off-nadir angle.

12.1.2 HOW FINE IS FINE?

When dealing with radar images, we often face the resolution problem. Recent radar resolutions are in the range of 1–10 m for spaceborne SAR and less than 1 m for airborne SAR. Many people and many applications would request "finer resolution!" A 3-m resolution may be better than 10-m resolution, and 1.5-m resolution seems better than 3-m resolution. But how fine is fine? For a realization of 10 cm, we need a bandwidth = 1500 MHz according to the definition of range resolution.

A 10-cm resolution is theoretically realizable, but it is impossible to design such a wide-band radar system including transmitting and receiving antenna in the microwave frequency (Figure 12.14).

In addition, a bandwidth of 1500 MHz is not allotted for any radar applications in International Telecommunication Union (ITU) regulation. Each country has its own regulation for frequency allocation. If the airborne radar operates in its country under the frequency allocation regulation, it may be possible to use a certain bandwidth. For the PiSAR-X3 system, 15-cm resolution can be achieved. A 20-cm resolution is available for the F-SAR system.

The advantage of high resolution is its quality in the image. Sometimes the PolSAR image is better than the optical one if the resolution is 20–30 cm. We cannot judge the superiority of the 20–30 cm resolution radar image and the optical one. Each has its own merits. The disadvantage is a big volume of data size. The data size becomes as much as several GB for a small area, which is not suitable for file transfer. If the radar resolution becomes too high, the scattering mechanism of

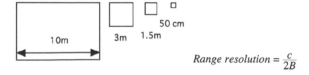

$$Range\ resolution = \frac{c}{2B}$$

FIGURE 12.14 Radar resolution and corresponding image pixel size.

the neighboring pixel becomes different. This sometimes makes interpretation of whole image difficult. Therefore, resolution requirements, resolution expectations, and related problems are dependent on its application.

Ensemble averaging is frequently carried out in the polarimetric data analysis. The total number of pixels usually exceeds 20 to endure the statistical property. If the number of samples is big, the resolution in the average becomes blurring, just like boxcar filtering. If 5×5 pixels are used, the corresponding area becomes $50 \times 50 \ m^2$ for 10-m resolution data, whereas it's $2.5 \times 2.5 \ m^2$ for 50-cm resolution data. Hence, the actual imaging window size is an important factor when applied object size. If the target is a man-made structure, a $50 \times 50 \ m$ area is too big; on the other hand, if the target is a forest area, the size may be acceptable. For a uniformly distributed target, very high resolution is not necessary. For complex man-made objects, finer resolution may be desired.

12.1.3 FREQUENCY RESPONSE

Let's take an example of radar images as functions of frequency. At first, scattering-power decomposition images around Niigata City are shown in Figure 12.15.

These images are obtained by PiSAR-X2 (NICT) on August 25, 2013 and PiSAR-L2 (JAXA) on August 3, 2016. The two images clearly show the difference in scattering mechanisms by frequency. The lower-half image is a rice paddy field, where green is dominant for the X-band, and the blue/dark color is dominant for the L-band. The scattering behaviors of the X- and L-bands are well represented with respect to wavelength against object size. For a shorter wavelength of 3 cm, rice stems are complex scattering objects yielding volume scattering. For 24-cm wavelength, rice stems are rather transparent objects and reflect weak signals. On the other hand, man-made structures are comparable with or larger than 24 cm, which leads to distinct and big double-bounce scattering (red) in the L-band.

PiSAR-L-band (3-m resolution) and X-band (1.5-m resolution) data sets were also used for the scattering-power decomposition image with a 5×5 window for comparison in Figure 12.15. We can see the difference in frequency response, that is, red (double bounce) and green (volume scattering) can be identified easily in the L-band. The dark area corresponds to the water surface

FIGURE 12.15 PiSAR-X2 and -L2 scattering power images of the Niigata area.

(river, sea, rice paddy field). On the other hand, the X-band image looks finer but covered with blue (surface scattering) on whole image. This blue color is characteristic of X-band scattering, because of the shorter wavelength to object. Therefore, L-band decomposition image looks vivid due to the dynamic scattering mechanism responses.

12.1.4 WINDOW SIZE RESPONSE

Ensemble averaging is frequently carried out in the polarimetric data analysis. The total number of pixels usually exceeds 20 to endure the statistical property. The imaging window size was changed from 5 × 5 to 7 × 7, and to 9 × 9 to check the resolution dependency. Figure 12.17 shows the decomposition images. As the window size increases (Figures 12.16 and 12.17), the statistical values become fixed, and color contrast becomes vivid. On the other hand, the resolution becomes blurring. Similar phenomena occur in other data sets. For a high-resolution image in the X-band, a window size of more than 7 × 7 is preferred.

For L-band data by ALOS2 [5], we have selected an image of San Francisco and picked up vegetation and oriented urban areas for testing. The window size was chosen as 2 × 4, 3 × 6, 4 × 8, and 5 × 10 so that ground area becomes close to square. The color-coded decomposition images are shown in Figure 12.18. It is seen from the comparison figures that a window size of more than 4 × 8 is preferred. If we choose 5 × 10, the total number of pixels is almost the same as that (7 × 7) of the X-band.

For quantitative comparison, the decomposition power and ratio are displayed in Figure 12.19 for the vegetation area, and in Figure 12.20 for the oriented urban area.

It is seen that the volume-scattering power increases as the window size increases. Other powers decrease as shown in the figure. Since the vegetation area is selected, the window size should be larger than 4 × 8 for the volume-scattering power dominant. For urban areas, all powers are stable except for the volume scattering as shown in Figure 12.20. Since the man-made structure is dominant in this region, the double-bounce scattering power P_d or the surface-scattering power P_s should be dominant. In this sense, the window size 5 × 10 is big enough.

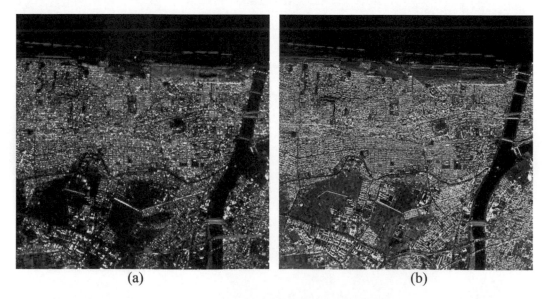

(a)   (b)

FIGURE 12.16 Frequency characteristics of scattering-power decomposition image over Kobari, Niigata, Japan. The data acquisition was carried out by PiSAR-X/L simultaneous observation: (a) PiSAR-L (5 × 5 window) and (b) PiSAR-X (5 × 5 window).

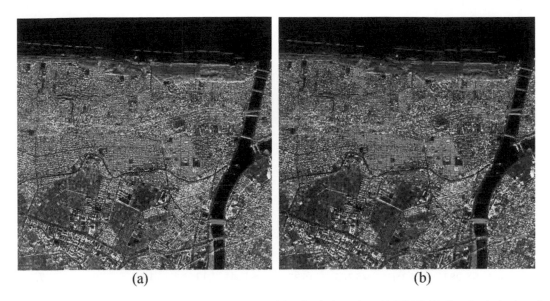

FIGURE 12.17 Decomposition images for 7 × 7 and 9 × 9 window size: (a) PiSAR-X (7 × 7 window) and (b) PiSAR-X (9 × 9 window).

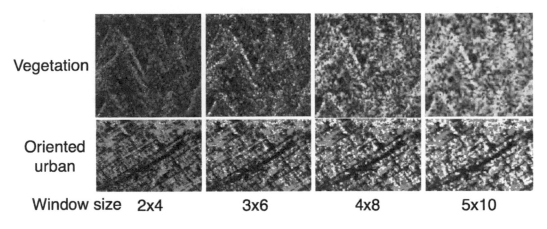

FIGURE 12.18 Window size vs. decomposition image in the L-band.

12.1.5 TIME SERIES DATA

Some targets change with time. For example, the lifetime of rice is short. Rice is a seasonally irrigated crop with three major growth phases in its life cycle: the vegetative phase (seedling, tillering, and stem elongation), the reproductive phase (booting, heading, and flowering), and the ripening phase (milk, dough, and mature grain). During each phase, the plant has a different aspect that changes the response of the radar signal. If we measure a paddy field in a time series, the response becomes as shown in Figure 12.21. Echigo Plains is a wide rice field of Japan. ALOS quad pol images in March, June, and August are compared in the figure. In March, the paddy field is just a bare soil surface, yielding surface scattering. In June, the paddy field is covered with water and small rice plants grow up to several centimeters. For L-band frequency, the small rice stem is transparent. Therefore, mirror reflection occurs in the paddy field. This is why we have dark colors in

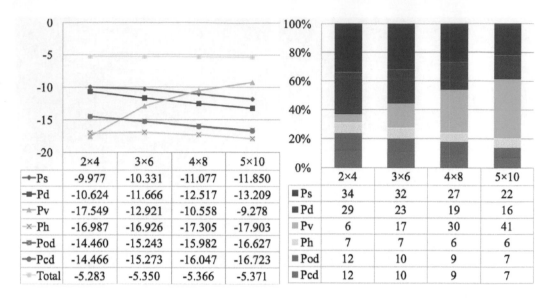

	2×4	3×6	4×8	5×10
Ps	-9.977	-10.331	-11.077	-11.850
Pd	-10.624	-11.666	-12.517	-13.209
Pv	-17.549	-12.921	-10.558	-9.278
Ph	-16.987	-16.926	-17.305	-17.903
Pod	-14.460	-15.243	-15.982	-16.627
Pcd	-14.466	-15.273	-16.047	-16.723
Total	-5.283	-5.350	-5.366	-5.371

	2×4	3×6	4×8	5×10
Ps	34	32	27	22
Pd	29	23	19	16
Pv	6	17	30	41
Ph	7	7	6	6
Pod	12	10	9	7
Pcd	12	10	9	7

FIGURE 12.19 Window size vs. decomposition powers and their ratio for vegetation.

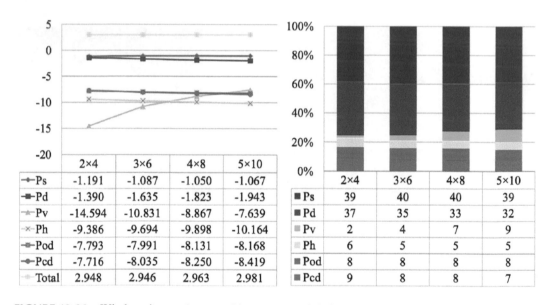

	2×4	3×6	4×8	5×10
Ps	-1.191	-1.087	-1.050	-1.067
Pd	-1.390	-1.635	-1.823	-1.943
Pv	-14.594	-10.831	-8.867	-7.639
Ph	-9.386	-9.694	-9.898	-10.164
Pod	-7.793	-7.991	-8.131	-8.168
Pcd	-7.716	-8.035	-8.250	-8.419
Total	2.948	2.946	2.963	2.981

	2×4	3×6	4×8	5×10
Ps	39	40	40	39
Pd	37	35	33	32
Pv	2	4	7	9
Ph	6	5	5	5
Pod	8	8	8	8
Pcd	9	8	8	7

FIGURE 12.20 Window size vs. decomposition powers and their ratio for oriented urban area.

the paddy field. In August, the rice stem grows up to approximately 50–60 cm. If the planting row is orthogonal to radar illumination, a right-angle structure by rice stems and the wet ground surface causes double-bounce reflection. We can see red areas in the paddy fields.

Similarly, deciduous (broad leaf) trees lose their leaves in fall. Therefore, the polarimetric response in summer and winter is different. Accumulated snow in mountainous seems stable. But its dielectric property changes drastically with the season, caused by water content.

Therefore, the radar scene captured by a certain time can be considered as one snapshot within a long period. We should recognize it as one of the samples taking into account the background

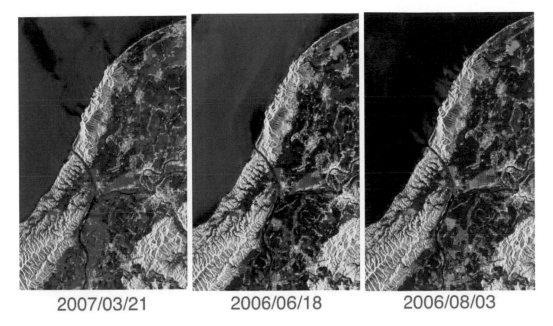

2007/03/21   2006/06/18   2006/08/03

FIGURE 12.21   Time series scattering power decomposition Y4R image of Echigo Plains, Niigata, Japan, by ALOS-PALSAR.

information. Time series data provide continuously changing phenomena and are helpful to detect the changes.

12.2   SUMMARY

In this chapter, the very basic properties of the SAR image, such as range direction, difference of incidence angle and off-nadir angle, and backscattering magnitude as a function of incidence angle are explained with illustrations. SAR image interpretation becomes quite difficult if the image is rotated by 180°. In most cases, the SAR image is displayed so that radar illumination is from top to bottom.

SAR images on frequency dependence and imaging window size are once again reviewed to understand the PolSAR image correctly. Even if the same area is used, L-band and X-band images are very different from each other. This is caused by the difference of scattering characteristics by frequency. It is important to interpret PolSAR images correctly based on these items.

REFERENCES

1. IPCC report on global warming. https://www.ipcc.ch/sr15/.
2. JAXA forest mapping. http://www.eorc.jaxa.jp.
3. F. M. Henderson and A. J. Lewis, *Principles & Applications of Imaging Radar, Manual of Remote Sensing*, 3d ed., vol. 2, ch. 5, pp. 271–357, John Wiley & Sons, Hoboken, NJ, 1998.
4. J. van Zyl and Y. Kim, *Synthetic Aperture Radar Polarimetry*, John Wiley & Sons, Hoboken, NJ, 2011. ISBN: 9781118115114.
5. M. Shimada, *Imaging from Spaceborne SARs, Calibration, and Applications*, CRC Press, 2019. ISBN: 978-1-138-19705-5.
6. K. S. Chen, S. B. Serpico, and J. A. Smith, eds., Special issue: Remote sensing of natural disasters, *Proceedings of the IEEE*, vol. 100, no. 10, 2012.

7. Y. Yamaguchi, "Disaster monitoring by fully polarimetric SAR data acquired with ALOS-PALSAR," *Proceedings of the IEEE*, vol. 100, no. 10, pp. 2851–2860, 2012.

8. G. Singh, Y. Yamaguchi, W.-M. Boerner, and S.-E. Park, "Monitoring of the March 11, 2011, Off-Tohoku 9.0 earthquake, with super-Tsunami disaster by implementing fully polarimetric high resolution POLSAR techniques," *Proceedings of the IEEE*, vol. 101. no. 3, pp. 831–846, March 2013.

13 Surface Scattering

In this chapter, we will see some examples of surface-scattering objects. Typical scatterers are bare soil surfaces, crop fields, rice paddy fields, volcano ash-covered areas, small grasslands, snow covered areas, snow and ice fields, etc. These scatterers have the same characteristic that the *HH* and *VV* component are in phase. Therefore, $|HH + VV|$ becomes a good indicator for surface scattering. $|HH + VV|$ and the surface-scattering power P_s are often assigned to "blue" in an RGB color-coded image. Some of surface scatterings are shown in the following figures. The images are RGB color-coded with red for the double-bounce scattering P_d, blue for the surface scattering P_s, and green for the volume scattering P_v (Figures 13.1 through 13.5).

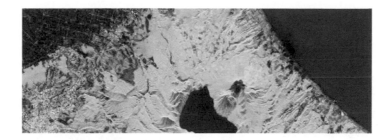

FIGURE 13.1 ALOS image of Shikotsu Lake, Hokkaido, Japan. Blue corresponds to the surface-scattering power, which locates the volcano summit, bare soil surface, crop field, and fallen tree area in a forest by a typhoon.

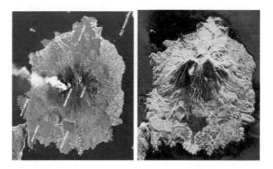

FIGURE 13.2 Sakura-jima volcano (left: ASTER optical image, right: ALOS-PALSAR). Ash and lava flow trace can be seen in blue.

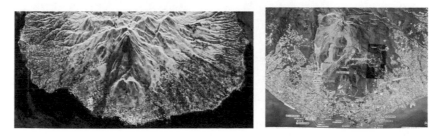

FIGURE 13.3 Unzen. Trace of pyroclastic flow can be identified by the dark surface scattering.

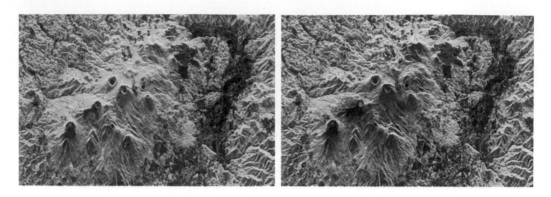

FIGURE 13.4 Shinmoe-dake eruption (left: before eruption on June 10, 2009, right: after March 16, 2011). Volcano ashes covered the surrounding area after the eruption. (From Yamaguchi, Y., *Proc. IEEE*, 100, 2851–2860, 2012.)

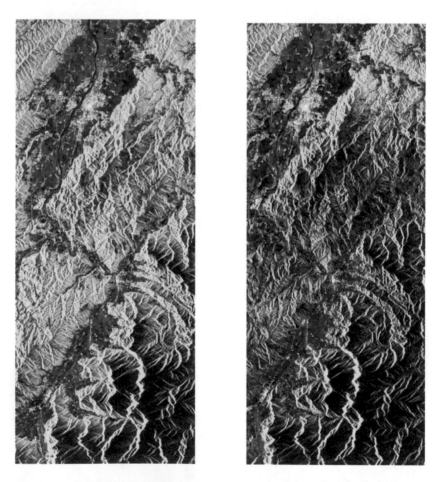

FIGURE 13.5 Accumulated snow in mountains. (left: March 26, 2009, right: March 29, 2010). The accumulated snow depth depends on the total amount of snowfall each year. We had heavy snow in 2009–2010 winter (right image). (From Yamaguchi, Y., *Proc. IEEE*, 100, 2851–2860, 2012.)

13.1 SCATTERING FROM ROUGH SURFACES

When a plane wave is impinging on a surface, the wave is scattered according to the roughness as shown in Figure 13.6. If the surface is flat (smooth surface), the scattering is similar to a mirror reflection, and hence no backscattering occurs. In this case, the surface looks dark in the radar image. This situation corresponds to scattering from the water surface, playground, and airport runway. As roughness increases (medium rough), the scattering wave tends to spread in wider range directions. Backscattering increases according to the roughness. A rice paddy field or crop field scattering in the L-band corresponds to this situation. For a completely rough surface, the scattered wave is radiated in omni directions (rough surface scattering).

Backscattering characteristics or the radar cross section (RCS) as a function of incidence angle for various rough surfaces are also depicted in Figure 13.6. For a smooth surface, strong RCS occurs only around normal incidence angles. A landslide brings rough bare soil on the ground or mountain slope. Due to wide angle characteristics of the rough surface scattering, it becomes easier for polarimetric radar to detect the landslide area. This landslide situation corresponds to rough surface RCS. The sea surface is also a good example of rough surfaces. The sea surface has various situations from smooth to very rough condition. For the L-band, the sea acts as a smooth to medium rough surface within 25° incidence angles.

In addition to the roughness, penetration capability of the wave plays an important role in backscattering. Backscattering is a complex function of frequency, dielectric constant, and conductivity. For P-band frequency, most of flat surfaces have small RCS and look dark.

The scattering from the rough surface has been a great topic by many researchers [2–20]. The details of the theoretical development are beyond the scope of this book. We just keep the general concept of scattering from the rough surface as shown in Figure 13.6 and see some basic properties for surface-scattering phenomena.

There are two surface parameters: mean height and correlation length. When we evaluate the surface, we have to take into account the size with respect to wavelength, that is, the ratio of mean height h with respect to wavelength λ in the vertical direction, and the correlation length in the horizontal direction. These are two major parameters describing surface roughness. For example, $h = 1$ cm is small for $\lambda = 30$ cm at L-band, but rather big for $\lambda = 3$ cm at X-band. Therefore, the ratio of the mean height by wavelength, $h/\lambda = 1/30$ or $1/3$, is an important factor for evaluating surface roughness.

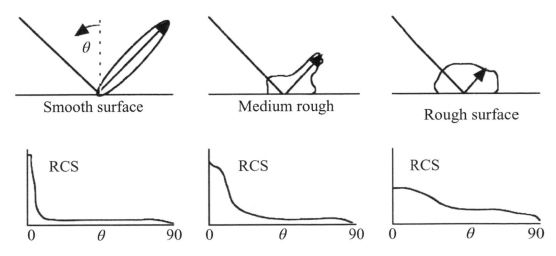

FIGURE 13.6   Scattering from smooth to rough surface.

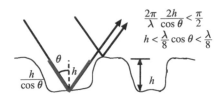

FIGURE 13.7   Rayleigh criterion: smooth surface corresponds to $h<\lambda/8$.

Whether the surface is rough or smooth is determined the Rayleigh criterion as shown in Figure 13.7. If the phase difference due to propagation paths 1 and 2 is less than $\pi/2$, the surface is said to be smooth. That is, $h < \lambda/8$.

According to $\lambda/8$ criterion, a smooth surface should have the mean height less than 3 cm for L-band, and 3 mm for X-band. There are not so many smooth surfaces for X-band on the actual scene except for a calm water surface. Even for a lake surface, small waves (ripples) appear caused by wind. In such a case, the lake surface is not necessarily smooth for X-band. Usually, rice paddy fields have rough surfaces for X-band and smooth surfaces for L-band. These characteristics can be well-recognized in Figure 12.15.

The correlation length l is defined as $$\rho(l) = \frac{1}{e}$$

such that the correlation function in the plane $\rho(\tau) = \dfrac{\langle z(x+\tau,y)\, z(x,y)\rangle}{\langle z^2(x,y)\rangle}$

becomes $1/e$.

If the correlation length is small, the surface is rough. On the other hand, if the correlation length is long, the surface is smooth. However, the length estimation is very difficult in actual situations because the correlation is not a direct measurable quantity.

13.2   BRAGG SCATTERING

Bragg scattering is a special resonance effect that occurs at periodical structures, such as sea surface waves or planted rows of crops. If the reflected waves from two points are in phase in the retro-directive direction, the backscattering wave becomes enhanced as shown in Figure 13.8. It is known that the Bragg scattering occurs at the VV component on the sea surface.

13.2.1   VALIDATION BY ALOS AND ALOS2 DATA

Bragg scattering in rice paddy fields is frequently observed by ALOS and ALOS2 observations. Among various data sets, special attention was paid to the Bragg scattering effect in paddy fields observed in Japan. It was found that the scattering is highly dependent on the direction of the planting row with respect to the illumination direction [21,22].

Figure 13.9 shows some model-based scattering power decomposition results acquired by ALOS/ALOS2 at various off-nadir angles. The locations are Aizu and Minami-Soma in Fukushima

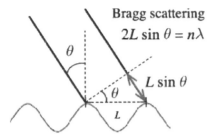

FIGURE 13.8 Bragg scattering.

Prefecture, Echigo Plains in Niigata Prefecture, and Joso City in Ibaraki Prefecture, Japan. Table 13.1 lists the scene information.

Observation results by fully polarimetric radar yielded a very strong surface scattering return from rice paddy fields. After the decomposition of the scattering power of these data, a bright blue (surface scattering P_s) area appeared in some parts of the paddy fields, but not necessarily in the whole field. The bright area depends on the location as shown in Figure 13.10. Even in the same paddy field area, bright blue areas are not uniformly distributed.

Bright blue color areas indicated by a red box can be seen in all paddy fields in Figure 13.9. These scattering phenomena are common to paddy fields, and this enhanced color pattern can be seen in almost everywhere in Japan. In addition, it seems that the squint angle larger than 10° does not show the enhancement effect (Figure 13.11). This bright pattern seems to be caused by Bragg scattering. Thus, the purpose is to confirm Bragg scattering caused by planting rows in paddy fields.

Bragg scattering is a well-known physical effect as shown in Figure 13.8. If the spacing L and incidence angle θ satisfy the equation, strong backscattering occurs by resonance. According to the specification of the ALOS systems, the off-nadir angle θ ranges from 21.5 to 34.9° for the polarimetric acquisition mode, and the interval spacing L in Figure 13.8 becomes 21–32 cm. The typical planting row in the paddy field spans approximately 24–30 cm and falls in the range of resonance intervals.

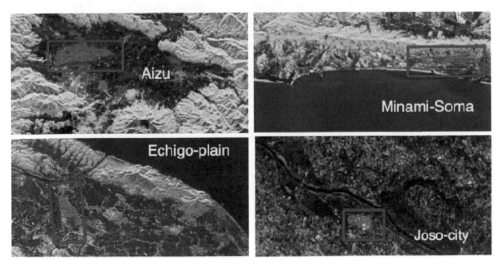

FIGURE 13.9 The four-component scattering power decomposition image. Red rectangles show rice paddy fields with enhanced backscattering.

TABLE 13.1

Scenes of L-Band Polarimetric Data

Area	Date	Off-Nadir Angle	Platform	Averaging Window
Aizu	April 7, 2009	21.5°	ALOS	3 × 18
Minami-Soma	October 12, 2006	21.5°	ALOS	3 × 18
Echigo Plains	September 27, 2015	32.7°	ALOS-2	5 × 10
Joso City	September 16, 2015	25°	ALOS-2	5 × 10

13.2.2 RCS MAGNITUDE

At first, we compared the surface-scattering power magnitude in the bright and non-bright spots in paddy fields. As an example of Aizu, numbered rectangular patches are depicted in Figure 13.10 with supporting Google Earth images.

If we look closely into the field, we notice planting rows in the rice paddy fields are aligned as shown in Figure 13.10 ① ⑤. The periodic structure of rows and its direction cause various scattering patterns. The anticipated scattering situation is depicted in Figure 13.11. If the squint angle is small and less than 10°, Bragg scattering may happen. If the squint angle φ becomes larger than 10°, the backscattering does not show a significant enhancement effect. Thus, the squint angle affects the occurrence of Bragg scattering. The scattering situation in Figure 13.11 seems common to other rice paddy fields. Some spots exhibit bright blue, and other fields look dark in the same paddy field area.

Table 13.2 compares the scattering magnitude. On average of these areas, the bright spot had −1.2 dB, whereas the dark spot exhibited −10.4 dB with the difference of approximately 9 dB.

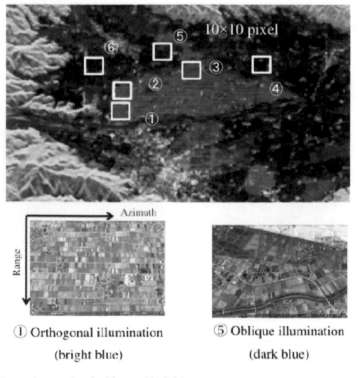

① Orthogonal illumination

(bright blue)

⑤ Oblique illumination

(dark blue)

FIGURE 13.10 Scattering patches in Aizu paddy fields.

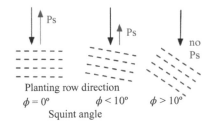

Planting row direction
$\phi = 0°$ $\phi < 10°$ $\phi > 10°$
Squint angle

FIGURE 13.11   Squint-angle characteristics of backscattering.

TABLE 13.2
Scattering Magnitude of Rice Paddy Field with and without Bragg Scattering

	Aizu		Minami-Soma		Echigo Plain		Joso City	
Bragg Scattering	**Yes (dB)**	**No (dB)**	**Yes (dB)**	**No (dB)**	**Yes (dB)**	**No (dB)**	**Yes (dB)**	**No (dB)**
Sample 1	0.4	−7	−1.5	−8.8	−3.7	−14.4	0.4	−13.4
Sample 2	−2	−8.2	−0.5	−7.4	−2.4	−13.8	−0.5	−11.2
Sample 3	−2.8	−8.7	−0.3	−8.6	0.9	−11.5	−0.5	−11.2
Average	−1.8	−8	−0.8	−8.3	−1.7	−13.2	−0.3	−11.9

13.2.3 POLARIZATION SIGNATURE OF BRAGG SCATTERING IN PADDY FIELD

Generally speaking, the scattering power of *VV* is greater than the *HH* power in Bragg scattering. To confirm this phenomenon, polarization signatures in these bright spots are depicted in Figure 13.12. The position of *VV* is in the center portion of Figure 13.12. Quite good agreement with a theoretical one [7] can be seen in the observation data (b–e). A similar polarization signature can be derived in

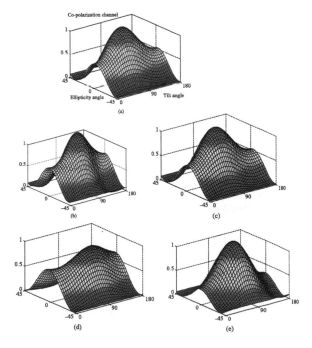

FIGURE 13.12   Polarization signatures of Bragg scattering in paddy fields: (a) Theoretical signature, (b) Aizu, (c) Minami-Soma, (d) Echigo Plains, and (e) Joso City.

FIGURE 13.13 Close-up images of Bragg scattering in paddy field (blue area), (left: Aizu, right: Echigo Plains, Niigata).

almost all bright surface-scattering regions. Therefore, it can be concluded that the bright blue area is caused by Bragg scattering (Figure 13.13).

13.3 MODELING OF THE SURFACE SCATTERING

When modeling the surface scattering, the surface is assumed to be smooth. From experimental results, it is well-known that: (1) *HH* and *VV* are in phase and (2) *HV* is small enough (−30 dB) to be neglected. The scattering matrix can be assumed to be (Figure 13.14)

$$[S] = \begin{bmatrix} R_h & 0 \\ 0 & R_v \end{bmatrix}$$

where

$$R_h = \frac{\cos\theta - \sqrt{\varepsilon_r - \sin^2\theta}}{\cos\theta + \sqrt{\varepsilon_r - \sin^2\theta}}, \quad R_v = \frac{(\varepsilon_r - 1)\{\sin^2\theta - \varepsilon_r(1+\sin^2\theta)\}}{(\varepsilon_r\cos\theta + \sqrt{\varepsilon_r - \sin^2\theta})^2}$$

ε_r is the relative dielectric constant and θ is the incidence angle.

Coherency matrix can be derived as

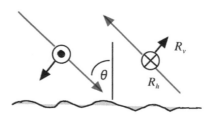

FIGURE 13.14 Surface-scattering model.

$$\mathbf{k}_p = \frac{1}{\sqrt{2}} \begin{bmatrix} R_h + R_v \\ R_h - R_v \\ 0 \end{bmatrix} = \frac{R_h + R_v}{\sqrt{2}} \begin{bmatrix} 1 \\ \beta \\ 0 \end{bmatrix}, \quad \beta = \frac{R_h - R_v}{R_h + R_v}, \quad |\beta| < 1$$

$$[T] = \langle \mathbf{k}_p \mathbf{k}_p^\dagger \rangle \Rightarrow [T]_{surface} = \begin{bmatrix} 1 & \beta^* & 0 \\ \beta & |\beta|^2 & 0 \\ 0 & 0 & 0 \end{bmatrix}$$

This model has the maximum value in T_{11}. The double-bounce component T_{22} is assumed to be small compared to T_{11}. The approximate value β can be calculated from the preceding equation. The cross-polarized HV term T_{33} is neglected.

The surface-scattering power P_s associated with this model appears in a flat surface area, such as in bare soil, paddy field, crop field, and sea surface at a small incidence angle in the L-band. It also appears in a disaster area, flooded area, tsunami-covered area, and landslide area. For high frequency above the X-band, the wavelength becomes less than 3 cm. In these higher frequencies, the surface scattering occurs for almost all targets larger than several centimeters.

13.4 APPLICATIONS OF SURFACE-SCATTERING POWER P_S

13.4.1 TSUNAMI

A great earthquake with magnitude 9.0 hit East Japan on March 11, 2011. This disaster was accompanied by a huge tsunami, which attacked the eastern seashore of the Tohoku area in Japan. ALOS-PALSAR had acquired fully polarimetric data over the Ishinomaki area before and after the earthquake on November 21, 2010 and April 8, 2011, respectively. The area was heavily destroyed not only by the earthquake but more so by the incoming and the retreating tsunami waves. The major part of Ishinomaki City and neighboring Onagawa-cho were completely destroyed and washed out by the tsunami. Figure 13.15 shows the corresponding ALOS-PALSAR polarimetric images of Ishinomaki City before and after the earthquake together with ground truth data [1]. Although the second data taken (April 8) was 28 days after the earthquake (March 11), it is possible to confirm several changes: red color (man-made) area turned into blue color (surface scattering caused by completely washed out area due to the impact of the tsunami) near by the seashore in Figure 13.15a and b. The ground truth was carried out by the Association of Japanese Geographers and the Geospatial Information Authority of Japan, respectively. Figure 13.15c shows the extent of the disaster area with blue indicating destruction by the tsunami and with orange indicating flooding by the tsunami. The legend color *"orange"* in Figure 13.15c denotes the flooded area where the tsunami hit. But there were still remaining some buildings/houses and man-made structures after the tsunami. The "blue" color in Figure 13.15c denotes the area where almost all buildings/houses and man-made structures had collapsed or were destroyed and washed away by the tsunami, leaving a bare surface on the ground. We can observe corresponding features very well in Figure 13.15b and c.

The Onagawa-cho image in Figure 13.16a lost the red color of the double-bounce scattering power in Figure 13.16b, indicating that almost all of the man-made structures were destroyed by the earthquake and the tsunami. Purple (blue + orange) color areas in the ground truth image of Figure 13.16c shows collapsed and destroyed areas by tsunami. The polarimetric decomposed image of Figure 13.16b seems to display exact correspondence to the ground truth image (c).

The final example of the disaster image is around the mouth of Kitakami River near Onagawa-cho as shown in Figure 13.17. Since this area is sparsely populated, there are not many houses or man-made structures, indicating fewer red spots in Figure 13.17a. Yet, the red spots

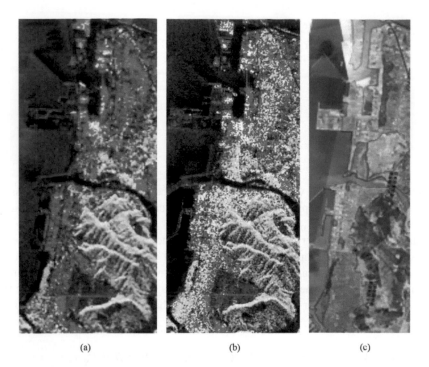

(a)   (b)   (c)

FIGURE 13.15 Ishinomaki area suffered from tsunami caused by great earthquake of East Japan 2011. (From Yamaguchi, Y., *Proc. IEEE*, 100, 2851–2860, 2012.): (a) before, (b) after the earthquake and tsunami, and (c) ground truth data. Blue color (surface scattering) areas show totally destroyed areas by tsunami. Orange color denotes the area flooded by tsunami. Ground truth provided by Association of Japanese Geographers and Geospatial Information Authority of Japan, respectively.

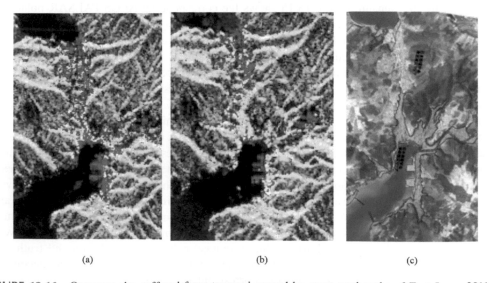

(a)   (b)   (c)

FIGURE 13.16 Onagawa-cho suffered from tsunami caused by great earthquake of East Japan 2011: (a) before and (b) after the earthquake and tsunami and (c) Ground truth data provided by Association of Japanese Geographics and Geospatial Information Authority of Japan, respectively. (From Yamaguchi, Y., *Proc. IEEE*, 100, 2851–2860, 2012.)

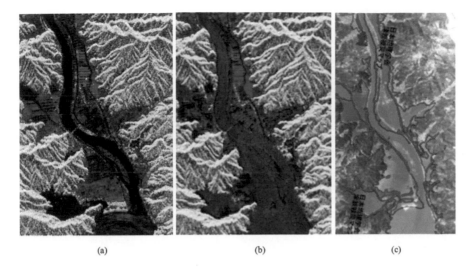

(a)   (b)   (c)

FIGURE 13.17 The mouth of Kitakami River suffered from tsunami caused by great earthquake of East Japan 2011: (a) before and (b) after the earthquake and tsunami and (c) Ground truth data provided by Association of Japanese Geographics and Geospatial Information Authority of Japan, respectively. (From Yamaguchi, Y., *Proc. IEEE*, 100, 2851–2860, 2012.)

(including Okawa Elementary School) are still reduced in Figure 13.17b by the tsunami disaster. In addition, it is possible to see at the center bottom of Figure 13.17 that a small green area has been completely wiped out by the tsunami. The river as well as its surrounding area shows the surface scattering to be predominant after the tsunami tide.

13.4.2 LANDSLIDE

When a landslide occurs, the bare soil causes the surface scattering. The scattering power P_s is rather easily recognized by polarimetric SAR (PolSAR) because it is scattered in wide-angle directions. Figure 13.18 shows one example of a huge landslide area, Aratozawa, observed by ALOS on June 14, 2008. The blue area corresponds to bare soil surface, which can be recognized easily by the comparison with optical image.

If a landslide occurs, the covering mud/soil on the upper cliff flows down, making the surface slope gentle as indicated by the black straight line in Figure 13.19 If radar observation is carried out from the same position (just like ALOS), the incidence angles in the upper/middle position decreases after the landslide. Since the decrease of the incidence angle brings an increase of the surface-scattering magnitude as shown in Figure 13.19, we see a brighter color on the upper side of the cliff in Figure 13.18. On the other hand, the lower portion of the landslide has opposite behavior and becomes dark. These characteristics apply not only to PolSAR observation, but also to general radar observations.

The next image shows landslide areas caused by typhoon no.12 in 2012 and observed by airborne PiSAR-L2 (data no. L204006) in 2014. The color-coding of the scattering power decomposition is red for surface scattering, blue for double-bounce scattering, and green for volume scattering. The reason for changing red and blue is just for contrast enhancement. By looking at red-colored areas, we can recognize they are corresponding well to the Google Earth optical image (not shown here). Hence, polarimetric data can identify the landslide area very effectively and serve as a disaster monitoring tool.

Airborne PiSAR-X2 was observed at the Kumamoto earthquake disaster area on April 17, 2016. In a similar way, the high-resolution data was processed and color-coded with red = surface

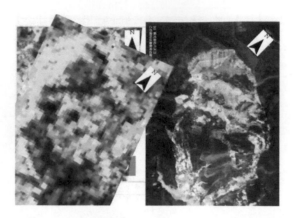

FIGURE 13.18   Aratozawa landslide (PolSAR and Optical image).

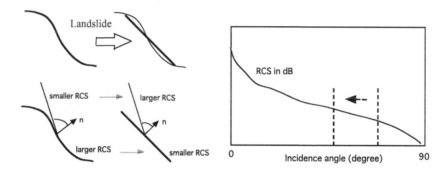

FIGURE 13.19   Landslide, incidence angle, and backscattering magnitude.

scattering and blue = double bounce, as shown in Figures 13.20 and 13.21 [22]. The image looks like an actual photo, where we can identify the landslide area very easily. From this figure, high-resolution PolSAR is very effective for detecting landslides.

13.4.3   VOLCANO ACTIVITY MONITORING

As depicted on the world map in Figures 13.22 through 13.24, the surrounding area of the Pacific Ocean is called the "Ring of Fire," where volcano activity is very strong. Volcano summits are covered with ashes, rocks, lava, etc. Sometimes toxic gases erupted. There is no vegetation. Since the place near the volcano summit is dangerous for human activity, radar plays a very important role in monitoring. The main scattering mechanism is the surface scattering.

The preceding figure shows the Miyake-jima volcano image before and after SO_2 gas eruption during the observation period November 6, 2008 to November 24, 2008. The island is located at 34.08N, 139.53E, Tokyo, Japan. Compared to the left image (20081009), we can see more blue caused by the surface scattering in the right image (20081124). This implies that surface scattering increased due to deposition of volcanic ash and that SO_2 gas at least partially defoliated the vegetation, making it more transparent at L-band frequencies, which also created barer surface scattering. Therefore, the change in the blue color area indicates the effect caused by this SO_2 gas eruption. Man-made structures such as houses are easily recognized by the red color and remain the same in these two images.

FIGURE 13.20 Landslides at Totsukawa village, observed by PiSAR-L2 in 2014 (red-colored areas correspond to landslide).

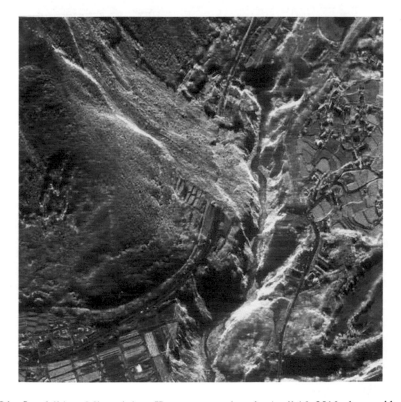

FIGURE 13.21 Landslide at Minami-Aso, Kumamoto earthquake April 16, 2016, observed by PiSAR-X2.

FIGURE 13.22 World volcano map.

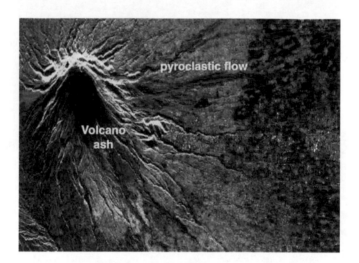

FIGURE 13.23 Volcano Merapi, Indonesia by ALPSRP262863760-20101231.

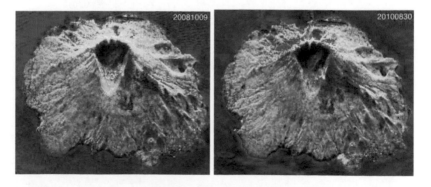

FIGURE 13.24 Miyake-jima, Japan by PiSAR-X (left: October 9, 2008, right: November 24, 2008). (From Yamaguchi, Y., *Proc. IEEE*, 100, 2851–2860, 2012.)

13.4.4 OIL SLICK DETECTION

An oil spill from a tanker or throwing away wasted oil from an engine, for example, are reported everywhere in the world. In addition, the collision of an oil tanker releases a huge amount of oil in the sea and causes a severe environmental disaster and ecological damages. On the other hand, a natural oil spill sometimes occurs from under the sea, such as in the Gulf of Mexico or west Africa. These cases may serve to discover new oil wells. Hence, oil slick detection is attracting attention both for environmental and economic purposes for exploring new oil wells. Oil slick detection by SAR has been carried out for many years. It is known that:

1. The *VV* return is stronger than the *HH* return from a rough sea surface. Bragg scattering occurs at the *VV* radar channel.
2. An oil-covered sea surface becomes smoother compared to the surrounding surface. The reflection magnitude becomes small.
3. The magnitude depends on the sea state.
4. The thickness of the oil spill is usually in the order of millimeters.
5. The effective parameter to detecting an oil spill is entropy or the correlation coefficient for the frequency above the C-band.

Clutter exists whose scattering is similar to an oil slick. The clutter has a localized patch with a small backscattering value. When the sea surface is rather calm, the cluttered area looks dark compared to its surroundings. This clutter may be floating skin, sea ice, etc., which reduces sea wave motion. Therefore, it is difficult to detect an oil slick accurately. In that sense, PolSAR will serve to pick up "an oil spill look-a-like" effective polarimetric parameters are proposed.

- Correlation coefficients $|\gamma_{HH-VV}| = \dfrac{|C_{13}|}{\sqrt{C_{11}C_{33}}} = \dfrac{|\langle S_{HH}S_{VV}^* \rangle|}{\sqrt{\langle |S_{HH}|^2 \rangle \langle |S_{VV}|^2 \rangle}}$ and $\gamma_{HH+VV,HH-VV} = \dfrac{T_{12}}{\sqrt{T_{11}T_{22}}}$

- $\text{Re}\{T_{12}\} = \dfrac{1}{2}\left(|S_{HH}|^2 - |S_{VV}|^2\right) < 0 \,(\text{Bragg scattering})$

- $\rho_B = \tan^{-1}\left(\dfrac{T_{22}}{T_{11}}\right) = \tan^{-1}\left(\dfrac{|S_{HH}-S_{VV}|^2}{|S_{HH}+S_{VV}|^2}\right)$

- Entropy H
- Roughness indicator
- $\gamma_{XX-YY}(0) = \dfrac{T_{11}-T_{33}}{T_{11}+T_{33}} = \dfrac{\langle |S_{HH}+S_{VV}|^2 - 4|S_{HV}|^2 \rangle}{\langle |S_{HH}+S_{VV}|^2 + 4|S_{HV}|^2 \rangle}, 0 = \text{very rough}, 1 = \text{very smooth}$

- $M = \dfrac{T_{22}+T_{33}}{T_{11}} = \dfrac{\langle |S_{HH}-S_{VV}|^2 + 4|S_{HV}|^2 \rangle}{\langle |S_{HH}+S_{VV}|^2 \rangle},$ is sensitive to small roughness.

13.5 REMARKS

In this chapter, some decomposition images are displayed to show how the surface-scattering power is useful for disaster monitoring, such as volcano eruption, tsunami disaster, accumulated snow in mountains, and landslide detection. It may be applicable to rice crop monitoring. Deforestation or detection of clear-cut open area is another important application. Typical scatterers are bare soil surfaces, mud and landslide areas, crop fields, rice paddy fields, volcano ash-covered areas, small grasslands, snow-covered areas, snow and ice fields, etc. According to target characteristics, the surface-scattering power and auxiliary parameters will serve the purpose.

REFERENCES

1. Y. Yamaguchi, "Disaster monitoring by fully polarimetric SAR data acquired with ALOS-PALSAR," *Proc. IEEE*, vol. 100, no. 10, pp. 2851–2860, 2012.
2. A. K. Fung and K.-S. *Chen, Microwave Emission and Scattering Models for Users*, 480 p., Artech House, 2010.
3. K.-S. Chen, *Principle of Synthetic Aperture Radar: A System Simulation Approach*, 200 p., CRC Press, FL, 2015.
4. K.-S. Chen, *Microwave Scattering and Imaging of Rough Surfaces*, CRC Press, 2019.
5. D. Masonnett and J.-C. Souyris, *Imaging with Synthetic Aperture Radar*, EPFL/CRC-Press, 2008, ISBN 978-0-8493-8239-4.
6. K. S. Chen, T. D. Wu, L. Tsang, Q. Li, J. C. Shi, and A. K. Fung, "The emission of rough surfaces calculated by the Integral Equation Method with a comparison to a three-dimensional moment method simulations," *IEEE Trans. Geosci. Remote Sens.*, vol. 41, no. 1, pp. 1–12, 2003.
7. T. D. Wu, K. S. Chen, J. C. Shi, H. W. Lee, and A. K. Fung, "A study of an AIEM model for bistatic scattering from randomly rough surfaces," *IEEE Trans. Geosci. Remote Sens.*, vol. 46, no. 9, pp. 2584–2598, 2008.
8. S. Huang, L. Tsang, E. Njoku, and K.-S. Chen, "Backscattering coefficients, coherent reflectivities, and emissivities of randomly rough soil surfaces at L-Band for SMAP applications based on numerical solutions of Maxwell equations in three-dimensional simulations," *IEEE Trans. Geosci. Remote Sens.*, vol. 48, no. 6, pp. 2557–2568, 2010.
9. T. D. Wu, K. S. Chen, A. K. Fung, and M. K. Tsay, "A transition model for the reflection coefficient in surface scattering," *IEEE Trans. Geosci. Remote Sens.*, vol. 39, no. 9, pp. 2040–2050, 2001.
10. J. J. VanZyl and Y. J. Kim, *Synthetic Aperture Radar Polarimetry*, Wiley, New Jersey, 2010.
11. I. Hajnsek, E. Pottier, and S. R. Cloude, "Inversion of surface parameters from polarimetric SAR," *IEEE Trans. Geosci. Remote Sens.*, vol. 41, pp. 727–745, 2003.
12. I. Hajnsek, T. Jagdhuber, H. Schon, and K. P. Papathanassiou, "Potential of estimating soil moisture under vegetation cover by means of PolSAR," *IEEE Trans. Geosci. Remote Sens.*, vol. 47, no. 2, pp. 442–454, 2009.
13. T. Jagdhuber, I. Hajnsek, A. Bronstert, and K. P. Papathanassiou, "Soil moisture estimation under low vegetation cover using a multi-angular polarimetric decomposition," *IEEE Trans. Geosci. Remote Sens.*, vol. 51, no. 4, pp. 2201–2214, 2012.
14. Y. Oh, K. Sarabandi, and F. T. Ulaby, "An empirical model and an inversion technique for radar scattering from bare soil surfaces," *IEEE Trans. Geosci. Remote Sens.*, vol. 30, pp. 370–381, 1992.
15. Y. Oh, "Quantitative retrieval of soil moisture content and surface roughness from multi-polarized radar observations of bare soil surfaces," *IEEE Trans. Geosci. Remote Sens.*, vol. 42, no. 3, pp. 596–601, 2004.
16. S. Huang, L. Tsang, E. G. Njoku, and K. S. Chen, "Backscattering coefficients, coherent reflectivities, emissivities of randomly rough soil surfaces at L-band for SMAP applications based on numerical solutions of Maxwell equations in three dimensional simulations," *IEEE Trans. Geosci. Remote Sens.*, vol. 48, pp. 2557–2567, 2010.
17. S. Huang and L. Tsang, "Electromagnetic scattering of randomly rough soil surfaces based on numerical solutions of Maxwell equations in 3 dimensional simulations using hybrid UV/PBTG/SMCG method," *IEEE Trans. Geosci. Remote Sens.*, vol. 50, pp. 4025–4035, 2012.
18. L. Tsang, K. H. Ding, S. H. Huang, and X. Xu, "Electromagnetic computation in scattering of electromagnetic waves by random rough surface and dense media in microwave remote sensing of land surfaces," *Proc. IEEE*, vol. 101, pp. 255–279, 2013.

19. J.-P. Wigneron, L. Laguerre, and Y. H. Kerr, "A simple parameterization of the L band microwave emission from rough agricultural soil," *IEEE Trans. Geosci. Remote Sens.*, vol. 39, no. 8, pp. 1697–1707, 2001.
20. A. K. Fung and K. S. Chen, *Microwave Scattering and Emission Models for Users*, Artech House, Massachusetts, 2010.
21. K. Ouchi, H. Wang, I. Ishitsuka, G. Saito, and K. Mohri, "On the Bragg scattering observed in L-band synthetic aperture radar images of flooded rice field," *IEICE Trans. Commun.*, vol. E89-B, no. 8, pp. 2218–2225, 2006.
22. Y. Yamaguchi, G. Singh, and H. Yamada, "On the model-based scattering power of fully polarimetric SAR data," *Trans. IEICE*, vol. J101-B, no. 9, pp. 638–647, 2018. doi:10.14923/transcomj.2018API0001.

14 Double-Bounce Scattering

Generally, a double-bounce scatterer has a right-angle structure as shown in Figure 14.1. The structure is composed of two flat orthogonal surfaces and is called a "dihedral corner reflector" or simply "dihedral" or sometimes "diplane." We can see various right-angle structures in the real world such as roads and building walls, water surfaces and bridge side structures, sea surfaces and ship side bodies, or even planted tall vegetation in crop fields and rice paddy fields. In this chapter, we investigate the double-bounce scattering phenomena by dihedrals in detail, and we see the double-bounce scattering power in actual PolSAR images.

14.1 DIHEDRAL

The specific features are:

- Retro-directive characteristics.
- Large RCS at 0° squint angle but fades away for large squint angles.
- Phase difference between HH and VV is 180° for metallic material.

This characteristic is suitable for polarimetric calibration targets.

- Phase difference between HH and VV is 180° for metal, and 130 to 140° for dielectric materials.

These characteristics can be understood through a simple theoretical model. Figure 14.1 shows scattering matrix derived from the Fresnel reflection coefficient. When a dihedral structure is placed orthogonal to radar illumination, we have strong HH and VV components. The cross-polarized HV component can be neglected since the magnitude becomes very small. Approximate scattering matrix can be written as

$$[S] = \begin{bmatrix} R_{h1}R_{h2} & 0 \\ 0 & -R_{v1}R_{v2} \end{bmatrix} \qquad (14.1.1)$$

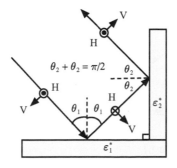

FIGURE 14.1 Double bounce from right-angle structure.

$$R_h = \frac{\cos\theta - \sqrt{\varepsilon_r^* - \sin^2\theta}}{\cos\theta + \sqrt{\varepsilon_r^* - \sin^2\theta}}, \quad R_v = \frac{\varepsilon_r^* \cos\theta - \sqrt{\varepsilon_r^* - \sin^2\theta}}{\varepsilon_r^* \cos\theta + \sqrt{\varepsilon_r^* - \sin^2\theta}} \qquad : \text{Fresnel reflection coefficient}$$

where $\varepsilon_r^* = \varepsilon_r - j\,60\sigma\lambda$, ε_r is the relative dielectric constant, σ is the conductivity, λ is the wavelength, and θ is the incidence angle.

For simplicity, we assume the wall and the ground have the same material $\varepsilon_1^* = \varepsilon_2^*$,

$$\text{then } [S] \Rightarrow \begin{bmatrix} S_{HH} & 0 \\ 0 & S_{VV} \end{bmatrix} = \begin{bmatrix} R_{h1}R_{h2} & 0 \\ 0 & -R_{v1}R_{v2} \end{bmatrix} \Rightarrow \begin{bmatrix} 1 & 0 \\ 0 & -1 \end{bmatrix} \text{ when the medium is metallic.}$$

For dielectric materials, the scattering matrix is in the next form.

$$[S] \Rightarrow \begin{bmatrix} S_{HH} & 0 \\ 0 & S_{VV} \end{bmatrix} \Rightarrow \begin{bmatrix} 1 & 0 \\ 0 & \rho \end{bmatrix} \text{ for dielectric materials.}$$

We define polarization ratio as $\qquad \rho = \dfrac{S_{VV}}{S_{HH}} = |\rho|\angle\phi \qquad\qquad$ (14.1.2)

and check the magnitude $|\rho|$ as a function of incidence angle. For example, right-angle structures composed of polyethylene and concrete alone have the value as shown in Figure 14.2. The relative dielectric constant is taken as

$$\varepsilon^* = 2.26 - j\,0.003 \text{ for polyethylene} \qquad \text{and} \qquad \varepsilon^* = 5.8 - j\,0.01 \text{ for concrete.}$$

The value $|\rho|$ is less than 0.3 throughout the incidence angle range. Compared to the metallic dihedral, $|\rho|$ is much smaller than 1 for dielectrics. These characteristics come from the property of the Fresnel reflection that Rh (*HH* reflection) is always larger than Rv (*VV* reflection). Right-angle structures cause double reflections. The result derives that the *HH* component is much larger than the *VV* component in urban scattering, which can be seen as red areas in Figure 14.3.

In addition to the amplitude information, the incidence angle range where the phase retains $\angle\phi = 180°$ becomes narrow as relative the dielectric constant becomes small. This means that *HH* and *VV* arc out of phase in a limited incidence angle range where the double-bounce scattering occurs. The right-angle structure does not necessarily exhibit the double-bounce scattering feature for very large or very small incidence angles.

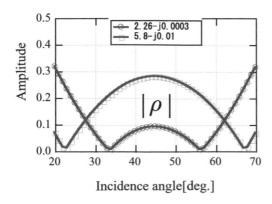
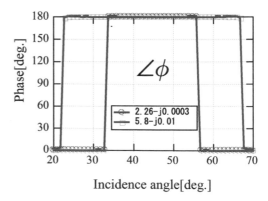

FIGURE 14.2   Polarization ratio as a function of incidence angle.

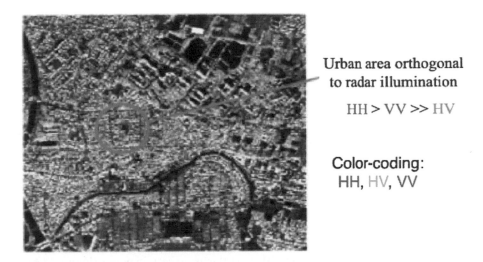

Urban area orthogonal
to radar illumination

HH > VV >> HV

Color-coding:
HH, HV, VV

FIGURE 14.3   Scattering in urban area.

14.1.1   DEPENDENCY ON INCIDENCE ANGLE

In order to check the scattering mechanism of the dihedral corner reflector on the incidence angle dependency, we measured both metallic and dielectric dihedrals in an anechoic chamber using X-band. Figure 14.4 shows the polarization signatures as a function of incidence angle at 15°, 30°, and 45°.

Since polarimetric calibration is conducted on 45° incidence angle, the signature of the metallic dihedral is the ideal one at 45°. The polarization signatures at 30° and 15° of metallic dihedral still keep the shape of the dihedral response. However, for the dielectric dihedral, the response of 45° looks like that of a horizontal dipole, which means the *HH* component is strongest, and no double-bounce feature is seen in the signature. When the incidence angle is 30°, the shape looks like those of dihedrals. The angular range of dihedral characteristics seems narrow as anticipated

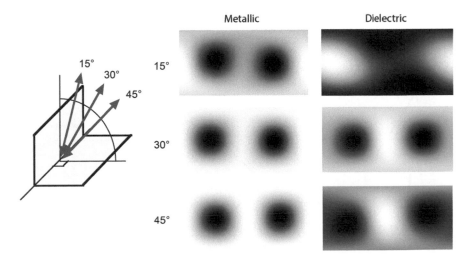

FIGURE 14.4   Incidence angle dependency of polarization signature of dihedrals.

in Figure 14.3. For the 15° incidence angle, the shape became similar to that of vertical dipoles. Therefore, for a metallic dihedral, a similar polarization signature can be seen at these incidence angles, but we can see strong angular dependency of the polarization signature of dielectric dihedrals.

14.1.2 Dependency on Material

In order to check the polarization signature by material constant, we have calculated the dihedral response by increasing conductivity in Equation (14.1.1). The approximate scattering matrix becomes as shown in Figure 14.5. As the conductivity of the medium increases, the shape of polarization signature changes from that of the H-dipole to a metallic dihedral.

14.1.3 Role as a Polarimetric Calibration Target

The dihedral corner reflector is used as a calibration target. It is a mandatory target for an accurate polarimetric calibration purpose. However, there still exist uncertain points (Figure 14.6).

How big is enough for the ideal polarimetric target with respect to wavelength? How much size is suitable for accurate scattering matrix? Squint angle dependency? The calibration target is set orthogonal to the radar beam. If the orientation of the calibrator is not normal or orthogonal, what kind of effect takes place? How is the scattering matrix?

These are frequently asked questions. To answer these questions, finite difference time domain (FDTD) analysis was carried out to examine scattering characteristics of dihedrals with respect to

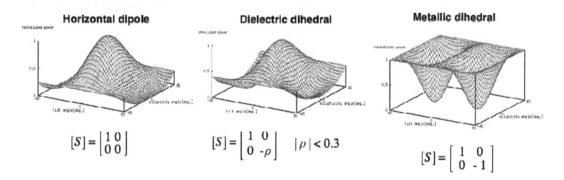

FIGURE 14.5 Conductivity and polarization signature (calculated).

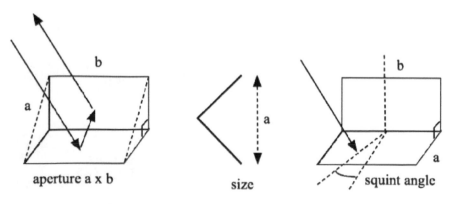

FIGURE 14.6 Dihedral as calibration target.

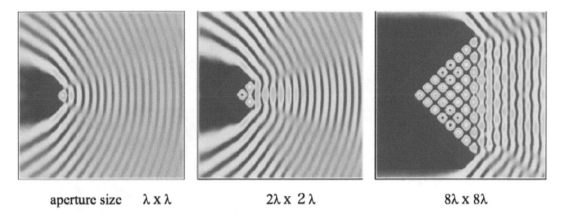

aperture size $\lambda \times \lambda$ $2\lambda \times 2\lambda$ $8\lambda \times 8\lambda$

FIGURE 14.7   Electric field distribution across the aperture (size: $\lambda \times \lambda$, $2 \times 2\lambda$, $8 \times 8\lambda$).

wavelength [1]. At first, the calculation was carried out on the reflected electric field distribution as a function of aperture size $\lambda \times \lambda$, $2\lambda \times 2\lambda$, and $8\lambda \times 8\lambda$. As the aperture size increases, the field pattern approaches that of a plane wave. The retro-directive wave pattern can be seen in the $8\lambda \times 8\lambda$ aperture (Figure 14.7).

14.1.4 DEPENDENCY ON SIZE WITH RESPECT TO WAVELENGTH

The calculation of polarization ratio $\rho = \frac{S_{VV}}{S_{HH}} = |\rho| \angle \phi$ indicates the necessary condition for actual calibration target size. Since this value directly represents the scattering matrix element, it is compared with the ideal case of the Fresnel reflection. From Figure 14.8, it is understood that the aperture size more than 8λ is required for realizing the dihedral corner reflector. This is a good indicator for achieving a dihedral polarimetric calibrator.

14.1.5 DEPENDENCY ON SQUINT ANGLE

We measured the dependency directly in an anechoic chamber using the same dihedral model in the previous section and tried to retrieve some tendency. The measured polarization signatures are displayed in Figure 14.9 for squint angles of 45° and 60°, with incidence angles of 15°, 30°, and 45°. It is understood that the shape of the polarization signature is distorted and that there is no special tendency among these signatures. Therefore, it is difficult to derive some conclusion from the experimental results.

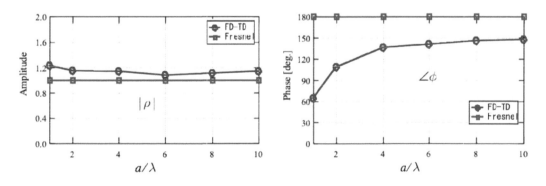

FIGURE 14.8   Polarization ratio as a function of aperture size in λ.

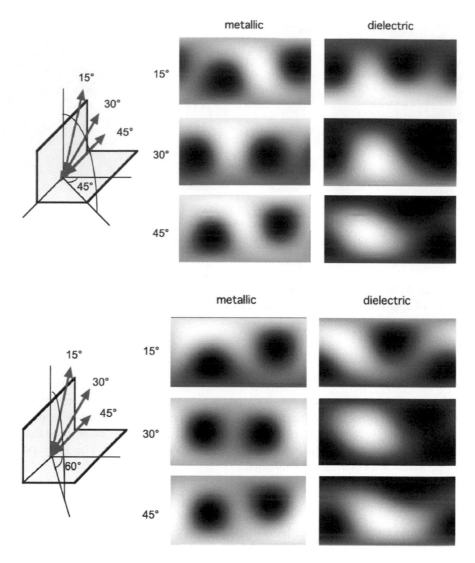

FIGURE 14.9 Polarization signature of dihedrals with squint-angle incidence (45°and 60°).

Squint-angle dependency is quite difficult to estimate theoretically. However, various experimental situations can be realized in FDTD analyses [2]. We have set the squint angle as variable and calculated various cases. From the calculation results, we came to a basic point on the radiation pattern from the aperture antenna. It is well-known in antenna theory that there is a Fourier transform relation between the aperture and radiation beam pattern. If the aperture size is small, the reradiated main beam becomes wide, which includes a double-bounce property. In this case, the main beam is wide and directed toward the specular direction. But it happens that the main beam includes the backscattering direction, and hence the double-bounce feature can be observed in the retro-directive direction. On the other hand, if the aperture size is wide enough compared with wavelength, the radiated main beam becomes very narrow. In this case, the main beam is sharp and directed toward the specular direction, and there is no radiation to the backscattering direction. In this case, there is no double-bounce scattering to off-main beam directions.

If we have an antenna with aperture size L, the radiation beam can be obtained by Fourier transform. If uniform distribution on the aperture is assumed, then the first null angle of the main beam is given by

$$\phi_c = 2 \sin^{-1}\left(\frac{\lambda}{2L \cos \theta}\right) \qquad (14.1.3)$$

where θ is the incidence angle. This situation is depicted for orthogonal incidence (zero squint) and oblique (squint) incidence in Figure 14.10. Both $L1$ and $L2$ reflect double-bounce scattering for the orthogonal incidence case, but large $L1$ does not reflect double bounce in the squint case because the main beam does not cover the backscattering direction.

The preceding considerations on dihedral scattering may serve actual PolSAR data analyses. The double-bounce scattering model is given as

$$[T]_{dihedral} = \begin{bmatrix} |\alpha|^2 & \alpha & 0 \\ \alpha^* & 1 & 0 \\ 0 & 0 & 0 \end{bmatrix}, \text{ with } |\alpha| < 1 \qquad (14.1.4)$$

where $\alpha = \dfrac{S_{HH} + S_{VV}}{S_{HH} - S_{VV}} = \dfrac{1 + \rho}{1 - \rho}, \quad \rho = \dfrac{S_{VV}}{S_{HH}}$

In typical values for radar observations, we have the following estimates, assuming the incidence angle is in a range 20°–50°, and concrete building blocks with relative dielectric constant 5–7.

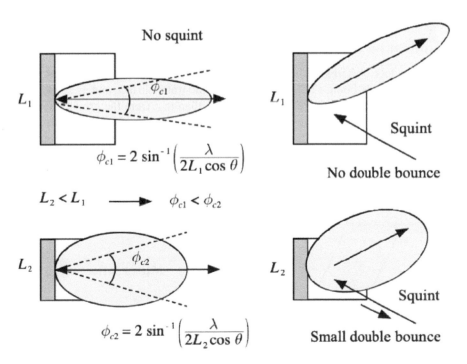

No squint

L_1 ϕ_{c1}

$$\phi_{c1} = 2 \sin^{-1}\left(\frac{\lambda}{2L_1 \cos \theta}\right)$$

$L_2 < L_1 \longrightarrow \phi_{c1} < \phi_{c2}$

L_2 ϕ_{c2}

$$\phi_{c2} = 2 \sin^{-1}\left(\frac{\lambda}{2L_2 \cos \theta}\right)$$

L_1 Squint

No double bounce

L_2 Squint

Small double bounce

FIGURE 14.10 The relation of double-bounce scattering observation and dihedral width for squint-angle incidence.

$$\rho = -0.3 \qquad \alpha = \frac{1+\rho}{1-\rho} = 0.53 \qquad \alpha^2 = 0.28$$

$$\rho = -0.25 \qquad \alpha = 0.6 \qquad \alpha^2 = 0.36$$

$$\rho = -0.2 \qquad \alpha = 0.666 \qquad \alpha^2 = 0.44$$

An example of concrete building becomes the following coherency matrix:

$$[S]_{dihedral}^{dielectric} \Rightarrow \begin{bmatrix} 0.255 - j0.041 & 0 \\ 0 & -0.059 + j\,0.021 \end{bmatrix},$$

$$\frac{[T]_{dihedral}}{T_{22}} = \begin{bmatrix} 0.377 & 0.612 + j0.055 & 0 \\ 0.612 - j0.055 & 1 & 0 \\ 0 & 0 & 0 \end{bmatrix}$$

T_{11} is smaller than T_{22}, and hence the matrix is a good model for concrete buildings.

For double-bounce scattering from vegetation, a tree trunk or stem to the ground surface or water surface constructs a right-angle structure. Especially, a strong double bounce occurs in rice paddy fields when the planting row is orthogonal to radar illumination, and rice stems are tall on water surface. We can formulate scattering model in the same way. The final form of the coherency matrix is the same as Equation (14.1.4).

14.2 EXAMPLES OF DOUBLE-BOUNCE SCATTERING

14.2.1 LOTUS IN THE LAKE

The wetland has lots of herbs such as lotus and reeds. They grow near the land as shown in Figure 14.11. The double-bounce reflection occurs on the water surface and its stems [3]. The magnitude of double-bounce scattering power P_d is larger than that from the ground. The annual herb grows up and fades down in a 1-year cycle. Therefore, at each stage of growing, these herbs exhibit distinct scattering behaviors. Figure 14.12 shows PiSAR-X/L observation of Sakata Lagoon, Niigata, Japan. The joint data collection simultaneously brought very interesting results as shown in the figure.

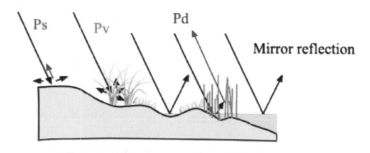

FIGURE 14.11   Wetland vegetation and scatterings.

L-band   X-band

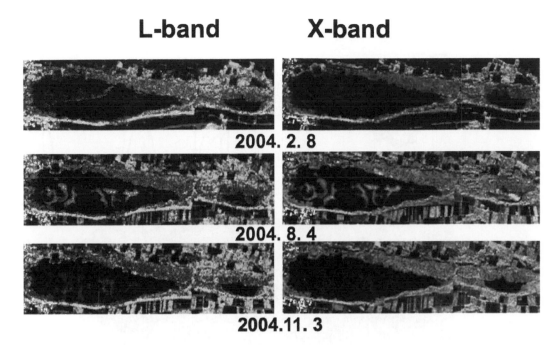

2004. 2. 8

2004. 8. 4

2004.11. 3

FIGURE 14.12 Pi-SAR X/L observation over Sakata Lagoon. (From Yajima, Y. et al., *IEEE Trans. Geosci. Remote Sens.*, 46, 1667–1673, 2008). Scattering-power decomposition image with red: double bounce caused by tall stems of vegetation and water surface, blue: surface scattering by bare soil surface, green: volume scattering from vegetation and trees.

This simultaneous measurement of X- and L-band observation revealed interesting scattering behavior as shown in Figure 14.13 [4]. For the L-band wave, it penetrates lotus leaves and reaches to the water surface of the lagoon. The lotus stem and water surface create double-bounce scattering for the L-band frequency. However, for the X-band wave, it cannot penetrate lotus leaves and is reflected from the leaf surface, leading to surface scattering. Therefore, the lotus acts as a surface-scattering object for X-band, and a double-bounce scatterer for L-band. A similar situation occurs for reed scattering near the lagoon. The situation is depicted in Figure 14.13. If we see the decomposition results, the scattering mechanisms in summer are clearly displayed by color-coding.

14.2.2 Rice Growth

Echigo Plain in Niigata is famous for rice production in Japan. Time series ALOS-PALSAR data over rice paddy fields are shown in Figure 14.14 to compare the growing stage on June 18 and August 3. The radar illumination direction is from left to right. Planting the row orthogonal to radar illumination exhibits a double-bounce scattering (red) feature. Red goes up with rice growth.

The next example is a rice paddy field in Bangladesh shown in Figure 14.15 where two data sets are compared. One is a bare soil surface in May, and the other is the ripening stage in November. The scattering powers P_s and P_d indicate the situation very clearly.

14.2.3 Detection of Building Collapse

The reduction of P_d is utilized to detect damage areas [5,6]. Color change from red to blue in the urban area or village indicates that houses are collapsed and the surface scattering becomes dominant. In Figure 14.16, both color changes of P_d and P_s show the damage by a mudslide caused by heavy rain in Mocoa, Colombia, on April 1, 2017. Blue areas on the right-hand side are mud.

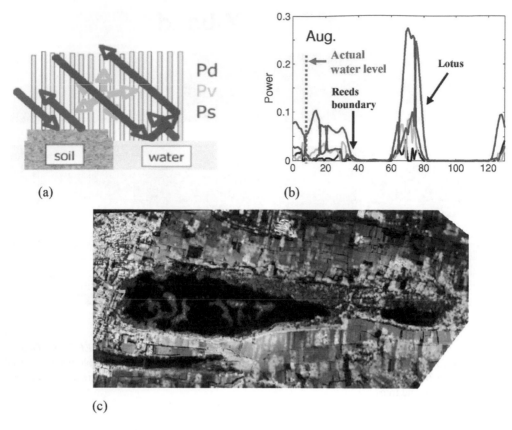

(a)   (b)

(c)

FIGURE 14.13 Scattering mechanisms and scattering powers in Sakata Lagoon. (From Yajima, Y. et al., *IEEE Trans. Geosci. Remote Sens.*, 46, 1667–1673, 2008): (a) scattering powers P_s, P_d, P_v; (b) range profile along a transect on lagoon; and (c) overlaid image of scattering mechanisms onto ground map.

FIGURE 14.14 Rice growth seen by ALOS-PALSAR (L-band).

Data no. ALOS2105220440-160504   ALOS2024490440-141105

FIGURE 14.15 Rice paddy field in Bangladesh.

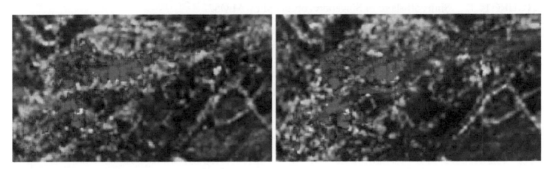

ALOS2103220010-160421   ALOS2154970010-170406

FIGURE 14.16 Mudslide April 1, 2017, Mocoa, Colombia.

As we have seen in Figure 13.15, the Ishinomaki area suffered from a tsunami caused by the great earthquake of east Japan in 2011; buildings were totally wiped out by the tsunami. The remaining was bare soil surface only. This tsunami damage was also recognized by the color change of fading down P_d and growing up P_s. The combination of P_s and P_d such as $(P_s - P_d)/(P_s + P_d)$ may serve to identify the disaster areas.

14.2.4 SHIP DETECTION

Ship detection is one of the main applications for ocean remote sensing. The scattering power decomposition image provides the direct result of the ship detection as shown in Figure 14.17 where a 6SD algorithm is applied. Various colors are seen in each ship depending on the orientation direction with respect to radar illumination.

14.2.5 OYSTER FARMING AND TIDAL LEVEL

An oyster farm in Taiwan is famous for a big tidal level change. Due to the tidal change, the vertical pole length in the air appearing from the sea water surface changes accordingly. This situation causes the double-bounce scattering power to change with the tidal level. RadarSAT2 data is shown in Figure 14.18 as a function of time series. It is anticipated that P_d becomes strong when the tidal

ALOS2106990020-160516

FIGURE 14.17 Ships off-shore of Singapore, observed by ALOS2.

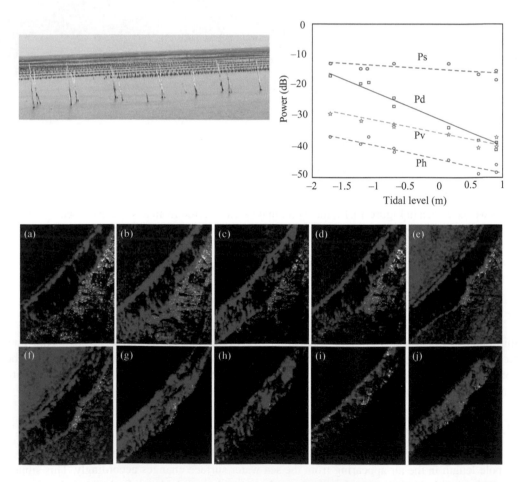

FIGURE 14.18 G4U image of oyster farming in Taiwan using RadarSAT2quad pol Data. (a)–(f) correspond to acquisition time with different tidal level. (From Cheng, T.Y., *IEICE Trans. Commun.*, E96-B, 2573–2579, 2013.)

FIGURE 14.19 Flooding in Joso City, Japan.

level is small since long vertical poles appear on the water surface, and that P_d becomes small for a big tidal level. The strong correlation between water level and double-bounce scattering power P_d can be recognized [7]. From Figure 14.19, we can see the P_d (red) gradually fades away when the tidal level becomes big.

14.2.6 FLOODING

Flooding causes devastating damage to the surrounding area. If inundated areas are detected by synthetic aperture radar (SAR) observation [8–19], it serves as a warning system as well as forecasting. Since the flooding situation changes time by time, it is necessary to acquire data and make it available as soon as possible (in quasi-real time). Figure 14.19 shows the flooding situation in Joso City, Japan, caused by typhoon 18 on September 11, 2015.

SAR is expected to play a key role in surveillance or monitoring flooding situations because optical sensors cannot be applied in heavy rainy conditions. SAR has advantages of its operation any time at any weather conditions. Now high-resolution spaceborne SAR, such as TerraSAR-X or CosmoSkyMed, has been operating to detect flooding. The resolution is in the order of 1–3 m. Time series data are also available. What is known from the observations:

- Inundated area looks dark compared to its surroundings.
- *HH* is better than *VV*.
- Larger incidence angle is preferred.
- Backscattering magnitude from inundated area depends on platform.
- Sensor parameters are different for each system.

From the observation results and theoretical considerations, we face the following two situations and associated problems.

14.2.6.1 Open Space Situations

Open space is just like a rural area where there is almost no big shadowing object for radar beams. The inundated area or water-covered area is called an open water area. The surface of the inundated area becomes flat like a mirror. The radar wave is mirror reflected, and hence no backscattering occurs. This situation causes a dark SAR image compared to the surroundings. If we look for the dark area, it is possible to find the inundated area. But there is a difficult situation for radar sensing. Is it possible to distinguish between an open water area and a road? Since both have flat surfaces, it is difficult to distinguish. How about standing water vs. open water? These are critical problems for open water and open space situations. Some radar cross section (RCS) threshold criteria obtained in the local area and statistical analysis may solve these problems.

14.2.6.2   Urban Area Situations

SAR is a side-looking radar, which causes a shadowing effect by the building. According to the orientation and density of buildings, various shadows are created for radar sensing. Figure 14.20 shows one typical example of a radar observation scenario in an urban area. The city center is assumed to be covered with water as indicated by the blue color. There are two main blind regions for radar, that is, shadowed by buildings and inundated surfaces. The area corresponds to (B–C) and (C–D) in Figure 14.20. No signal comes from these areas. It is difficult to distinguish these two.

Also, radar cannot distinguish radar echo from point G and point E, because they are in the same range from the radar. Since the scattering magnitude from point E is close to zero, the radar echo is masked by that of point G. The reflections from (E–F′) are also masked by the layover effect by the building. Therefore, the inundated region (B–F′) cannot be properly detected. Only a small portion (C–D) is illuminated by radar, which causes a mirror reflection. Most of the portions are not illuminated by radar. From this scattering mechanism, it is quite difficult to detect urban area flooding.

The double-bounce scattering created by the right-angle structure (G–F′–E) becomes the dominant echo in inundated area. The magnitude raises up due to strong reflection by the water surface in the inundated area. Usually, a 1–2 dB increase is often reported in actual urban flooding situations.

Now, we use two data sets before and after big flooding that occurred on September 11, 2015 over Joso City, Japan. This flooding was caused by localized heavy rain and initiated by a broken dike.

#ALOS2065720720-150811,	right looking,	off-nadir angle 25° (before flooding)
#ALOS2071040740-150916,	left looking,	off-nadir angle 25° (after flooding)

At first, we tried to confirm RCS-based detection. The water area with RCS < −11 dB was picked up in total power image as well as in the *HH* image. The results are shown in Figure 14.21. The threshold value −11 dB was chosen to match the actual photo situation with the inundated area.

Scattering-power decomposition was carried out on the data sets. The second one was taken 5 days after the flooding. Therefore, the flooding situation, including water level and extent of inundated area, changed from the original one. The elliptical circle A and B are rice paddy fields in the decomposition image. However, the harvesting has finished on A, not inundated, leaving bare soil in the field. Paddy field B was inundated before harvesting as shown in the photo. Rice stems are partially in the water similar to the oyster farming example. This rice situation caused strong double-bounce scattering as is indicated by the pink color (Figure 14.22).

14.2.6.3   Flooding Effect: Scattering from Riverbed

When heavy rain continues, we often suffer from flooding. As in the Amazon, there are riverbed and wetlands with grass or trees surrounding the river. When flooding occurs, the water surface level goes up, and the scattering mechanism changes as shown in Figure 14.23.

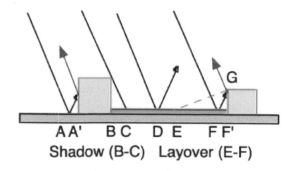

FIGURE 14.20   Scattering scenario in urban area. Radar shadow (B–C), layover (E–F), specular surface (C–D) in flooded urban street (B–F′), stronger double bounce FF′ > AA′.

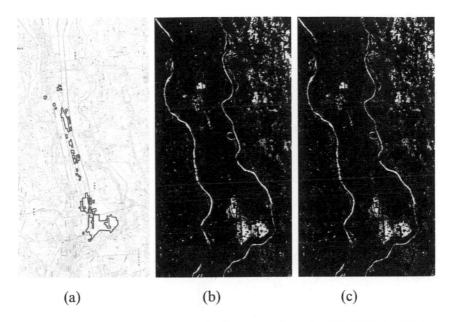

(a)   (b)   (c)

FIGURE 14.21   Detected flooding area: (a) ground truth on September 16, 2015, by Ministry of Land, Infrastructure, Transport and Tourism, Japanese Government; (b) total power; and (c) *HH* polarization.

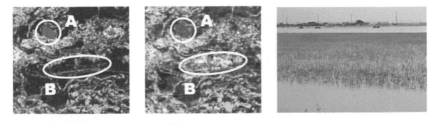

20150811 (Before) 20150916 (5 days after flooding)

FIGURE 14.22   Scattering-power decomposition images of Joso City. RGB color-code is assigned to the scattering mechanisms with R (double bounce), G (volume), and B (Surface).

Before flooding, the scattering mainly consists of volume scattering by vegetation or canopy. The radar wave penetrated onto the ground is scattered as surface scattering, combined with some herbaceous volume scattering. However, the contribution of surface scattering from ground surface is usually very small. For the rainforest area, the volume scattering becomes dominant, and hence, the decomposition image looks completely green.

On the other hand, the ground surface is covered with water for flooding case. The scatterer on the ground changes from a rough soil to a water surface. Since the reflection magnitude is larger than the normal ground, the reflected wave hits tree stems and returns to the radar. This causes strong double-bounce scattering. An increase of P_d (red) to P_v (green) results in a mixture of these two colors, yielding yellow or orange. Therefore, the inundated area looks yellow or orange, even red, due to increasing P_d. This situation is depicted in Figure 14.24 for the Amazon River.

14.2.7   Land Use

An amazing image was obtained after scattering-power decomposition as shown in Figure 14.25. The location is close to the border of Brazil and Argentina where the double-bounce scattering area

Scattering from riverbed:

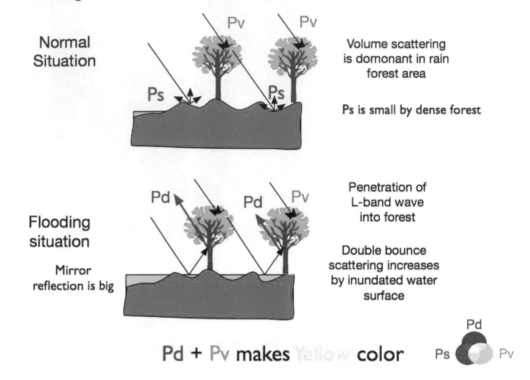

Normal Situation — Volume scattering is domonant in rain forest area

Ps is small by dense forest

Flooding situation — Penetration of L-band wave into forest

Mirror reflection is big

Double bounce scattering increases by inundated water surface

Pd + Pv makes Yellow color

FIGURE 14.23 Scattering mechanism change in riverbed due to flooding.

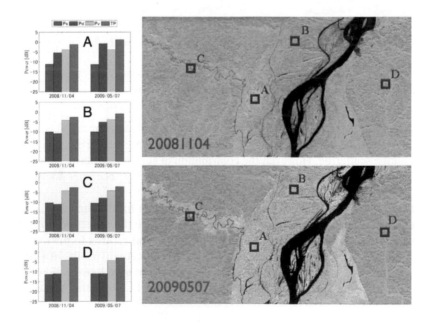

Along Amazon river

FIGURE 14.24 Scattering from riverbed after flooding along upper Amazon in Peru.   *(Continued)*

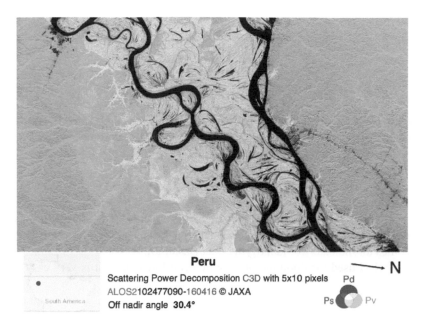

FIGURE 14.24 (Continued)   Scattering from riverbed after flooding along upper Amazon in Peru.

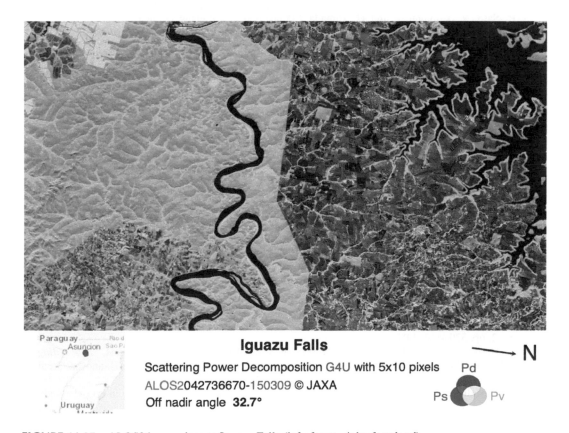

FIGURE 14.25   ALOS2 image close to Iguazu Falls (left: forest, right: farmland).

(red) and forest area (green) are distinctively identified. Red areas correspond to farmlands with tall crops such as corn. The light green area in the upper-left corner is an agricultural field with a lot of planting rows.

14.3 REMARKS

In this chapter, the double-bounce scattering phenomena by metal and dielectric dihedral structures are examined in detail. If the dihedral is placed normal to radar illumination, it acts as a double-bounce scatterer. However, it does not act as double-bounce reflector anymore when obliquely placed to the radar illumination.

We can see very strong double-bounce scattering power in actual PolSAR images in urban areas, buildings, man-made structures, tall vegetation such as corn, ships on the sea, etc. Since the power is strong compared to those of the surrounding area, it is easier to find man-made objects. If there is a reduction of the power after disaster, it means the area is damaged or collapsed. Therefore, reduction also becomes an indicator for identifying the disaster area.

If L-band SAR is used for flood monitoring in a rainforest area, it is very easy to identify the inundated place.

REFERENCES

1. H. Kobayashi, Y. Yamaguchi, and H. Yamada, "Scattering matrix from dielectric corner reflector," Technical Report of IEICE, AP2002-15, 2002-5.
2. K. Hayashi, R. Sato, Y. Yamaguchi, and H. Yamada, "Polarimetric scattering analysis for a finite dihedral corner reflector," *IEICE Trans. Commun.* vol. E89-B, no. 1, pp. 191–195, 2006.
3. R. Sato, Y. Yamaguchi, and H. Yamada, "Polarimetric scattering feature estimation for accurate wetland boundary classification," *Electronic Proceedings of IGARSS 2011*, Vancouver, Canada, July 2011.
4. Y. Yajima, Y. Yamaguchi, R. Sato, H. Yamada, and W.-M. Boerner, "POLSAR image analysis of wetlands using a modified four-component scattering power decomposition," *IEEE Trans. Geosci. Remote Sens.*, vol. 46, no. 6, pp. 1667–1673, 2008.
5. Y. Yamaguchi, "Disaster monitoring by fully polarimetric SAR data acquired with ALOS-PALSAR," *Proc. IEEE*, vol. 100, no. 10, pp. 2851–2860, 2012.
6. G. Singh, Y. Yamaguchi, S.-E. Park, W.-M. Boerner, "Monitoring of the 2011 March 11 off-Tohoku 9.0 earthquake with super-tsunami disaster by implementing fully polarimetric high resolution POLSAR techniques," *Proc. IEEE*, vol. 101, no. 3, pp. 831–846, 2013.
7. T. Y. Cheng, Y. Yamaguchi, K. S. Chen, J. S. Lee, and Y. Cui, "Sandbank and oyster farm monitoring with multi-temporal polarimetric SAR data using four component scattering power decomposition," *IEICE Trans. Commun.*, vol. E96-B, no. 10, pp. 2573–2579, 2013.

FLOODING REFERENCES

8. G. Boni, F. Castelli, L. Ferraris, N. Pierdicca, S. Serpico, and F. Siccardi, "High resolution COSMO/SkyMed SAR data analysis for civil protection from flooding events," *Proceedings of IGARSS'2007*, IEEE, 2007.
9. L. Pulvirenti, N. Pierdicca, M. Chini, and L. Guerriero, "An algorithm for operational flood mapping from Synthetic Aperture Radar (SAR) data using fuzzy logic," *Nat. Hazards Earth Syst. Sci.*, vol. 11, no. 2, pp. 529–540, 2011.
10. V. Herrera-Cruz and F. Koudogbo, "TerraSAR-X rapid mapping for flood events," *Proceedings of ISPRS Workshop on High-Resolution Earth Imaging for Geospatial Information*, Hannover, Germany, pp. 170–175, 2009.
11. R. T. Melrose, R.T. Kingsford, and A.K. Milne, "Using radar to detect flooding in arid wetlands and rivers," *Proceedings of IGARSS'2012*, IEEE, 2012.
12. S. Martinis, A. Twele, S. Voigt, and G. Strunz, "Towards a global SAR-based flood mapping service," *Proceedings of IGARSS'2014, IEEE*, 2014.

13. D. C. Mason, I. J. Davenport, J. C. Neal, G. J.-P. Schumann, and P. D. Bates, "Near real-time flood detection in urban and rural areas using high-resolution synthetic aperture radar images," *IEEE Trans. Geosci. Remote Sens.*, vol. 50, no. 8, pp. 3041–3052, 2012.

14. L. Giustarini, R. Hostache, P. Matgen, G. J.-P. Schumann, P. D. Bates, and D. C. Mason, "A change detection approach to flood mapping in urban areas using TerraSAR-X," *IEEE Trans. Geosci. Remote Sens.*, vol. 51, no. 4, pp. 2417–2430, 2013.

15. M. Watanabe, M. Shimada, M. Matsumoto, and M. Sato, "GB-SAR/PiSAR simultaneous experiment for a trial of flood area detection," *Proceedings of IGARSS'2008, IEEE*, 2008.

16. B. M. Tanguy, M. Bernier, K. Chokmani, Y. Gauthier, and J. Poulin, "Development of a methodology for flood hazard detection in urban areas from RadarSAT-2 imagery," *Proceedings of IGARSS'2014*, IEEE, 2014.

17. J. Lu, J. Li, G. Chen, L. Zhao, B. Xiong, and G. Kuang, "Improving pixel-based change detection accuracy using an object-based approach in multi-temporal SAR flood images," *IEEE JSTARS*, vol. 8, no. 7, pp. 3486–3496, 2015.

18. G. Boni, N. Pierdicca, L. Pulvirenti1, G. Squicciarino, and L. Candela, "Joint use of X- and C-band SAR images for flood monitoring: The 2014 PO river basin case study," *Proceedings of IGARSS'2015*, IEEE, 2015.

19. P. Iervolino, R. Guida, A. Iodice, and D. Riccio, "Flooding water depth estimation with high-resolution SAR," *IEEE Trans. Geosci. Remote Sens.*, vol. 53, no. 5, pp. 2295–2307, 2015.

15 Volume Scattering

This chapter is devoted to the volume scattering and its power applications. The typical volume-scattering object is vegetation, including forests, trees, bushes, branches, leaves, crops, grass, etc. There are numerous scattering points inside vegetation that cause "volume" scattering. The cross-polarized *HV* component is the main source for the volume scattering in polarimetric SAR (PolSAR) observation. We know that the *HV* component is created by vegetation, in addition to oriented surface and edges of man-made objects. The most important application of the *HV* component is forest/tree monitoring. Since forests absorb greenhouse gas and mitigate global warming, the preservation and monitoring of forests is very important and attracting attention [1]. The key parameter for monitoring forests is the amount of biomass. Therefore, biomass estimation becomes the most important application for radar remote sensing. Since the *HV* component, $|S_{HV}|$ or $|S_{HV}|^2$, is directly related to the forest volume, the measured value is utilized to create forest/non-forest map of the world as shown in Figure 15.1 [2].

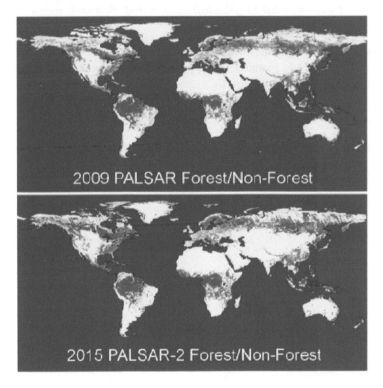

FIGURE 15.1 Forest/non-forest map by JAXA (2009&2015) [2]. This figure has been created by ALOS observation and has 84% accuracy compared with ground-based data. Forest areas are color-coded by green with the threshold level of 100 ton/ha.

15.1 FOREST MAPPING

The main application of the volume scattering is forest mapping related to biomass estimation. Radar observables are backscattering coefficients. Since biomass is not directly related to the backscattering, the estimation of biomass has been investigated based on these observables. Although some experimental equations relating biomass and backscattering coefficients have been presented [3–6], it is always necessary to update empirical equations for more accurate estimation. There are several methodologies in the forestry area that biomass is estimated through measurable height parameters. Tree height or diameter at breast height (DBH) can be measured directly and used for accurate estimates of biomass [6]. This information can also be used as sampling data. Therefore, the SAR measurement is further extended toward the PolInSAR or TomoSAR technique as shown in Figure 15.2 to measure tree height and volume structure. Since the detail of TomoSAR measurement [7] is out of scope of this book, it is omitted here.

An advanced polarimetric parameter is defined as the radar vegetation index (RVI) [3] to monitor forest as

$$\mathrm{RVI} = \frac{8\sigma_{hv}}{\sigma_{hh} + \sigma_{vv} + 2\sigma_{hv}} = \frac{4\min(\lambda_1, \lambda_2, \lambda_3)}{\lambda_1 + \lambda_2 + \lambda_3} \qquad 0 \le \mathrm{RVI} \le \qquad (15.1.1)$$

Based on this index, experimental equation has been derived for the radar cross section (RCS) by a polynomial of biomass expression.

$$\sigma_{f,pp}^0 = a_0 + a_1\beta + a_2\beta^2 + a_3\beta^3 \qquad \beta = \log \; \mathrm{Biomass}\left(\frac{\mathrm{tons}}{\mathrm{ha}}\right) \qquad (15.1.2)$$

The coefficients and other parameters are listed in Table 15.1. A good correlation can be achieved for L-band HV component and biomass. An approximate equation for L-band HV component is derived in terms of biomass up to 400 ton/ha.

$$\sigma_{L,HV}^0 = -21.898 + 5.806\beta - 0.336\beta^2 - 0.258\beta^3 \qquad (15.1.3)$$

The relation of radar parameter, forest parameter, and biomass is shown in Figure 15.2.

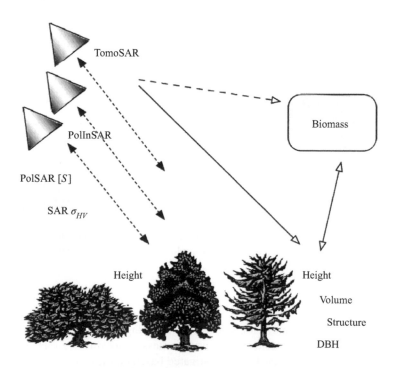

FIGURE 15.2 Advanced SAR observation of a forest to estimate biomass.

TABLE 15.1
Coefficients for Backscattering Equation (15.2)

Band	Pol	a_0	a_1	a_2	a_3	r^2
P-band 440 MHz	*HH*	−19.6	4.534	1.592	−0.4582	0.929
	VV	−19.5	4.037	0.826	−0.4381	0.904
	HV	−29.7	3.893	4.666	−1.5122	0.978
L-band 1.22 GHz	*HH*	−12.6	4.444	−1.545	0.2916	0.894
	VV	−14	4.144	−1.226	0.1815	0.916
	HV	−21.9	5.806	−0.336	−0.258	0.957
C-band 5.3 GHz	*HH*	−10.1	3.464	−2.085	0.4924	0.547
	VV	−10.1	1.629	−0.644	0.1698	0.511
	HV	−16.9	4.06	−1.781	0.4006	0.749

15.1.1 Frequency Characteristics of Forest Volume Scattering

To understand the frequency response of forests, scattering power decomposition images are prepared by PiSAR-L/X observation data over the Tomakomai National Forest area in Hokkaido, Japan. Figure 15.3 shows an RGB color-coded image with L-band (upper) and X-band (lower) data at different acquisition times. The forest has a lot of tree species and is managed by the government as can be seen in the figure.

These images display time series change by frequency. Data on February 13 was a snow-covered scene. The L-band data seems to have surface scattering from accumulated snow. The X-band images with a non-snow season have bluish color because the surface scattering is large, whereas L-band images do not have the surface scattering so much. In the winter season, leaves have fallen off from deciduous trees. Numerous branches and twigs induced complex double-bounce scattering in the X-band image as the higher-order scattering models in Chapter 7.

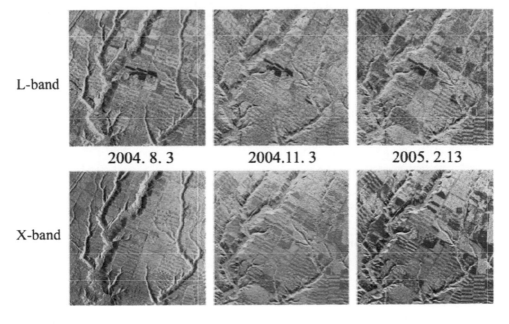

FIGURE 15.3   Tomakomai National Forest area observed by PiSAR-L/X.

15.1.2 SCATTERING CENTER

The scattering position in forests or trees is dependent on the penetration capability of the frequency band. Figure 15.4 shows a rough scattering position of each band. P-band has the deepest penetration capability, and hence the wave penetrates canopy, branches, stem, and finally reaches the ground surface or even in the dry underground medium. Then it is reflected back to the radar. L-band can penetrate canopy and sometimes go to the bottom. Most leaves are transparent for the L-band wave. The main scattering occurs in the branches or stems for L-band and S-band. C-band also penetrates ~80 cm in the canopy and is reflected from there. So it contains canopy information. X-band is high frequency and cannot penetrate the canopy so deeply (up to 20–30 cm), depending on the tree species. Since the wavelength is 3 cm, the leaf size is sometimes bigger than the wavelength. In this case, the surface scattering becomes large, although the volume scattering is dominant. The scattering mainly occurs in the canopy for the X-band. Table 15.2 lists some experimental penetration results.

In this chapter, based on the research works [8–14], we will see some useful examples of the volume-scattering power obtained from ALOS/ALOS2 and PolSAR.

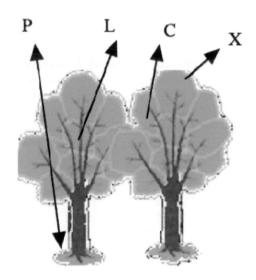

FIGURE 15.4   Main scattering location.

TABLE 15.2
Penetration Depth

Tree Species	S-band (cm)	C-band (cm)	X-band (cm)
Cedar	150	83	18
Cypress	170	77	32
Pine	280	57	25

15.1.3 Tree-Type Classification

Scattering-power decomposition provides various scattering powers in an imaging window. If tree species are different in neighboring windows, the power components become different. To see the details of tree species, six-component scattering power decomposition (Chapter 8) was applied to an Indonesia rainforest as shown in Figures 15.5 and 15.6.

Plantation, Indonesia

Scattering Power Decomposition 6SD with 5x10 pixels

ALOS2047697110-150411 © JAXA

Off nadir angle **32.7°**

FIGURE 15.5 Palm oil plantation in Sumatra, Indonesia. Dark purple corresponds to palm oil plantation area.

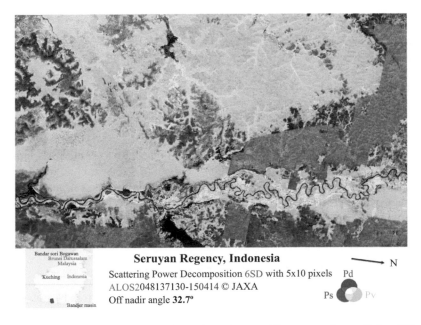

Seruyan Regency, Indonesia

Scattering Power Decomposition 6SD with 5x10 pixels

ALOS2048137130-150414 © JAXA

Off nadir angle **32.7°**

FIGURE 15.6 Plantation (dark) and natural forest (green) area. All areas are forest. Color difference is due to the difference of scattering mechanism by tree species. Dark purple corresponds to palm oil plantation area.

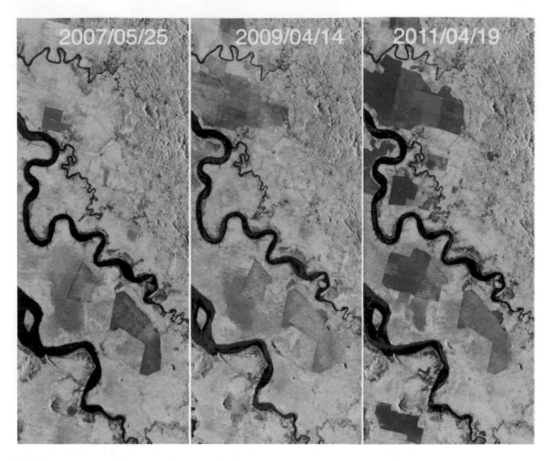

FIGURE 15.7 Time series deforestation in Malaysia. Blue area corresponds to clear-cut.

Surprisingly, it is possible to distinguish tree types easily by color, based on the scattering mechanisms. The natural forest area looks green, whereas the oil palm plantation areas look dark purple (combination of surface scattering and volume scattering). Tall trees reflect red color corresponding to the double-bounce scattering. Since these rainforest areas are usually covered with clouds more than 90% of the time, the situation cannot be seen by any optical image sensor. Even if an optical image is obtained, it is difficult to distinguish tree types. These figures show the advantage of L-band polarimetric SAR sensing.

Time series data in the rain forest provide us with an interesting feature [8]. Figure 15.7 shows ALOS images of Malaysia deforestation. Just after clear-cut, the color is blue, caused by the surface scattering. Blue (deforestation) area expands as time goes on, while the surface scattering tends to fade due to regrowth of trees. Regrowth causes new volume scattering P_v on the surface scattering P_s. The double bounce and the helix scattering are usually very small. We can see the forest change clearly in Figure 15.7.

15.2 CONIFER AND BROAD-LEAF TREE CLASSIFICATION

Is it possible to distinguish a conifer (needle-like) tree and broad-leaf tree by polarimetric radar? Since PiSAR-X2 has acquired data sets over the Tokyo area, we tried to make a G4U [15] scattering power decomposition image over Yoyogi Park in downtown Tokyo area. A normal RGB color-coded image was created; however, the forested area as a whole looked bluish due to X-band

Green: Broad leaf

Blue: Conifer

FIGURE 15.8   Conifer and broad-leaf tree classification by PiSAR-X2.

surface scattering. So we assigned blue for the volume scattering, and green for the surface scattering, just for our intuitive recognition of trees. The result is shown in Figure 15.8. Blue corresponds to conifer trees, and green corresponds to broad-leaf trees as evidenced by the Google Earth image.

X-band scattering is mainly accompanied with the surface scattering on the top surface of scatterer. If there is a broad-leaf tree, the surface scattering may be dominant because the leaf size is more than the wavelength. On the other hand, the volume scattering or HV component generation becomes dominant for conifer trees at X-band. Using the difference in the scattering mechanism, it is possible to distinguish conifer and broad-leaf trees at high-frequency radar above X-band [16].

15.2.1   Conifer and Broad-Leaf Tree Measurement in an Anechoic Chamber

In order to check the capability of PolSAR for distinguishing these trees, a polarimetric measurement was conducted using a network analyzer at Ku-band in a well-controlled anechoic chamber [16]. Test samples are shown in Figure 15.9.

The PolSAR measurement was conducted at:

Incidence angle was chosen as 30°, 45°, and 60°, for which tree the arrangement is shown in Figure 15.10. The center frequency is 15 GHz, resulting in the range resolution of 3.75 cm.

The final decomposition images are shown in Figure 15.10 corresponding to tree arrangement.
It is seen that blue (surface-scattering power) is dominant for the broad-leaf tree, whereas the conifer tree is characterized by green (volume scattering). This point is further investigated to show the scattering power contribution ratio in Figure 15.11.
Once scattering matrices are obtained, it is possible to carry out eigenvalue analysis (Chapter 5), which yields useful polarimetric indexes: entropy H, mean alpha-angle α, and anisotropy.

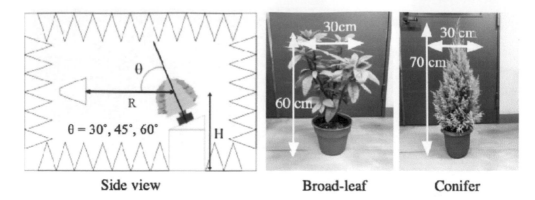

FIGURE 15.9 Polarimetric measurement of broad leaf and conifer trees.

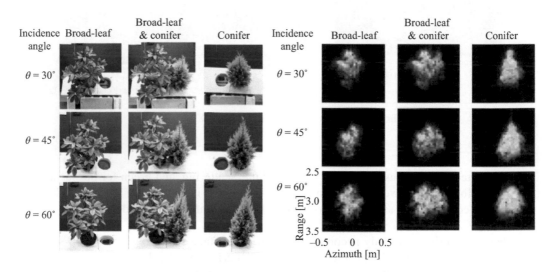

FIGURE 15.10 Tree arrangements and the scattering power decomposition images.

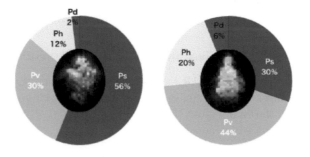

FIGURE 15.11 Scattering powers.

Using $\bar{\alpha}$ information, Figure 15.12 was calculated to show the possibility of tree-type discrimination. From Figure 15.12, it seems possible to classify tree types by a simple algorithm shown in Figure 15.13. A threshold level of $\bar{\alpha} \geq 30°$ was selected to classify conifer and broad-leaf trees. A median filter is applied to the classification result in Figure 15.14. It seems that high-frequency PolSAR has a great potential to distinguish conifer and broad-leaf trees. The average coherency matrices for conifer and broad-leaf trees were calculated as

$$\text{conifer } \langle[T]\rangle = \begin{bmatrix} 1 & -0.07-0.03j & 0.05-0.07j \\ -0.07+0.03j & 0.44 & -0.01+0.05j \\ 0.05+0.07j & -0.01-0.05j & 0.45 \end{bmatrix}$$

$$\text{broad-leaf } \langle[T]\rangle = \begin{bmatrix} 1 & -0.07-0.17j & 0.05j \\ -0.07+0.17j & 0.09 & 0.01j \\ -0.05j & -0.01j & 0.08 \end{bmatrix}$$

These values may serve for correlation analysis or similarity analysis.

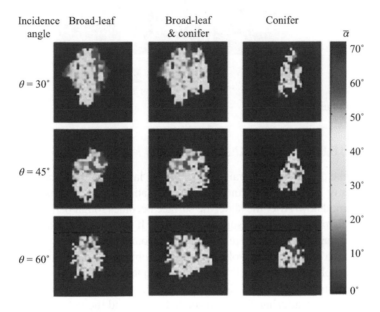

FIGURE 15.12 Alpha-angle distribution.

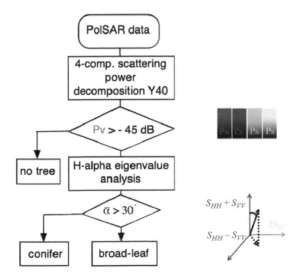

FIGURE 15.13 Classification algorithm.

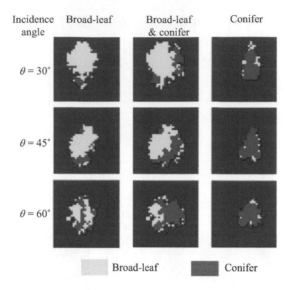

FIGURE 15.14   Classification result.

15.3   LANDSLIDE DETECTION IN VEGETATED MOUNTAINS

When a big earthquake occurs or heavy rain fall continues, landslides frequently happen in mountainous areas. Landslides or mudslides tear surface trees, vegetations, soils, houses, and bring them downward all together as mud flow. After the slide, the surface becomes rough bare soil. When observed by PolSAR, the area is characterized by the surface scattering. If the area is surrounded by trees or vegetations, the scattering nature is totally different from the surrounding. The area may be characterized by a surface-scattering area surrounded by volume-scattering objects from the polarimetric point of view.

Let's take an example of Hokkaido Iburi Tobu earthquake on September 6, 2018. ALOS2 had acquired data on August 26, 2017 and September 8, 2018. For comparison and also for contrast enhancement, the surface scattering P_s is color-coded as red and the volume-scattering P_v as green in Figure 15.15. It is seen that the red area has increased in the mountainous area by the earthquake.

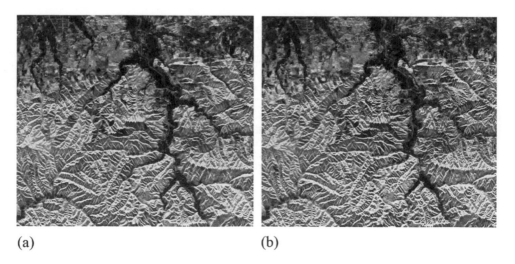

(a)   (b)

FIGURE 15.15   Hokkaido Iburi Tobu earthquake (before and after): (a) August 26, 2017: before and (b) September 8, 2018: after.

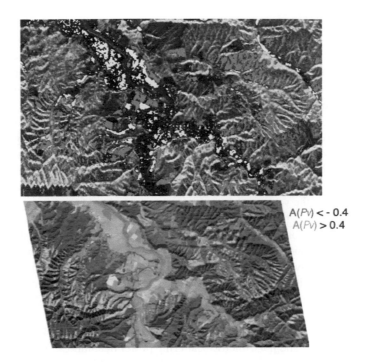

$A(Pv) < - 0.4$
$A(Pv) > 0.4$

FIGURE 15.16   Landslide by earthquake at Atsumacho, Hokkaido, Japan. $A(P_v) < -0.4$ corresponds to landslide area (bare soil). $A(P_v) > 0.4$ corresponds to area where debris emerged.

For detection of the landslide area, we may think of using "increasing P_s" or "decreasing P_v" as a polarimetric index. However, the following polarimetric index served the best performance to identify the landslide location after many trials.

$$\left| A(P_v) \right| > 0.4, \qquad A(P_v) = \frac{P_v^{after} - P_v^{before}}{P_v^{after} + P_v^{before}}$$

$$\left| A(TP) \right| > 0.35 \qquad TP = P_s + P_d + P_v + P_h$$

In this case, the information of P_v was useful, rather than P_s. Figure 15.16 shows the landslide area detected by the preceding parameters together with an air photo [17].

15.4   WETLAND MONITORING

Wetlands cover at least 6% of the earth's surface. Wetland ecosystems play a key role in hydrological and biogeochemical cycles and comprise a large part of the world's biodiversity and resources [18]. They provide a critical habitat for a wide variety of plant and animal species, including the larval stages of many fish and insects, a resort for migrating birds, forage for cattle grazing, and bee flora. Wetlands also deliver a wide range of important services, including water supply, water purification, carbon sequestration, coastal protection, and outdoor recreation. Intact wetlands perform as buffers in the hydrological cycle and as sinks for organic carbon, counteracting the effects of the increase in atmospheric CO_2. Thus, their sustainable use ensures human and economic development and quality of life.

However, depending on the region, 30%–90% of the world's wetlands have already been destroyed or strongly modified mainly due to agriculture and urban development in many countries. Climate-change scenarios predict additional stresses on wetlands, because of changes in hydrology, temperature increases, and a rise in sea level. In addition, global warming causes organic carbon stored in permafrost (peatland) release back into the air. This increases an abrupt change in the arctic region.

Understanding the roles of wetlands correctly and keeping healthy condition are essential for sustaining the environment. One of the inherent features for wetland monitoring is the distribution of emergent species. Vegetation patterns are considered an emergent feature of the local hydrological regime. They can be used as indicators of environmental conditions. Since the plant communities represent the amount of biomass accumulated above and below ground, it shares one of the earth's carbon stocks. If the vegetation is correctly mapped, the information will serve to monitor the wetland environment correctly.

As far as radar observation is concerned, the review of frequency characteristics and polarization channel effects are presented in [19,20]. Emergent plant communities cover more than 90% of the wetland area and are usually dominated by one or very few species. It is known that higher frequency above C-band may be sensitive to this vegetation, that is, suitable for emergent species monitoring.

This section presents time series results of X-band fully polarimetric radar observation of wetland "Sakata Lagoon" in Niigata, Japan, during 2013–2015. The second generation of airborne polarimetric and interferometric synthetic aperture system (PiSAR-2) took flight over the Niigata area in Japan, three times. The polarimetric scattering power decomposition has applied to the data sets, and the wetland region was successfully imaged with very high resolution, which revealed the detailed response of the emergent plants, such as lotus and reed.

15.4.1 STUDY AREA

Sakata Lagoon is a marsh, located in 37.48N, 138.52E, in Niigata Prefecture, Japan. It has been registered as a RAMSAR site according to the Convention on Wetland of International Importance especially as a migratory waterfowl habitat as shown in Figure 15.17. It is a famous local place where more than 3,000 migrating cranes stay in the winter.

During the flight campaign of airborne PiSAR-2, the area was observed three times on August 25, 2013, October 17, 2013, and March 3, 2015. The PiSAR-2 has a capability of very high resolution and fully polarimetric data take function [21], that is, 30 cm range resolution and 30 cm azimuth resolution. For exploring the possibility of the PiSAR-2 system, some experimental flights have been conducted by the NICT research team.

FIGURE 15.17 Sakata Lagoon and lotus.

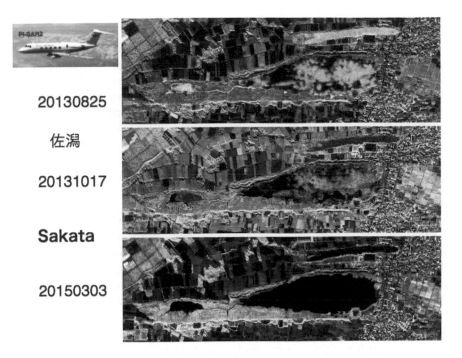

FIGURE 15.18 Scattering power decomposition images of Sakata with red for the double bounce, green for the volume scattering, and blue for the surface scattering.

15.4.2 SCATTERING POWER DECOMPOSITION IMAGE

All the data sets after polarimetric calibration were processed by the general four-component scattering power decomposition with unitary transformation: G4U [15]. Time series of the polarimetric scenes are shown in Figure 15.18, where the RGB color-coding is used. In the decomposition, we used a 9×9 window size to derive each of the scattering powers, resulting in approximately 3×3 m resolution on the ground.

15.4.3 DISCUSSION

It is seen in Figure 15.18 that color on the surface of Sakata completely changes from blue to red, and to black. This color change reflects the scattering mechanism change from surface scattering to double-bounce scattering and to mirror reflection. On the August image, the main contribution of the surface scattering on the mid-lake is the broad leaf of lotus in its most active period. The thick broad leaves act as dishes for X-band waves, reflecting surface scattering. This scattering causes the blue color. On the October image, the leaves are faded and stems become apparent. The water surface and stems generate double-bounce scattering power, yielding a red color. In the March image, all vegetation is gone, and the surface becomes flat. This situation causes a mirror reflection.

The upper left areas of Sakata Lagoon are sandy crop fields, and the right corner is a residential area and small rice paddy field. Reeds and other vegetation look green in the boundary region of the lagoon. A very high-resolution image helps us to identify the objects in the scene [22].

15.5 GLACIER IN ANTARCTICA

ALOS2-PALSAR2 acquired quad pol data over the whole Shirase Glacier in Antarctica as shown in Figure 15.19. The image is composed of two time series data sets. From the beginning to the end (to ocean), the glacier flow can be clearly seen by a green color (the *HV* component). It is approximately 100 km long. Although this *HV* component is not created by vegetation, the volume scattering is dominant in the whole image. This is caused by compressed ice particles of which size is dependent on the position just like in the mouth–stomach–intestines. It is rather surprising that the surface scattering is not so strong.

15.6 REMARKS

Volume-scattering objects and the volume-scattering power P_v are main topics of this chapter. Among volume scatterers, tree species including height, volume, and biomass will be the most important quantities to be measured or estimated by remote sensing data. PolSAR and TomoSAR at L- to Ku-band have potential to provide an accurate estimation because of abundant information. As we have seen in L-band data by ALOS2, there is a possibility to distinguish tree types in the rain forest. Since scattering points inside forest changes by frequency band, a combination of frequency bands will increase the accuracy of desired values.

The main source of the volume scattering is the cross-polarized *HV* component. Appropriate use of this component serves for crop monitoring, wetland monitoring, etc. Further combination of other scattering powers helps to identify landslide/mudslide areas in vegetated regions, deforestation, or even vegetation species classification. A combined use of other parameters will be helpful to achieve that purpose.

FIGURE 15.19 Shirase Glacier, Antarctica.

REFERENCES

1. IPCC report on Global warming. https://www.ipcc.ch/sr15/
2. JAXA forest mapping. http://www.eorc.jaxa.jp
3. Y. Kim and J. van Zyl, "Comparison of forest estimation techniques using SAR data," IEEE *Proceeding of IGARSS2001*, Sydney, Australia, 2001.
4. M. Watanabe et al., "Forest structure dependency of the relation between L-band sigma zero and biophysical parameters," *IEEE Trans. Geosci. Remote Sens.*, vol. 44, pp. 3154–3156, 2006.
5. C. Dobson, F. T. Ulaby, T. L. Toan, A. Beaudoin, E. S. Kasischke, and N. Christernsen, "Dependence of radar backscatter on coniferous forest biomass," *IEEE Trans. Geosci. Remote Sens.*, vol. 30, no. 2, pp. 412–415, 1992.
6. H. Wang and K. Ouchi, "Accuracy of the K-distribution regression model for forest biomass estimation by high resolution polarimetric SAR: Comparison of model estimation and field data," *IEEE Trans. Geosci. Remote Sens.*, vol. 46, no. 4, pp. 1058–1064, 2010.
7. A. Moreia, P. P. Iraola, M. Younis, G. Krieger, I. Hajnsek, and K. Papathanassiou, "A tutorial on synthetic aperture radar," *IEEE GRSS Magazine*, March 2013.
8. S. Kobayashi et al., "Comparing polarimetric decomposition and in-situ data on forest growth of industrial plantation in Indonesia," *Proceedings of the 54th Conference of the Remote Sensing Society of Japan*, pp. 782–785, 2013.
9. S.-E. Park, W. Moon, and E. Pottier, "Assessment of scattering mechanism of polarimetric SAR signal from mountainous forest areas," *IEEE Trans. Geosci. Remote Sens.*, vol. 50, no. 11, pp. 4711–4719, November 2012.
10. E. Lehmann, P. Caccetta, Z.-S. Zhou, S. McNeill, X. Wu, and A. Mitchell, "Joint processing of Landsat and ALOSPALSAR data for forest mapping and monitoring," *IEEE Trans. Geosci. Remote Sens.*, vol. 50, no. 1, pp. 55–67, Jan. 2012.
11. T. Shiraishi, T. Motohka, R. B. Thapa, M. Watanabe, and M. Shimada, "Comparative assessment of supervised classifiers for land use–land cover classification in a tropical region using time-series PALSAR mosaic data," *IEEE JSTARS*, vol. 7, no. 4, pp. 1186–1199, 2014.
12. J. Reiche, C. M. Souzax, D. H. Hoekman, J. Verbesselt, H. Persaud, and M. Herold, "Feature level fusion of multi-temporal ALOS PALSAR and Landsat data for mapping and monitoring of tropical deforestation and forest degradation," *IEEE Journal of Selected Topics in Applied Earth Observations and Remote Sensing (JSTARS)*, vol. 6, no. 5, pp. 2159–2173, 2013.
13. O. Antropov, Y. Rauste, H. Ahola, and H. Hame, "Stand-level stem volume of boreal forests from spaceborne SAR imagery at *L*-band," *IEEE JSTARS*, vol. 6, no. 1, pp. 35–44, 2013.
14. N. Wenjian, S. Guoqing, G. Zhifeng, Z. Zhiyu, H. Yating, and H. Wenli, "Retrieval of forest biomass from ALOS PALSAR data using a lookup table method," *IEEE JSTARS*, vol. 6, no. 2, pp. 875–886, 2013.
15. G. Singh, Y. Yamaguchi, and S.-E. Park, "General four-component scattering power decomposition with unitary transformation of coherency matrix," *IEEE Trans. Geosci. Remote Sens.*, vol. 51, no. 5, pp. 3014–3022, 2013.
16. Y. Yamaguchi, Y. Minetani, M. Umemura, and H. Yamada, "Experimental validation of conifer and broad-leaf tree classification using high-resolution PolSAR data above X-band," *IEICE Trans. Communications*, vol. E102-B, no. 7, pp. 1345-1350, 2019.
17. Y. Yamaguchi, M. Umemura, D. Kanai, K. Miyazaki, "ALOS-2 Polarimetric SAR Observation of Hokkaido-Iburi-Tobu Earthquake 2018," *IEICE Communications Express (ComEX)*, vol. 8, no. 2, pp. 26-31, 2019. DOI: https://doi.org/10.1587/comex.2018XBL0131
18. W. J. Junk, S. An, C. M. Finlayson, B. Gopal, J. Květ, S. A. Mitchell, W. J. Mitsch, and R. D. Robarts, "Current state of knowledge regarding the world's wetlands and their future under global climate change: A synthesis," *Aquat. Sci.*, vol. 75, pp. 151–167, 2013.
19. F. M. Henderson and A. J. Lewis, "Radar detection of wetland ecosystems: A review," *Int. J. Remote Sens.*, vol. 29, no. 20, pp. 5809–5835, 2008.
20. R. Touzi, A. Deschamps, and G. Rother, "Wetland characterization using polarimetric RADARSAT-2 capability," *Can. J. Remote Sens.*, vol. 33, no. 1, pp. S56–S67, 2007.
21. M. Satake, T. Kobayashi, J. Uemoto, T. Umehara, S. Kojima, T. Matsuoka, A. Nadai, and S. Uratsuka, "New NICT airborne *X*-band SAR system, PISAR-2: Current status and future plan," *3rd International Polarimetric SAR WS in Niigata*, Japan, 2012.
22. Y. Yamaguchi, H. Yamada, and S. Kojima, "Time series observation of wetland 'Sakata—Ramsar site' by PiSAR-2," IEICE *Electronic Proceedings of ISAP 2016*, Okinawa, Japan, 2016.

Index

Note: Page numbers in italic and bold refer to figures and tables, respectively.

T - #0177 - 111024 - C350 - 254/178/16 - PB - 9780367503109 - Gloss Lamination